Engineering Analysis with ANSYS Workbench 19

GUANGMING ZHANG

Department of Mechanical Engineering
The University of Maryland @ College Park

COLLEGE HOUSE ENTERPRISES, LLC
Knoxville, TN

THANKS FOR PERMISSIONS

The author and the publisher thank the copyright and trademark owners providing permission for use of copyrighted materials and trademarks. We apologize for any errors or omissions in obtaining permissions. Where appropriate, we referenced similar treatments and gave credit for prior work. Errors or omissions in obtaining permissions or in giving proper references are not intentional. We will correct them at the earliest opportunity after the error or omission is brought to our attention. Please contact the publisher at the address given below.

The manuscript was prepared in Microsoft Word 2013 using 11 point Times New Roman font. The book was printed from camera ready copy by Publishing and Printing Inc, Knoxville, TN.

College House Enterprises, LLC.
5713 Glen Cove Drive
Knoxville, TN 37919-8611, U.S.A.
Phone: (865) 558 6111
FAX (865) 558 6111

ISBN 978-1-935673-50_7

**To my wife Jinyu
my children Zumei and Haowei, and my
grandchildren Qianqian, Jiajia and Lele**

who have always been with me especially at those difficult times. Their love and
understanding have supported me in a journey for survival and success.

ABOUT THE AUTHOR

Guangming Zhang obtained a bachelor degree and a master degree both in mechanical engineering from Tianjin University, the People's Republic of China. He obtained a master degree and a Ph. D. degree in mechanical engineering from the University of Illinois at Urbana-Champaign. He is currently an associate professor in the Department of Mechanical Engineering.

Professor Zhang worked at the Northwest Medical Surgical Instruments Factory in China where he served as a principal engineer to design surgical instruments and dental equipment. He also taught at the Beijing Institute of Printing and received the National Award for Outstanding Teaching from the Press and Publication Administration of the People's Republic of China in 1987. In 1992 he was selected by his peers at the University of Maryland to receive the Outstanding Systems Engineering Faculty Award. He was the recipient of the E. Robert Kent Outstanding Teaching Award of the College of Engineering in 1993, and the Poole and Kent Company Senior Faculty Teaching Award in 2004. He received the Professional Master of Engineering Program Award from the A. James Clark School of Engineering in 2006. He received the Pi Tau Sigma 2006 Faculty/Staff Appreciation Award, and the Pi Tau Sigma 2015 Purple Camshaft Award, He was a recipient of the 1992 Blackall Machine Tool & Gage Award of the American Society of Mechanical Engineering. In 1995, 1997, 1998 and 1999, he received the Award of Commendation, Award of Member of the Year, and the Award of Appreciation from the Society of Manufacturing Engineers, Region 3, for his outstanding service as the faculty advisor to the SME Student Chapter at College Park. In 2006, he was named as one of the 6 Keystone Professors, the Clark School of Academy of Distinguished Professors.

Professor Zhang actively participated in the NSF sponsored ECSEL grant between 1990 and 2000. He served as the principal investigator for this grant at the University of Maryland between 1997 and 2000, and coordinated the ECSEL sponsored projects on integration of design, on active learning and hands-on experiences, and on developing methods for team learning. He also participated in the NSF sponsored ENGAGE program. He has written 4 textbooks, co-authored 3 textbooks, about 70 technical papers, and holds one patent.

PREFACE

This book presents an introduction to the ANSYS Workbench, an integrated simulation software system for product development. ANSYS has been widely used in industry, not only nationally, but worldwide as well. With capabilities of computers expanding at an unthinkable pace, engineers have already realized the importance of using digital technology and product development in their daily work as technology and its benefits are becoming more sophisticated and pervasive. The book is written as an introductory text for undergraduate students in engineering. The book should also be useful to those engaged in the engineering design and engineering analysis.

The book is organized into 7 chapters. The first chapter provides a fundamental coverage of the ANSYS Workbench platform. It stresses the importance and effectiveness of combining geometrical modeling with mathematical and mechanical fundamentals to carry out finite element analysis. Chapter 2 focuses on the static structural analysis. The six case studies at the component level provide readers with a step-by-step procedure, starting from geometric modeling using DesignModeler to FEA using Mechanical. Chapter 3 focused on FEA with assemblies, illustrating the importance of using feature-based modeling and the uniqueness of DesignModeler in creating assembly models. Chapter 4 presents 6 case studies related to the steady-state thermal analysis at both component and assembly levels. Chapter 5 focuses on frequency and vibration analysis. The six case studies cover modal analysis, pre-stressed modal analysis, eigenvalue buckling analysis, transient structural analysis, harmonic response analysis, combined static structural analysis, modal analysis and harmonic response analysis. Chapter 6 focuses on 2D geometry modeling. There are 2 case studies dealing with two truss structures. There is a study with a cantilever beam to illustrate the procedure of obtaining the reaction forces at the support(s) and constructing shear-moment diagrams. Chapter 6 also presents the plane stress analysis using a 2D model to simplifying the FEA process. Chapter 7 presents FEA applications. Concepts of contact mechanics are introduced such as large deflection and modeling with contact regions. Chapter 7 also presents the ANSYS CAD integration focusing on importing a CAD file to DesignModeler. The entire book provides readers with thirty-four case studies and twenty-six questions in the exercise sections. The hands-on practice should help the readers to enhance his/her digital skills, and to gain new digital skills.

The material assembled in this book is an outgrowth of a senior-level design elective course taught by the author at the University of Maryland at College Pars. Moreover, this book is written in a style adaptable for self-study and reference. Suggestions for improving the contents are welcome and the author deeply appreciate the efforts made by the readers in this regard.

It is a pleasure to extend special thanks to Dr. James W. Dally, President of the College House Enterprises, LLC. His support has been invaluable to the author, not only in the academic area, but also in many aspects of the author's life and career development.

Guangming Zhang
College Park, MD
August 2019

CONTENTS

CHAPTER 5 Frequency and Vibration Analysis

CHAPTER 6 Concept Modeling: Line and Surface Bodies

CHAPTER 7 FEA Applications

CHAPTER 1

Introduction to ANSYS Workbench

1.1   Introduction

Engineering design is an activity which demands the responsibility of the designer or the design team to meet the customer needs in terms of the design specifications. When an initial design is completed, a critical step to follow is to evaluate the performance of the designed component under a specified working environment. Results obtained from the evaluation lead to accepting the initial design or directing to an improvement through a redesign process. Finite element analysis is such a process, which has been used to carry out those evaluations of the initial designs through simulation under a computer based environment.

      To meet such a need in the design process, most computer aided design software systems are equipped with simulation modules. It has become a common practice for the design engineers to create 3D solid models of parts and assemblies using the CAD software systems, and simulate the performance of the created 3D solid models under a defined operational environment. It is well known that ANSYS is a finite element analysis software system developed in 1970 for structural physics that could simulate static, dynamic and thermal problems. With the rapid advancement of computer technology, the integration of the CAD systems with the ANSYS finite element analysis system has already become a reality to meet the pressing need of industry to revolutionize its product development process. The ANSYS Workbench platform is now the backbone for delivering a comprehensive and integrated simulation system to meet the need of the design engineers in workplace and the need of the research engineers in the scientific field of exploring the new and innovative technology.

      This book focuses on the ANSYS Workbench system, which originates from the development of the ANSYS Parametric Design Language System or ANSYS APDL system. The ANSYS APDL system is the scripting language allows users to automate common tasks or build the physical models in terms of parameters (variables). While APDL is the foundation for ANSYS, containing the core simulation tools, to provide the FEA tools design users in their day-to-day analyses, its lack of the model visualization prevents from its widespread use in the engineering community. It is not uncommon that only a few senior engineers take the fullest responsibility of running FEA when using ANSYS APDL. The introduction of the ANSYS Workbench system just breaks such a barrier by providing the design modeling tool, which follows the common practice in 3D solid modeling widely used by the CAD community in engineering. Furthermore, the ANSYS Workbench system combines the strength of the ANSYS core simulation tools with its new design modeling tool to create a unique environment where the integration of design and evaluation is naturally and interactively implemented. In Section 1.2, we decompose the ANSYS Workbench system into two major parts: DesignModeler and Analysis/Component Systems. A brief introduction to the Analysis and Component Systems is covered in Section 1.2. In Section 1.3, a brief introduction to the DesignModeler is presented with focus on the feature-base modeling approach.

1.2 Introduction to Analysis Systems and Component Systems

The ANSYS Workbench Platform can be viewed as two parts: DesignModeler and Analysis/Component Systems. . When a user opens Workbench, the Project screen is on display, as shown below. On the left side. **Toolbox** is on display. Next to **Toolbox** is **Project Schematic**.

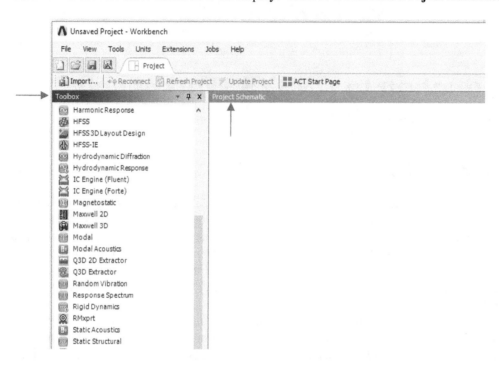

As shown below, the **Analysis Systems** and the **Component Systems** are listed in the **Toolbox**. The **Analysis Systems** contains the different types of systems, such as Explicit Dynamics, Modal, Static Structural, etc. Users can select one of the listed analysis systems to work on.

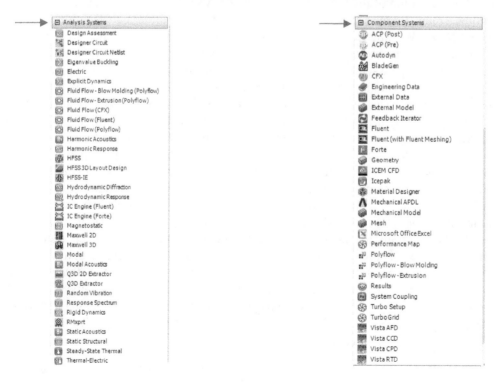

For example, a user wants to perform a static structural analysis. The first step is to highlight the Static Structural analysis system, drag Static Structural from the Toolbox and drop it into the Project Schematic, as shown below. By highlighting the name: Static Structural, a user may give the analysis system a specific name, for example, Torsion of a Tube.

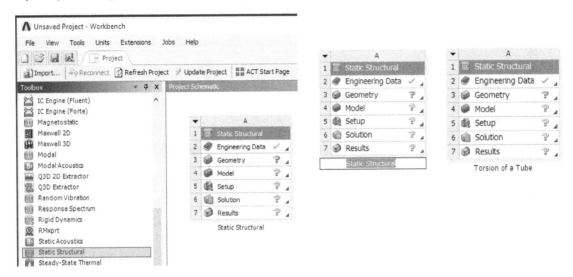

As illustrated above, an analysis system consists of 7 cells, namely, A1 Static Structural, A2 Engineering Data, A3 Geometry, A4 Model, A5 Setup, A6 Solution and A7 Results. These 7 cells are all pre-defined and ready to be populated for the analysis, starting from Engineering Data through Results. For example, double click A2 Engineering Data. We enter the Engineering Data Sources, which is a material library. Click the icon of Engineering Data Sources, Structural Steel is listed as No. 4 type of material with a check (activated) symbol, indicating Structural Steel is active, and is assigned to the geometric model automatically. In the Engineering Data Sources, material properties, such as Young's Modulus and Poisson's Ratio are presented in the **SI** units.

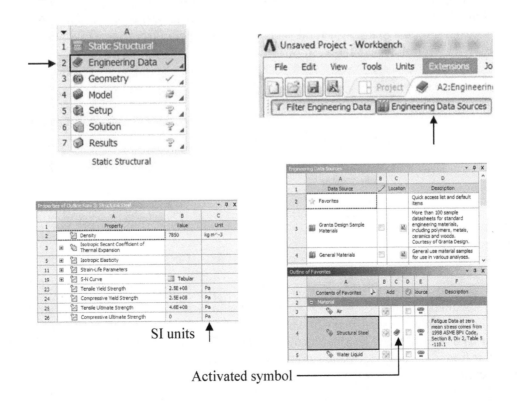

A component system is different from an analysis system because a component system represents only a subset of a complete analysis system. For example, we highlight the **Geometry** component system listed under **Component Systems**, drag **Geometry** from the **Toolbox** and drop it into the **Project Schematic**, as shown below.

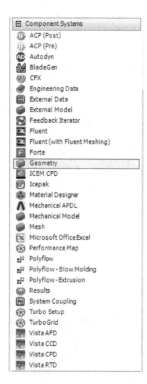
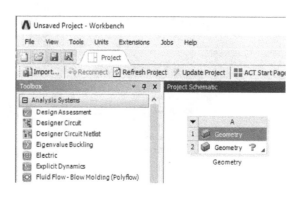

A user may work with this Geometry component system to create a 3D solid model for a part being designed. Afterwards, the user may select the Static Structural analysis system and drag it from the Toolbox. Before dropping it into the Project Schematic, look for the phrase "Share A2". When the phrase appears, drop the Static Structural analysis system. Both Geometry and Static Structural are displayed in the Project Schematic, as shown below.

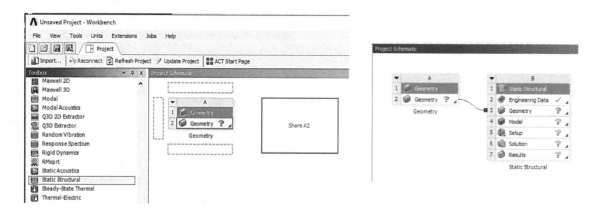

This unique feature of ANSYS Workbench allows the user to create and link to a second system. For example, the same geometric model created under the Geometry component system can be linked to a Steady-State Thermal analysis system. By highlighting Steady-State Thermal, and dragging it from the Toolbox. Make sure the phrase "Share A2" is on display before dropping it to the Project Schematic, as shown below.

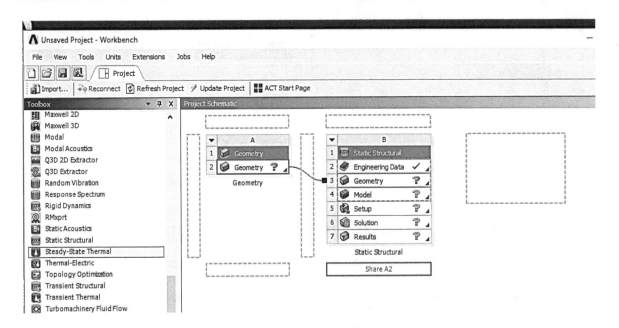

As shown below, the Geometry component system is shared by both Static Structural and Steady-State Thermal. Such a unique feature of ANSYS Workbench allows users to perform two (2) different system analyses. One of such examples presented in this book is a combined thermal and mechanical analysis discussed in Chapter 3. It is important to note that systems are added from left-to-right, and from top-to-bottom. When running an FEA analysis, all data transfer occurs from left (upstream) to right (downstream).

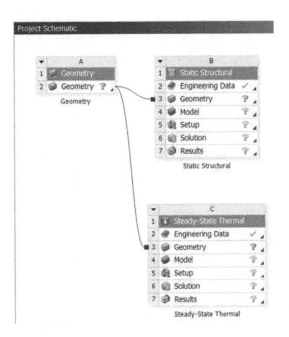

Another unique feature of ANSYS Workbench is "Duplicating Systems." To create a duplicate system of an existing system, in which all cells can be edited independently of the original system, a user may just click the system header and select Duplicate, as shown below. The duplicated system is named *Copy of Static Structural*, as shown. All data in the 7 cells in the duplicated system are shared with the 7 cells in the original system to allow users to investigate an alternative design from Engineering Data to Results. All modifications to the duplicated system can be made independently.

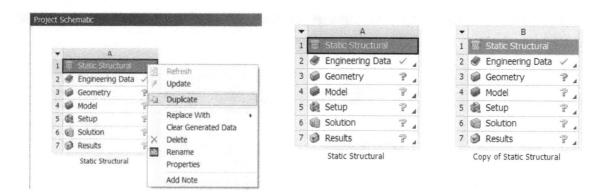

1.3 Introduction to the DesignModeler Geometry Editor

The second major part of the ANSYS Workbench platform is the DesignModeler. A CAD model is a must-to-have component to perform an FEA run. Because of this reason, all analysis systems and several component systems, such as Geometry, begin with a geometry-definition step. In ANSYS Workbench, users can define the geometry using three different methods. The first method is to create a geometry model from scratch. In the Project Schematic, a user needs to highlight the Geometry cell, right-click mouse to display the context menu, and select New DesignModeler Geometry. Afterwards, the DesignModeler window will be on display. The default unit system is Meter. The user may select Millimeter or Inch systems, based on his/her needs. This textbook uses this method to initiate the geometry creation.

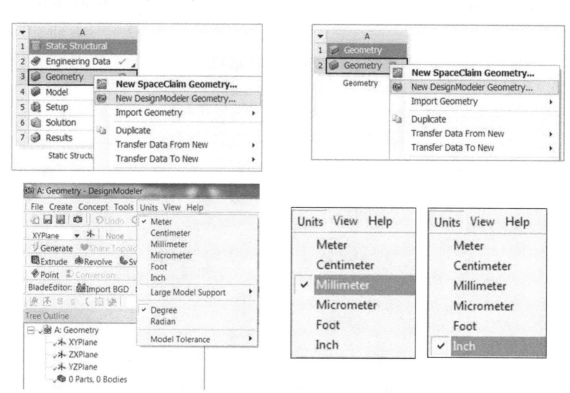

The second method is to import a CAD model that is already created by a CAD system to the DesignModeler. It is important to note the create CAD model has to be saved in a format accepted by ANSYS Workbench. One of the popular formats is IGES (Initial Graphics Exchange Specification). In the Project Schematic, a user needs to highlight the Geometry cell, right-click mouse to display the context menu, and select Import Geometry, and use Browse to locate the CAD file.

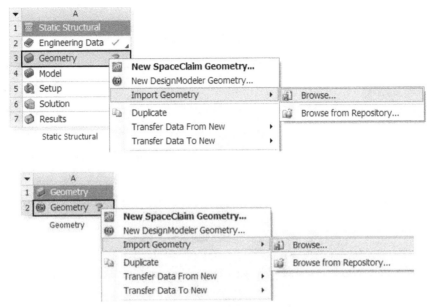

The third method is to import the CAD model directly from an active CAD session that is already running on the same machine. As shown below, we assume that Creo Parametric CAD session is running on a machine where ANSYS Workbench is installed. On the top menu of Creo Parametric, ANSYS 19.2 is on display. Click ANSYS 19.2 will lead to another window where Workbench is on display. Click the icon of Workbench will lead to start loading ANSYS Workbench, as shown below. In this way, the CAD model is directly imported to the ANSYS Workbench. Certainly, this requires several Plug-Ins when the ANSYS Workbench is installed on the same machine. We will discuss the process of adding Plug-Ins in Chapter 7.

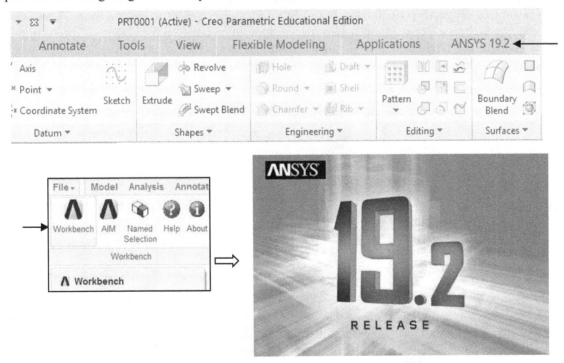

As illustrated above, users can launch ANSYS Workbench directly from a CAD system, such as Creo Parametric, SolidWorks, Autodesk Inventor and NX when using method 3. The unique feature of data integration of CAD and FEA systems is displayed in the following figure.

Initial Graphics Exchange Specification (IGES)

Up to this point of the introduction to the ANSYS Workbench, users may have already noticed that there are two (2) geometry editors in ANSYS Workbench for users to choose from. These two (2) geometry editors are DesignModeler Geometry Editor and SpaceClaim Geometry Editor. In this textbook, we use the DesignModeler Geometry Editor. Users need to pay special attention when picking the New DesignModeler from the context menu, as shown below.

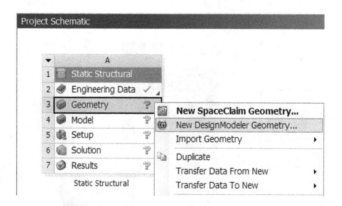

Most users of a CAD system use the feature-based modeling process to create the geometry. When a designer creates a model representing a part, he/she views the model as composed by a set features, such as blocks, cylinders, holes, etc. and an assembly of those features forms a 3D solid model. The ANSYS DesignModeler follows the same approach as the approach used by most CAD systems. We use the creation of a gearbox as an example, illustrating the feature-based approach.

As shown below, we start with creating a hollow-shaped solid called Feature 1. Afterwards, we create a plate solid feature called Feature 2. The ANSYS Workbench will automatically combine or add/merge these 2 features together to form a new feature called Feature 1+2.

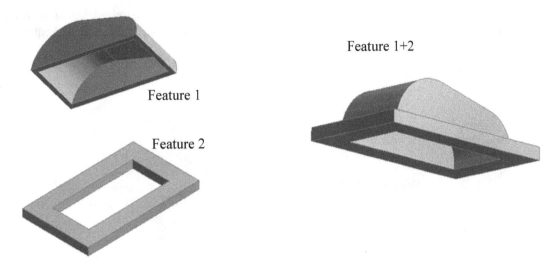

Feature 1

Feature 2

Feature 1+2

Now let use create a box-shaped solid called it Feature 3. The ANSYS Workbench will automatically add/merge Feature 3 to the feature called Feature 1+2 to form a new feature called Feature 1+2+3.

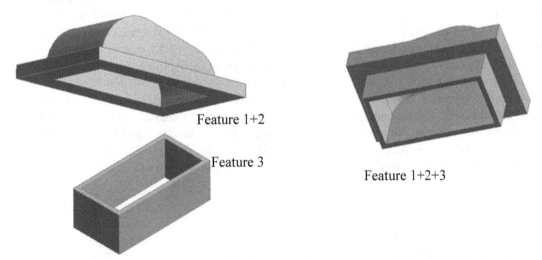

Feature 1+2

Feature 3

Feature 1+2+3

Now let us create a plate solid feature called Feature 4. The ANSYS Workbench will automatically add/merge Feature 4 to the feature called Feature 1+2+3 to form a new feature called Feature 1+2+3+4, which looks like a gearbox.

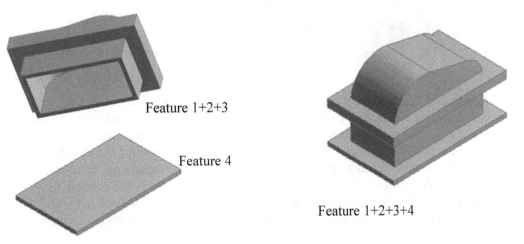

Feature 1+2+3

Feature 4

Feature 1+2+3+4

In the following, we create a new feature called Feature 5, which is a cylindrical-shaped solid, serving as the bearing supports. The ANSYS Workbench will automatically add/merge Feature 5 to Feature 1+2+3+4 to form a new feature called Feature 1+2+3+4+5.

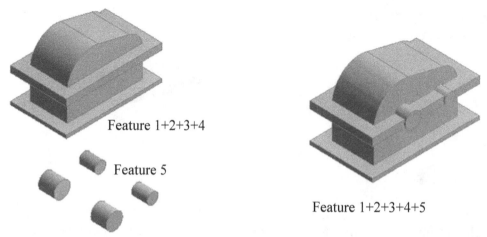

Feature 1+2+3+4

Feature 5

Feature 1+2+3+4+5

Now we need to create two through-all holes to reserve the space for inserting the shaft components. Let us call the hole feature as Feature 6. The ANSYS Workbench will use a subtraction operation to remove the material representing Feature 6 from Feature 1+2+3+4+5 to form a new feature called F[(1+2+3+4+5)-6], as shown in the following figure. For the readers of this textbook, who have learned the feature-based modeling approach from Creo Parametric, or Solidworks or Autodesk Inventor or NX, the process of creating the gearbox in the ANSYS Workbench is not new. For those readers of this textbook, who are new to the feature-based modeling approach, they should feel comfortable to create the geometry using the feature-based modeling approach because the graphic visualization is so clear and powerful to understand how to create a 3D solid model in the design and analysis process. As a conclusion, the DesignModeler shares a great deal of similarities with most CAD systems in the process of creating 3D solid models. This is also one of the major reasons why there are so many engineers, who are using the DesignModeler to create 3D solid models to run FEA.

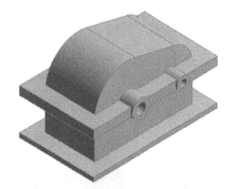

Feature [(1+2+3+4+5)-6]

Besides the feature-based modeling approach, the DesignModeler also offers certain operations to facilitate the feature-based modeling process. Take the gearbox as an example, the gearbox should have two portions, namely, the upper portion and the lower portion. The gearbox, as a real-life product, is an assembly of these two portions. When using most CAD systems, the upper portion and lower portion should be created as two independent components. Afterwards, these two portions are assembled to form the gearbox model. Such a process is not needed when using the ANSYS Workbench. The Slice operation offered by the ANSYS Workbench separates the created Feature [(1+2+3+4+5)-6] to form two solid bodies just through one operation of slicing, as shown below.

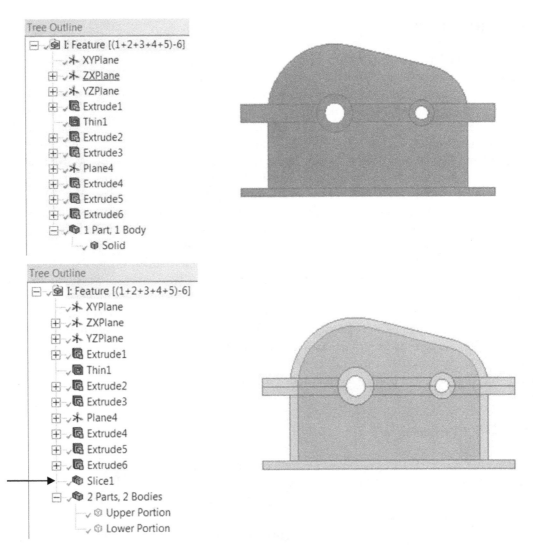

Examining the information shown in the above figure. The two Tree Outlines list the six features represented by the 6 Extrudes, which were created in order. The differences between the 2 Tree Outlines are the Slice1 operation listed in the second Tree Outline, and 1 Part, 1 Body listed in the first Tree Outline and 2 Parts, 2 Bodies listed in the second Tree Line. The 2 pictures clearly show that one is a single solid or a component model, and the other is an assembly of 2 solids or an assembly model. The DesignModeler capabilities in creating the geometry are clearly demonstrated. The 3D solid model created using the DesignModeler also enjoys the fullest utilization of the core simulation tools well developed in the ANSYS Workbench platform. The following figure shows the mesh generation for the gearbox model. With the versatile meshing capabilities, users can easily choose the mesh types, such as the Tetrahedra mesh and the Hexahedra mesh.

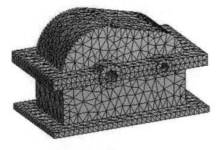

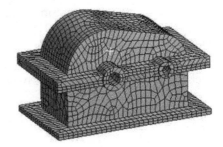

All-Tetrahedra Mesh   Hexahedra Dominant Mesh

1.4   Summary

In Chapter 1, we have presented a comprehensive picture on the ANSYS Workbench platform. As an FEA application software system, the ANSYS Workbench integrates the strength of the ANSYS core simulation tools with the modeling capabilities of the CAD software system. Such an integrated system provides engineers with a unique platform to meet the need of industry to revolutionize the product development process. Such an integrated system also provides the students and educators with a unique teaching/learning environment, under which academic projects can be effectively carried out and their results can be vividly displayed. There is no doubt that the ANSYS Workbench platform represents a much-needed software system for people in the engineering community to learn and apply the new knowledge to solve technical problems presented in the real-life environment.

Learning a new software application, such as the ANSYS Workbench, is exciting because we will gain the knowledge and skills. Such a learning process also demands the consistent efforts of a user to follow a set of guided instructions. The two most valuable references are ANSYS Parametric Design Language Guide and ANSYS Workbench User's Guide published by ANSYS, Inc. Several important references are also listed in Section 1.5 References. The author of this textbook hopes the examples and case studies presented in this book help the readers in their learning process. Remember the phrase: "Practice, Practice and Practice!" The process of learning a new software application is the process of accumulating the experience gained from the repeated click-by-click process. Do not be afraid of making mistakes while clicking. Very often, a user may learn more from making those mistakes since the process of making corrections enforces the user to think more and deeper, thus gaining the knowledge of how to use the ANSYS Workbench platform.

1.5   References

1. ANSYS Parametric Design Language Guide, Release 15.0, Nov. 2013, ANSTS, Inc.
2. ANSYS Workbench User's Guide, Release 15.0, Nov. 2013, ANSYS, Inc.
3. X. L. Chen and Y. J. Liu, Finite Element Modeling and Simulation with ANSYS Workbench, 1st edition, Barnes & Noble, January 2004.
4. J. W. Dally and R. J. Bonnenberger, Problems: Statics and Mechanics of Materials, College House Enterprises, LLC, 2010.
5. J. W. Dally and R. J. Bonnenberger, Mechanics II Mechanics of Materials, College House Enterprises, LLC, 2010.
6. X. S. Ding and G. L. Lin, ANSYS Workbench 14.5 Case Studies (in Chinese), Tsinghua University Publisher, Feb. 2014.
7. G. L. Lin, ANSYS Workbench 15.0 Case Studies (in Chinese), Tsinghua University Publisher, October 2014.
8. K. L. Lawrence. ANSYS Workbench Tutorial Release 13, SDC Publications, 2011.
9. K. L. Lawrence. ANSYS Workbench Tutorial Release 14, SDC Publications, 2012
10. Huei-Huang Lee, Finite Element Simulations with ANSYS Workbench 14, Theory, Applications, Case Studies, SDC Publications, 2012.
11. Huei-Huang Lee, Finite Element Simulations with ANSYS Workbench 16, Theory, Applications, Case Studies, SDC Publications, 2015.
12. E. H. Dill, Finite Element Method for Mechanics of Solids with ANSYS Applications, Barnes & Noble, September 2011.
13. Jack Zecher, ANSYS Workbench Software Tutorial with Multimedia CD Release 12, Barnes & Noble, 2009.
14. G. M. Zhang, Engineering Analysis with Pro/Mechanics and ANSYS, College House Enterprises, LLC, 2011.

1.6   Exercises

1.   How to launch the ANSYS Workbench 19.2.
     Method 1: On the computer screen, look for the logo Workbench 19.2. If this logo is on display, make a double click on it to launch Workbench 19.2.

Method 2: If this logo is not on display, click the start menu of your computer. Look for Workbench 19.2. If Workbench 19.2 is on the list, make a left-click to launch it.

Click Workbench 19.2 ⟶

Click the start menu ⟶

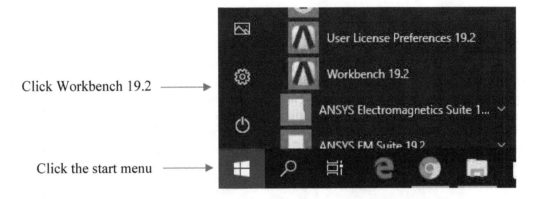

Method 3: If Workbench 19.2 is not on the list, type ANSYS Workbench in the search box. The best match box should list ANSTS Workbench as the search result. Just press the Enter key to start Workbench 19.2.

Search Result ⟶

Type ANSYS Workbench

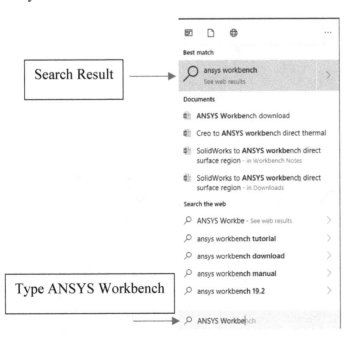

2.   Launch Workbench 19 first. The Toolbox and Project Schematic are on display. Assume that you want to perform a Steady-State Thermal analysis. Highlight Steady-State Thermal listed in Analysis Systems, drag it from the Toolbox and drop it into the Project Schematic.

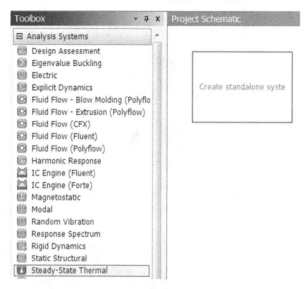

3.   Launch Workbench 19 first. The Toolbox and Project Schematic are on display. Assume that you want to perform two analyses. The first analysis is Static Structural and the second analysis is Modal. Do the following: highlight Static Structural, drag it and drop it into the Project Schematic. Afterwards, highlight Modal, drag it to the Solution of Static Structural and wait for Share A2 to A4. Afterwards, drop it into the Project Schematic. Your work should look like the one shown below.

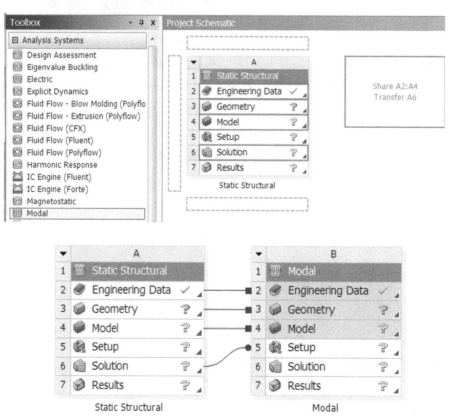

4.   Launch Workbench 19 first. The Toolbox and Project Schematic are on display. Assume that you want to perform the Static Structural analysis. Highlight Static Structural, drag it and drop it into the Project Schematic. Afterwards, double click the Engineering Data cell. You enter the material library. Click Engineering Data Sources > General Materials. Fill the table with the information copied from the material library. Make sure the units are also filled in the table.

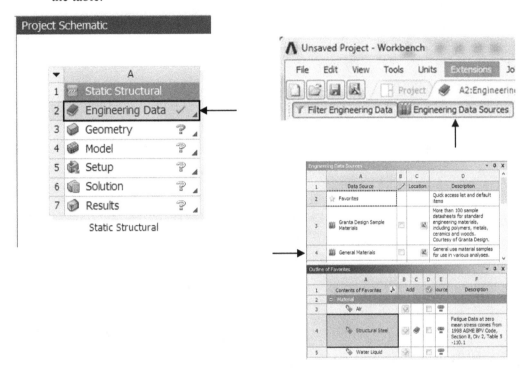

Static Structural

Material Type	Young's Modulus	Poisson's Ratio
Structural Steel		
Gray Cast Iron		
Magnesium Alloy		
Concrete		

CHAPTER 2

Static Structural Analysis

2.1 Introduction

As discussed in Chapter 1, to perform finite element analysis, we need a geometric model which is a representation of the physical object under study. When the geometric model is available, finite element analysis can be carried out. In this chapter, we present six case studies using static structural analysis. Section 2.2 presents the first case study using a cantilever beam subjected to a pressure load. We start with the creation of a 3D solid model representing the beam, assign material properties, create mesh, add the displacement constraint, define the pressure load, perform FEA, and display the results obtained from the FEA run. Section 2.2 reviews the method of creating imprint faces to be used as surface regions to define load conditions. The geometric model in Section 2.3 represents a welding structure. We use the U.S. Customary Unit system (inch). The geometric model in Section 2.4 provides basic steps in the process of geometry creation. We also introduce the procedure to assign a new type of material, besides the default material assignment of Structural Steel. Section 2.5 presents the revolving operation to create the tube geometry where a torque load is applied. Section 2.6 presents the sweeping operation to create an Allen wrench where pressure loads act on two surface regions. Section 2.7 presents the design of a connecting rod, illustrating the application of bearing loads. The material covered by these six case studies provide the readers with a comprehensive understanding of the fundamentals related to static structural analysis in ANSYS Workbench. Section 2.8 summarize the information covered by the 6 case studies.

2.2 Cantilever Beam Subjected to a Pressure Load

In this case study, we perform FEA on the cantilever beam shown below. We use the millimeter unit system to create the 3D solid model. The material type is structural steel (system default, Young's modules: 2e+11 Pascal and Poisson's ratio: 0.30). The left end of the beam is fixed to the wall. The right end is set to free. There is a pressure load acting on the top surface of the cantilever beam. Its magnitude is 0.5 MPa in the downward direction. We use the Design Modeler of Workbench to create a 3D solid model and perform FEA.

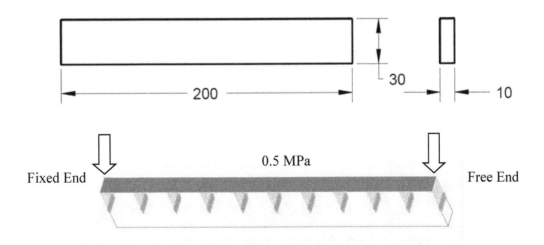

To start a design process using ANSYS Workbench, select the "**Workbench 19**" icon. This icon is displayed on the start menu. Click ANSYS 19 > Workbench 19. Users may directly click Workbench 19 when the symbol of Workbench 19 is displayed on the desktop of the computer.

 Workbench 19.2

Step 1: Become familiar with the Workbench Application Window.

The following figure presents the window on display after a user launches Workbench 19. The window offers the following features:

1. Provides a Main Toolbar, from which a new file can be opened for initiating the application process, such as creating a new project or save a project.
2. Provides a Toolbox window where users may select a specific application, such as Static Structural or Steady-State Thermal, etc.
3. Provides a Project Schematic Area where the selected application(s) are on display.

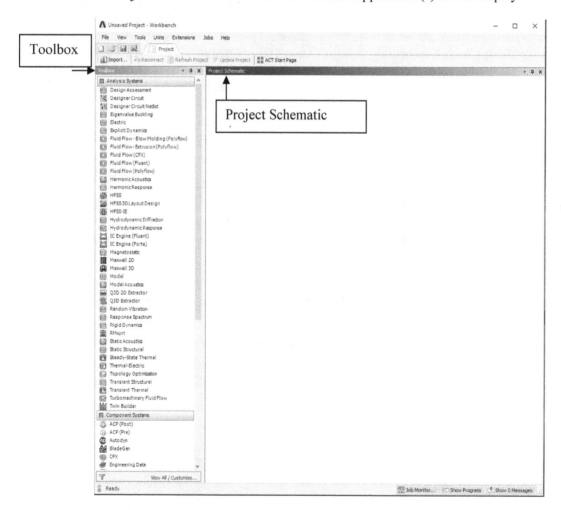

Step 2: From **Toolbox > Analysis Systems**, highlight **Static Structural**, drag it and drop it to the **Project Schematic** area.

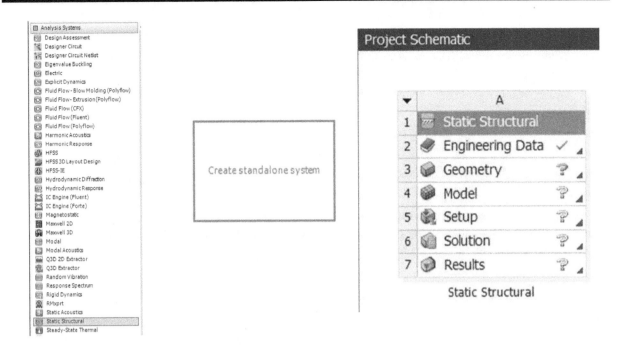

Step 3: Start Design Modeler and create a 2D sketch.

Right-click the **Geometry** cell and pick **New DesignModeler Geometry** to start **Design Modeler**. Users may double click **Geometry** to directly enter **Design Modeler** if **New SpaceClaim Geometry** is not present. Click the **Units** drop down on the top menu, and select **Millimeter**.

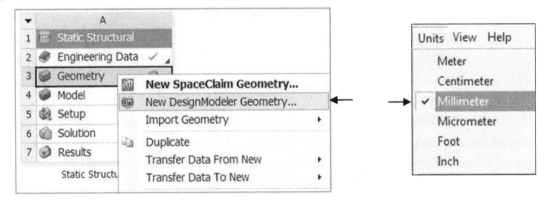

In **Tree Outline**, select **XYPlane** as the sketching plane.

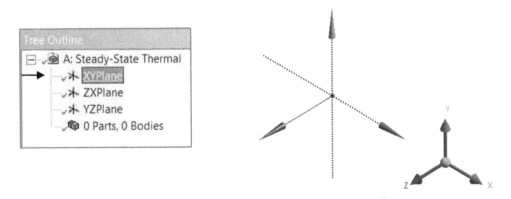

To orient the sketching plane, from the top menu, click the icon of **Look At** so that the selected sketching plane is oriented to be parallel to the screen and ready for preparing a sketch.

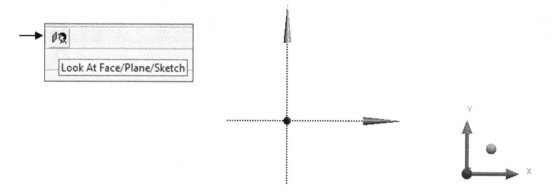

At the bottom of **Tree Outline**, click the icon of **Sketching**. The Sketching Toolboxes window appears. Click the box called **Draw**. Select **Rectangle**. Sketch a rectangle and set its left and lower corner at the origin of the displayed coordinate system, as shown below.

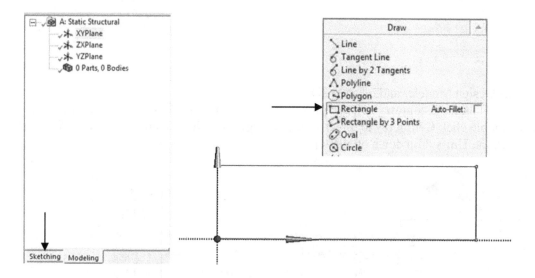

To add dimensions, click the box called **Dimensions.** In **Dimensions**, select **General**. Click the top edge and a left click to position the horizontal size dimension. Click the left edge and a left click to position the vertical size dimension. Specify 200 and 30, respectively. These 2 values are displayed in the **Details View** window.

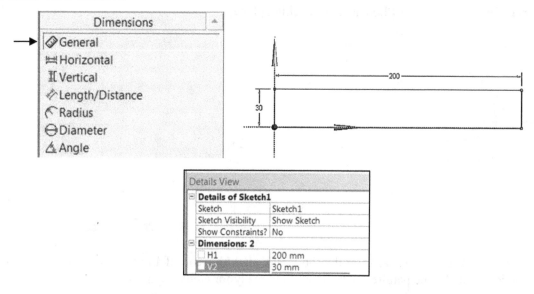

If the numerical values of 200 and 30 are not on display, and instead, two symbols of H1 and V2 are on display, the user needs to check **Value** instead of **Name** from the Display section, as shown below.

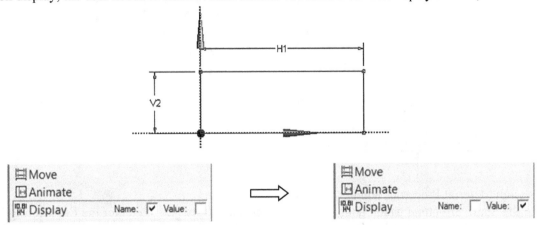

Upon completion of the sketch, click the icon of **Extrude**. Click the created sketch (Sketch1) from **Tree Outline**, and click **Apply** in **Details View**. Specify 10 as the extrude thickness. Click the **Generate** icon to obtain the 3D solid model of the beam as shown.

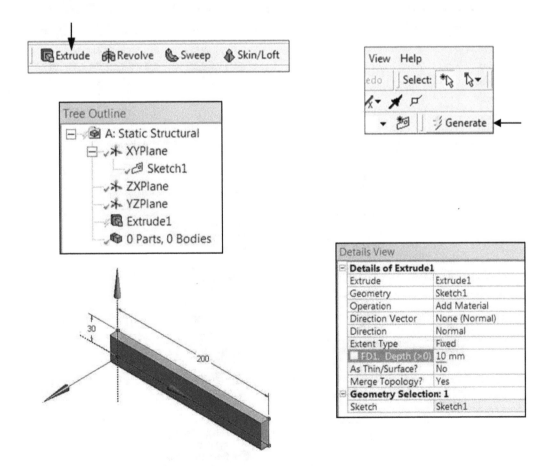

Upon completing the creation of this 3D solid model, we need to close Design Modeler. Click **File** > Close **DesignModeler**.

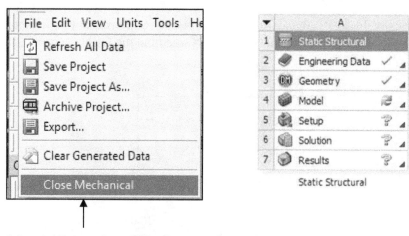

Step 4: Assign Material Properties and Perform FEA

In the Static Structural panel, double click **Engineering Data A2**. We enter the Engineering Data Mode. Click the icon of **Engineering Data Sources**. In Outline of Favorites, Structural Steel is listed and activated with the activated symbol, listed in the column C location by system default.

The material properties of **Structural Steel** are listed in the following table. The Young's Modules value is 2E11 Pascal and the Poisson's Ratio is 0.3. Users may notice that all the listed units are in the SI unit system.

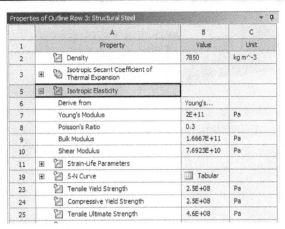

Afterwards, click the icon of **Engineering Data Sources** and click the icon of **Project** to go back to Project Schematic. Double-click the **Model A4** cell. We enter the **Mechanical Mode**. Check the unit system to be: Metric (mm, t, N, s, mV, mA).

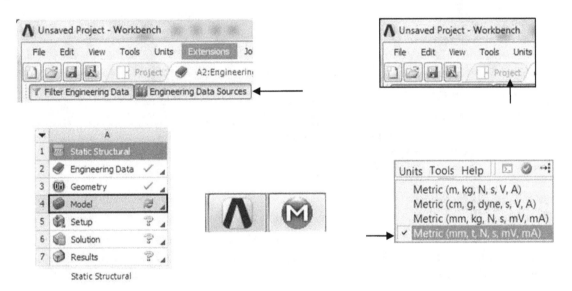

From the Project tree, highlight **Mesh.** In the Details of "Mesh" window, expand **Sizing**, and set Resolution to 4. Right-click **Mesh** to pick **Generate Mesh**.

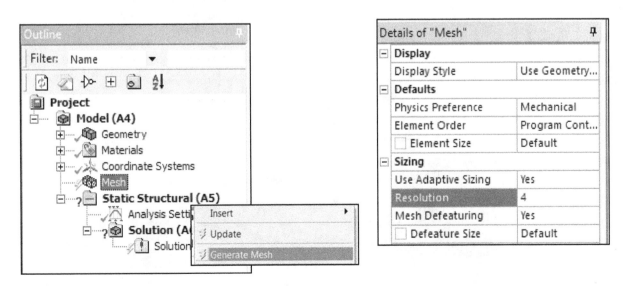

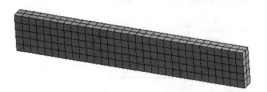

From the Project tree, highlight **Static Structural (A5)** and right-click to pick **Insert**. Afterwards, select **Fixed Support**. From this screen, pick the surface on the left end of the beam, as shown. As a result, this 3D solid model has a fixed support on the left end.

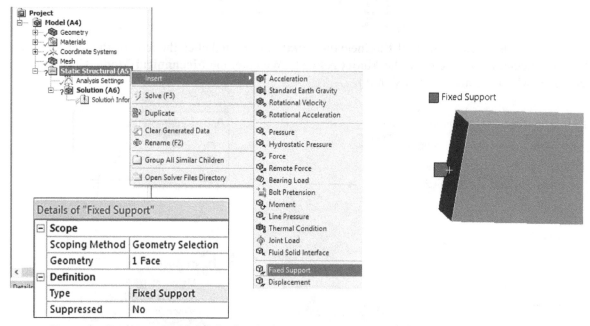

From the Project tree, highlight **Static Structural (A5)** and right-click to pick **Insert**. Afterwards, select **Pressure**. From this screen, pick the top surface of the rectangular block, as shown, and then select **Apply** for Geometry. Specify 0.5 MPa as the magnitude of the pressure load. The direction is downwards.

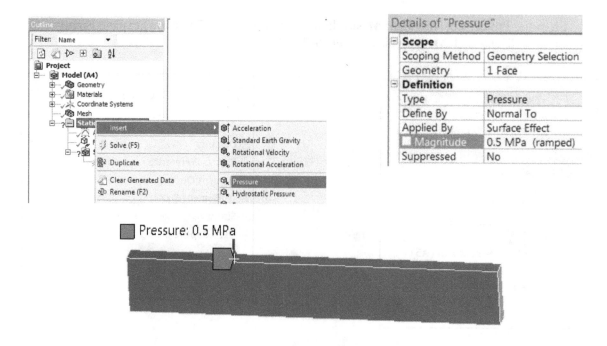

From the Project tree, highlight **Solution (A6)** and right-click to pick **Solve**.

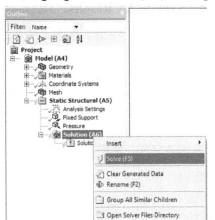

To plot the normal stress in the X direction, highlight **Solution (A6)** and pick **Insert**. Right click on **Solution (A6)**. Afterwards, highlight **Stress** and pick **Normal** (select the X direction). Click **Evaluate All Results.** The maximum value is 83.8 MPa located at the fixed end. Users may need to click the icon of Maximum to let the Max symbol on display.

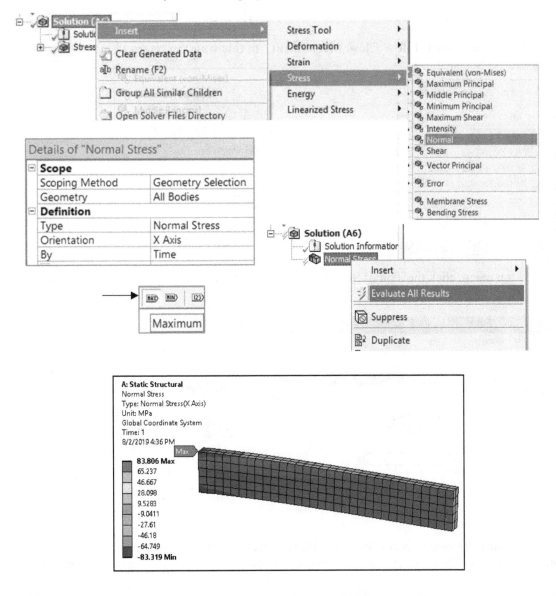

To plot the displacement distribution, highlight **Solution (A6)** and pick **Insert**. Afterwards, highlight **Deformation** and pick **Total.** Click **Evaluate All Results.** The maximum value is 0.227 mm, located at the free end.

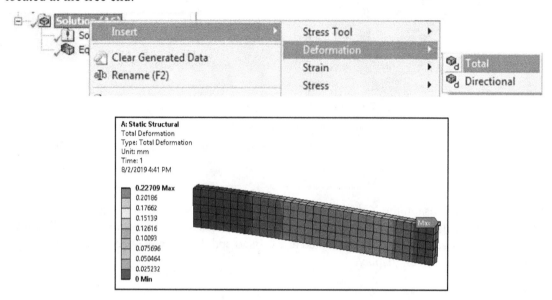

From the top menu, click **File > Close Mechanical**. In this way, we return to the main page of Workbench, or Project Schematic.

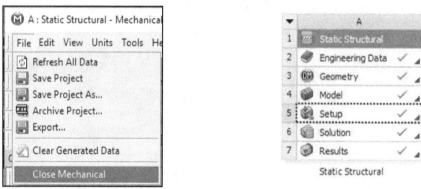

From the top menu, click the icon of **Save Project**. Specify Cantilever Beam as the file name and click **Save**.

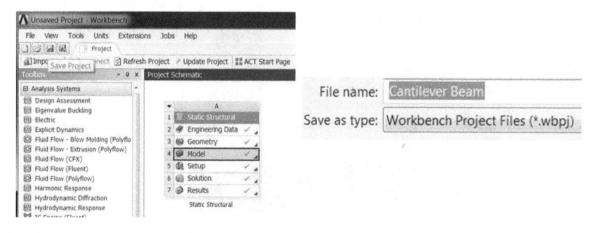

Step 5: Create a rectangular surface region by imprinting a rectangular area to the top surface of the beam. A pressure load will be only acted on the imprinted rectangular area of the cantilever beam.

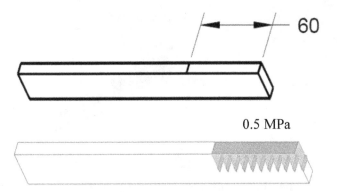

0.5 MPa

In **Project Schematic**, highlight **Static Structural**, change it to **Cantilever Beam**.

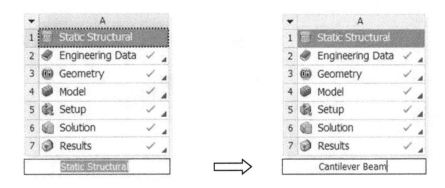

Highlight **Static Structural (A1)**, and right click and select **Duplicate** from the pop up window. Change Copy of Cantilever Beam to Surface Region 1.

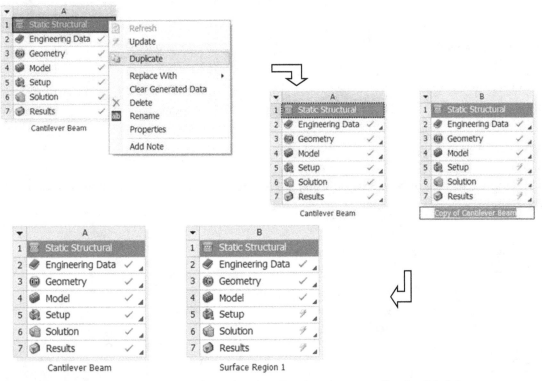

Now, double click **Geometry** from **Surface Region 1** to enter **DesignModeler**.

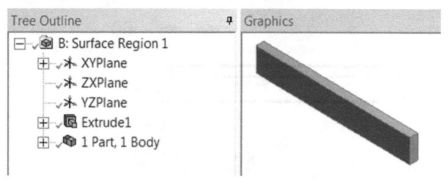

Click the icon of **New Plane**. In Details View, select **From Face** in the Type field. In the Base Plane field, left click to show **Apply** and pick the top surface of the beam model. Click **Apply**. Click **Generate** and Plane4 is created.

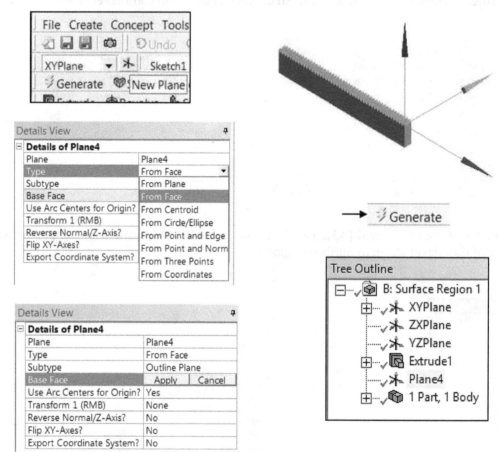

Now, click the icon of **New Sketch**. Click **Sketching > Draw**. Select **Rectangle** to sketch a corner rectangle as shown.

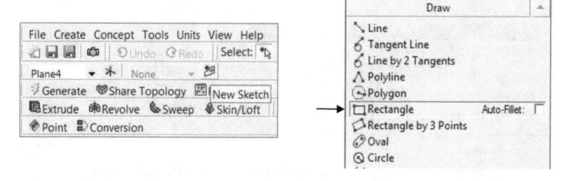

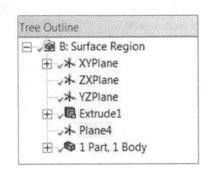

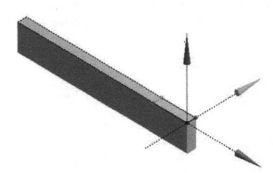

From Sketching Toolboxes, click **Dimensions.** And select **General**. Click the edge and a left click to position the size dimension. Specify as 60.

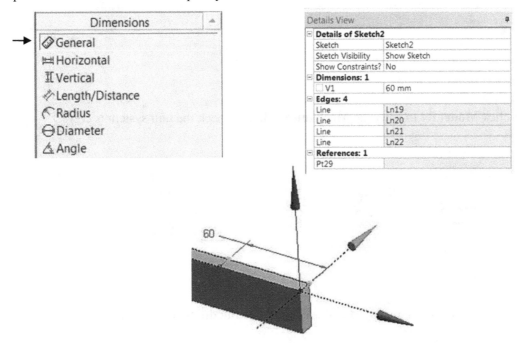

Click **Modeling > Extrude**. Click Sketch2 listed below Plane 4. Click **Apply**. Select **Imprint Faces** from the **Operation** field. Click **Generate**.

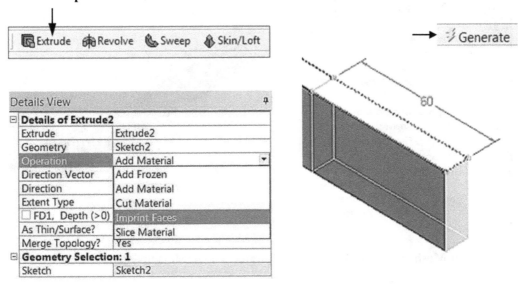

Upon completion of adding an imprinted surface to the beam model, we need to close **Design Modeler**. Click **File** > Close **Design Modeler**, to return to **Project Schematic**.

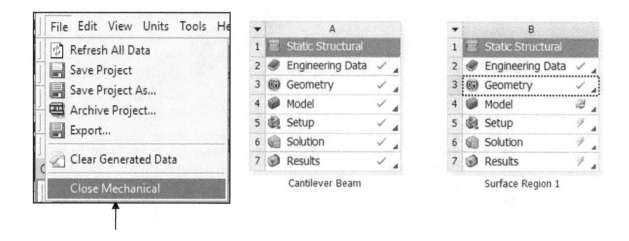

Step 6: Perform FEA
Double-click **Model B4** to enter the **Mechanical Mode**. Check the unit system is correct.

Upon examining the items displayed in the **Project** tree, modifications are needed.

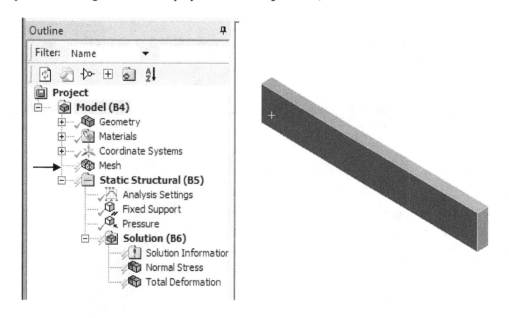

First highlight **Mesh** and right-click to pick **Update**. The mesh is updated.

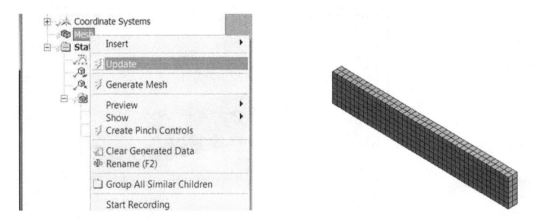

Highlight the **Pressure** load item. In Details of "Pressure", Click **1 Face** listed in Geometry. Pick the imprinted surface, and Click **Apply**.

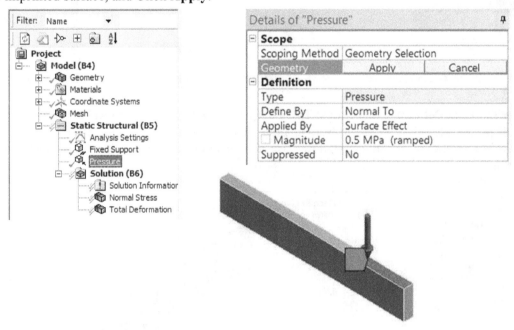

Highlight **Solution (B5)** and right-click to pick **Solve**. Workbench system is running FEA.

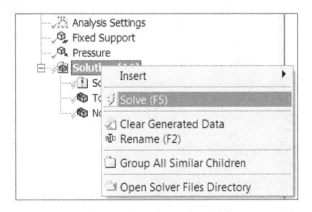

To plot the normal stress in the X-axis direction, highlight **Normal Stress**. The maximum value is 41.9 MPa. Click Max to displace its location. Click Probe to identify the coordinates of the location: X = 0, Y = 30, Z = 10, shown in Graphics Annotations window. Use integer numbers, considering the scale factor to display the deformation.

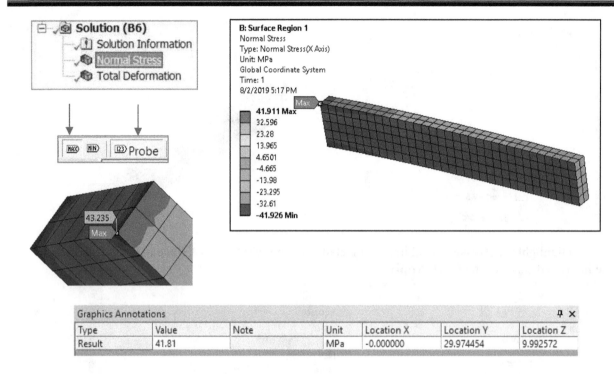

Type	Value	Note	Unit	Location X	Location Y	Location Z
Result	41.81		MPa	-0.000000	29.974454	9.992572

To plot the Total Deformation, highlight **Total Deformation**. The maximum value is 0.14086 mm. Click Max to displace its location. Click Probe to identify the coordinates of the location: X = 200, Y = 20, Z = 10, shown in Graphics Annotations window. Use integer numbers, considering the scale factor to display the deformation.

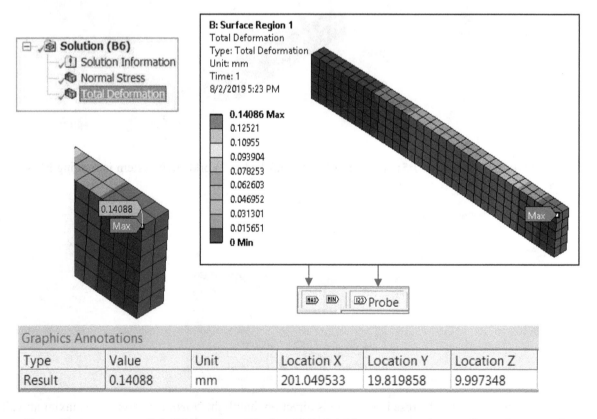

Type	Value	Unit	Location X	Location Y	Location Z
Result	0.14088	mm	201.049533	19.819858	9.997348

From the top menu, click **File** > **Close Mechanical**. In this way, we return to the main page of Workbench, or Project Schematic. From the top menu, click the icon of **Save Project**.

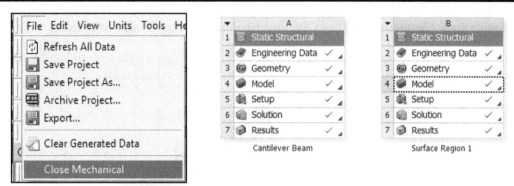

Step 7: Modify the created surface region (60x10) to a surface region (30x10). Afterwards, perform FEA.

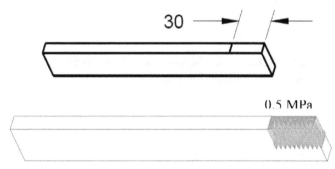

Highlight **Static Structural (B1)**, and right click to pick **Duplicate** from the pop up window. Change Copy of Surface Region 1 to Surface Region 2.

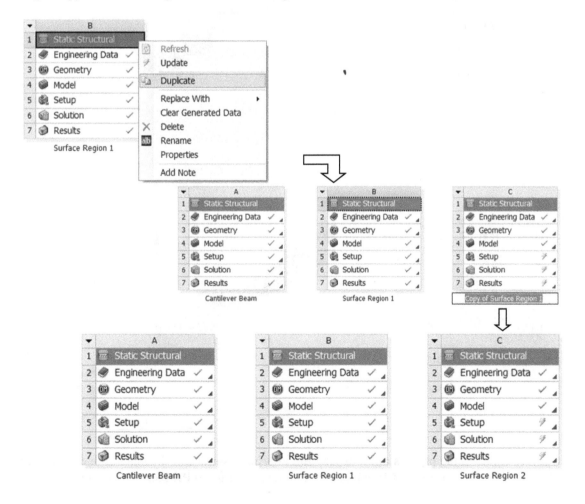

Now double click **Geometry (C2)** from **Surface Region 2** to enter **Design Modeler**.

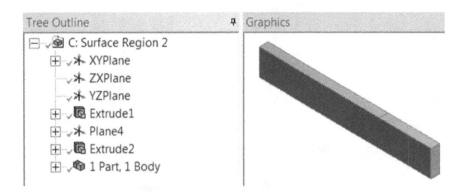

In Tree Outline, highlight Sketch2. The dimension of 60 is on display. In Details View, modify 60 to 30. Click **Generate** to complete the modification process.

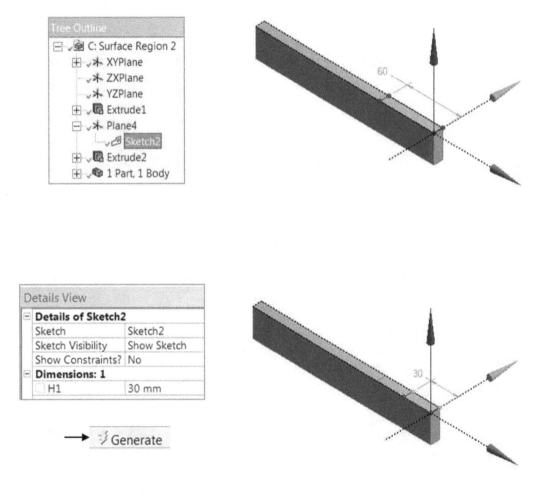

Upon completing the process of modifying the imprinted surface to the beam model, we need to close **Design Modeler**. Click **File** > **Close Design Modeler**, to return to **Project Schematic**.

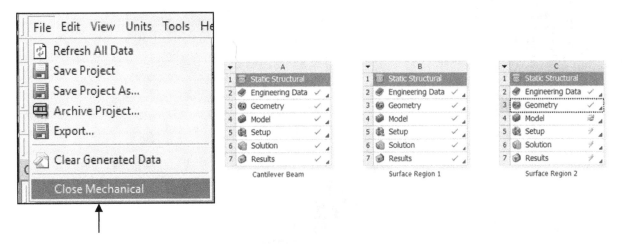

Step 8: Perform FEA

Double-click **Model (C4)**. We enter the **Mechanical Mode**. Check that the unit system is correct.

Upon examining the items displayed in the **Project** tree, modifications are needed.

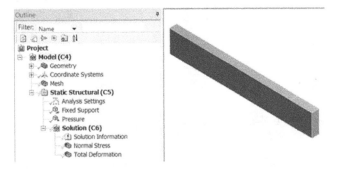

First highlight **Mesh** and right-click to pick **Update**. The mesh is updated.

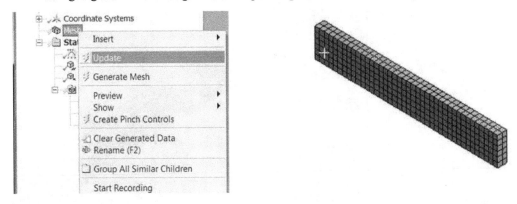

Highlight the **Pressure** load item. In Details of "Pressure", Click **1 Face** listed in Geometry. Pick the imprinted surface, and Click **Apply**.

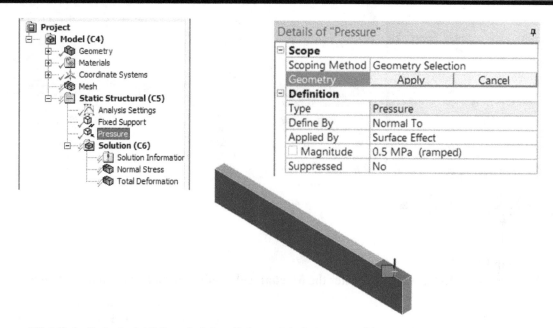

Highlight **Solution (C6)** and right-click to pick **Solve**. Workbench system is running FEA.

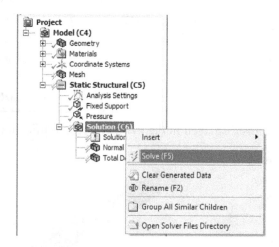

To plot the normal stress in the X-axis direction, highlight **Normal Stress**. The maximum value is 22.7 MPa.

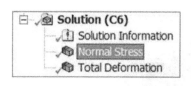

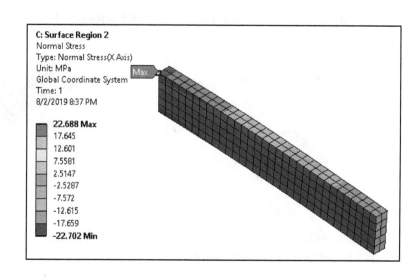

To plot the Total Deformation, highlight **Total Deformation**. The maximum value is 0.080 mm.

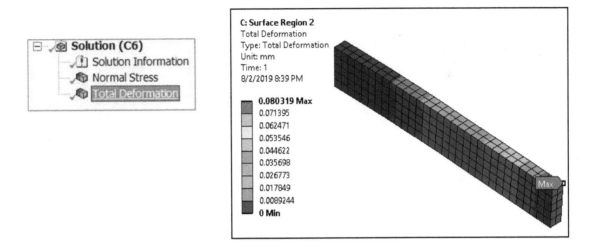

From the top menu, click **File > Close Mechanical**. In this way, we return to the main page of Workbench, or Project Schematic. From the top menu, click the icon of **Save Project.**

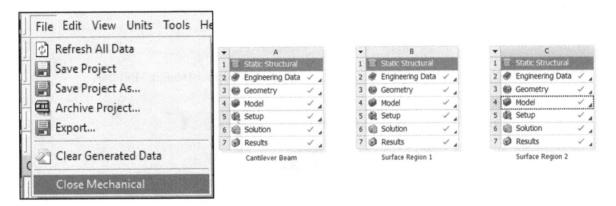

Step 9: Keep the surface region (60 X 10) and add a new rectangular surface region (20 x 10) with a position dimension of 70 with respect to the fixed end surface. Determine the maximum value of total deformation and the maximum value of normal stress in the X direction.

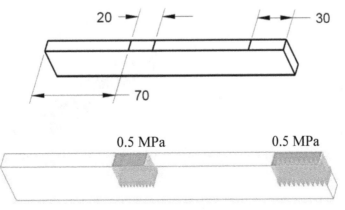

Highlight **Static Structural (C1)**, and right click to pick **Duplicate** from the pop up window. Change **Copy of Surface Region 2** to **Surface Region 3**.

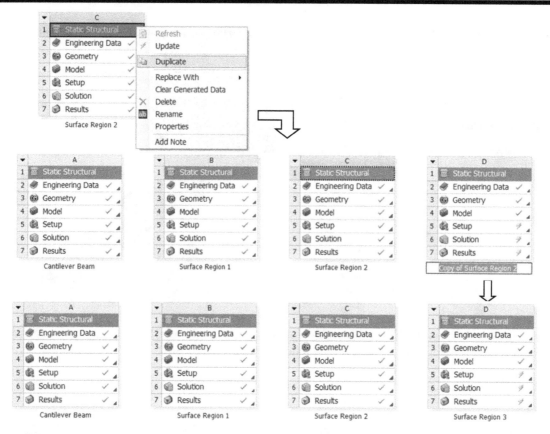

Now double click **Geometry (D2)** from **Surface Region 3** to enter **Design Modeler**.

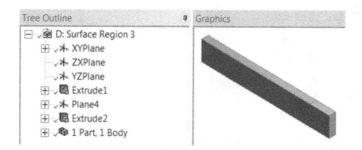

In Tree Outline, highlight Sketch2. The dimension of 30 is on display. This value is correct.

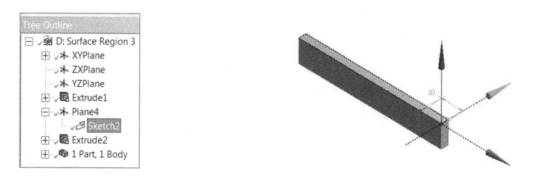

Now add the second surface region (20x10). Highlight Plane4 listed in Tree Outline. Click the icon of **New Sketch**. Click Sketching and select **Rectangle** to sketch a corner rectangle, as shown.

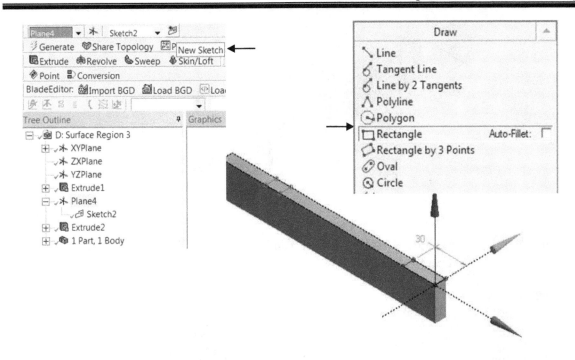

From Sketching Toolboxes, click **Dimensions.** And select **General**. Click the edge and a left click to position the size dimension. Specify 20. Afterwards.

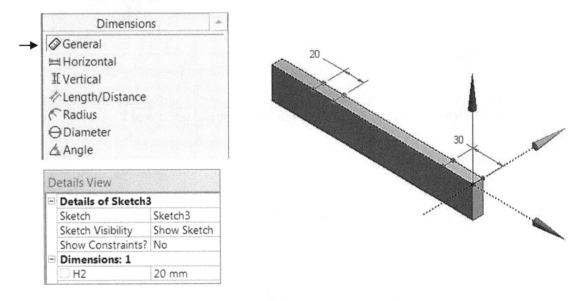

From Sketching Toolboxes, click **Dimensions.** And select **Horizontal**. Click the 2 edges to add a position dimension of 70.

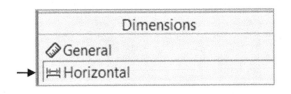

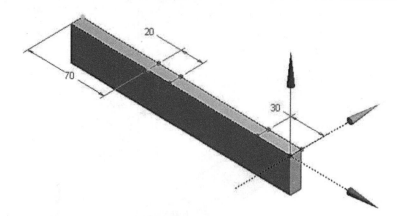

Click **Modeling** > **Extrude**. Highlight **Sketch3** and click **Apply**. Select **Imprint Faces** from the **Operation** field. Click **Generate**.

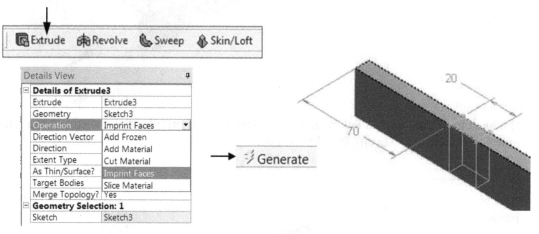

Upon completing the process of adding the new second imprinted surface to the beam model, we need to close **Design Modeler**. Click **File** > Close **Design Modeler**, and go back to **Project Schematic**.

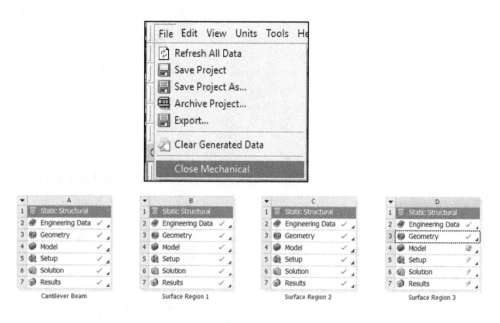

Step 8: Perform FEA

Double-click **Model (D4)**. We enter the **Mechanical Mode**. Check the unit system is correct.

Surface Region 3

Upon examining the items displayed in the **Project** tree, modifications are needed.

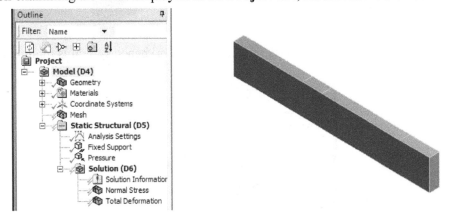

First highlight **Mesh** and right-click to pick **Update**. The mesh is updated.

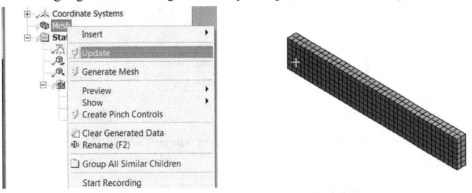

To add a new second pressure load. Highlight **Static Structural (D5)** > **Insert**. Pick **Pressure.** Click **Apply.** While holding down the **Ctrl** key, pick the two surface regions.

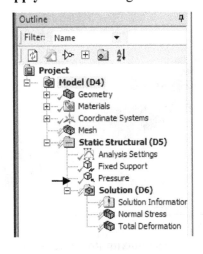

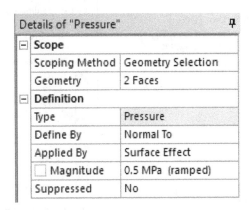

Details of "Pressure"	
Scope	
Scoping Method	Geometry Selection
Geometry	2 Faces
Definition	
Type	Pressure
Define By	Normal To
Applied By	Surface Effect
Magnitude	0.5 MPa (ramped)
Suppressed	No

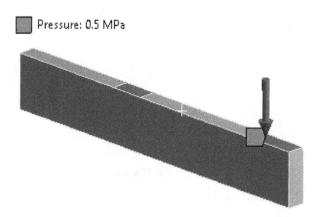

Highlight **Solution (D6)** and right-click to pick **Solve**. Workbench system is running FEA.

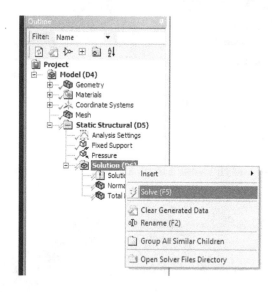

To plot the normal stress in the X-axis direction, highlight **Normal Stress**. The maximum value is 29.4 **MPa.**

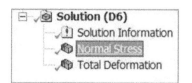

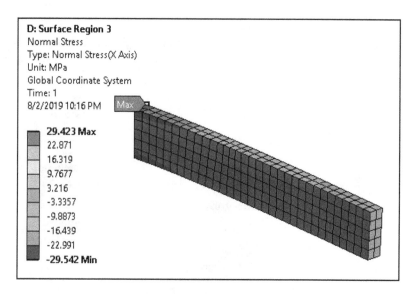

To plot the Total Deformation, highlight **Total Deformation**. The maximum value is 0.093 mm.

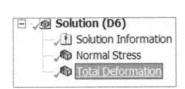

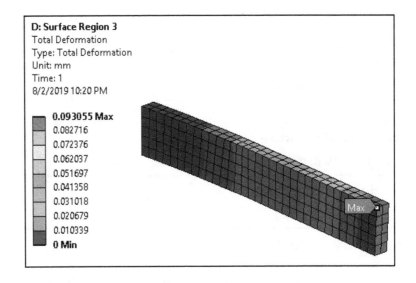

From the top menu, click **File > Close Mechanical**. In this way, we return to the main page of Workbench, or Project Schematic. From the top menu, click the icon of **Save Project**

From the top menu, click the icon of **Save Project**. From the top menu, click **File > Exit**.

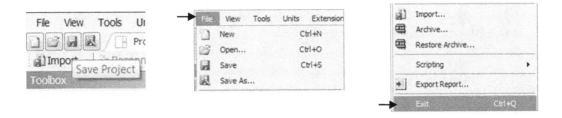

The data listed in the following table summarizes the results obtained from the 4 FEA runs.

Case Study	Max Displacement (mm)	Max Normal Stress X (MPa)
Pressure load on the top surface	0.23	83.8
Pressure load on surface region (60x10 mm^2)	0.14	41.9
Pressure load on surface region (30x10 mm^2)	0.08	22.7
Pressure load on 2 surface regions	0.09	29.4

Let us verify the accuracy of the results obtained from the FEA runs. Take the case study of the pressure load acting on the entire top surface. From the table data, the obtained magnitude of maximum displacement is 0.23 mm. The formula used in this verification process is taken from a textbook on Mechanics of Materials [1].

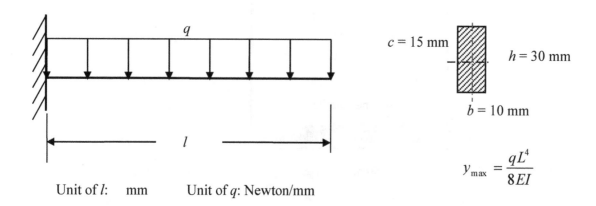

Unit of l: mm   Unit of q: Newton/mm

$$y_{max} = \frac{qL^4}{8EI}$$

The Young's modulus value is 200 GPa or 200x10^3 MPa. The section moment of inertia is 22500 mm^4. Pay attention to the units used when evaluating the displacement from the uniform load.

$$I = \frac{1}{12}bh^3 = \frac{1}{12}(10)(30)^3 = 22500 \ (mm^4) = 2.25\text{x}10^{-8} \ (m^4)$$

$$q = (0.5\frac{N}{mm^2})(10mm) = 5\frac{N}{mm} = 5000\frac{N}{m}$$

$$L = 200mm = 0.2m$$

$$E = 200GPa = 200\text{x}10^3 \ MPa = 2\text{x}10^5 \ \frac{N}{mm^2} = 200\text{x}10^9 \ Pa = 2\text{x}10^{11} \ \frac{N}{m^2}$$

$$y_{max} = \frac{qL^4}{8EI} = \frac{(5\frac{N}{mm})(200mm)^4}{8(2x10^5 \frac{N}{mm^2})(22500mm^4)} = 0.22 \ (mm)$$

This calculated value of 0.22 mm is very close to the maximum displacement value obtained from the FEA run, which is 0.23 mm. Remember: FEA is a numerical method for finding approximate solutions.

2.3 Welding Structure Subjected to a Pressure Load

In this case study, we perform FEA for the welding structure shown below. The units in the engineering drawing are in inches. The material type is structural steel (system default, Young's modules: 2e+11 Pascal or 29e6 psi, and Poisson's Ratio: 0.30. The two holes are fixed to ground. A pressure load of 10 psi is acting on the top surface of the welding structure. Use the Design Modeler of Workbench to create a 3D solid model and perform FEA.

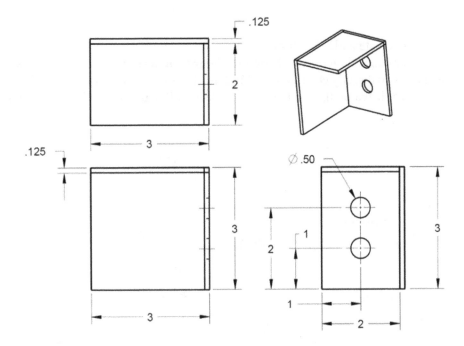

Step 1: Start Workbench.
To start a design process using ANSYS Workbench, select the icon, called "**Workbench 19**". This icon is displayed on the start menu. Click **ANSYS 19** > **Workbench 19**. Users may directly click Workbench 19 when the symbol of Workbench 19 is on display.

Workbench 19.2

Step 2: From **Toolbox**, highlight **Static Structural** and drag it to the **Project Schematic** area.

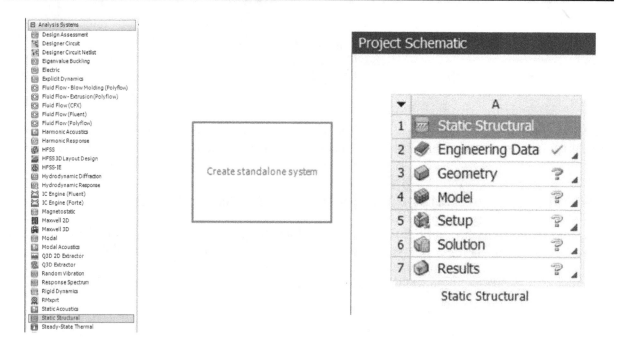

Step 3: Start Design Modeler and create a 2D sketch first.

Right-click the **Geometry** cell and pick **New DesignModeler Geometry** to start **Design Modeler**. Users may double click **Geometry** to directly enter **Design Modeler** if New **SpaceClaim Geometry** is not present. Click the **Units** drop down listed on the top menu, and select **U. S. Customary Units (Inch, lbm, lbf, ºF, s, V, A)**.

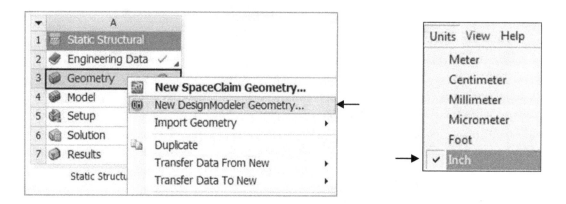

In **Tree Outline**, select **XYPlane** as the sketching plane.

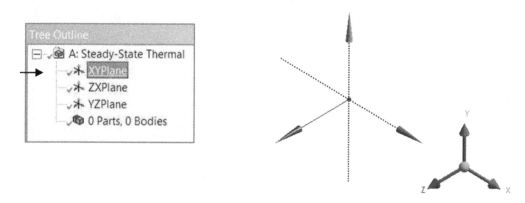

To orient the sketching plane, from the top menu, click the icon of **Look At** so that the selected sketching plane is oriented to be parallel to the screen and ready for preparing a sketch.

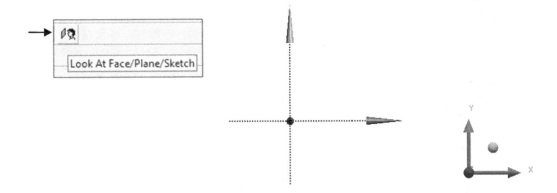

At the bottom of **Tree Outline**, click the icon of **Sketching**. The Sketching Toolboxes window appears. Click the box called **Draw**. Select **Rectangle**. Sketch a rectangle, as shown.

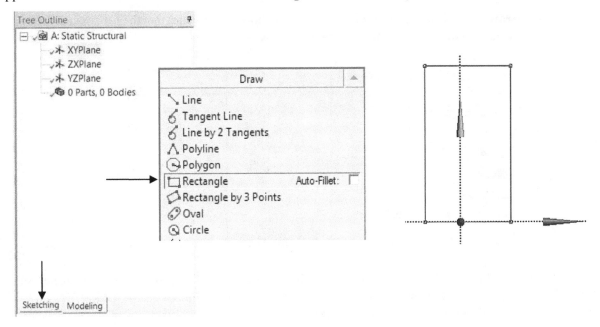

Add a symmetry constraint. In **Constraints**, pick **Symmetry** and pick the vertical axis first. Afterwards, pick the left side and the right side. The sketched rectangle is symmetric about the vertical axis.

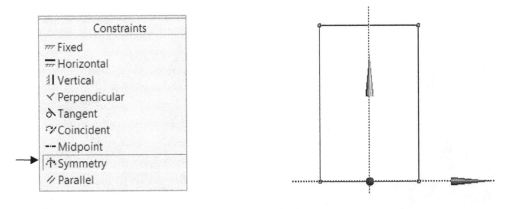

From Sketching Toolboxes, click the box called **Dimensions.** From the **Dimensions** panel, select **General**. Define 2 dimensions: 2 and 3 as shown.

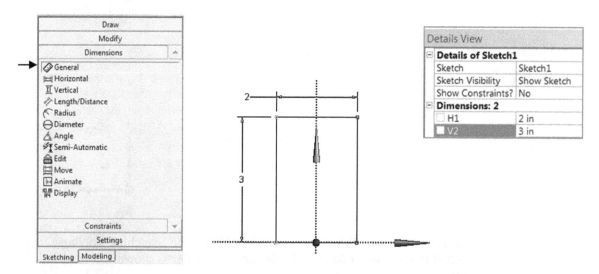

Click **Draw**, and select **Circle**. Sketch 2 circles, as shown. Make sure that their diameters are equal. If not, users may need to define the diameter dimension twice.

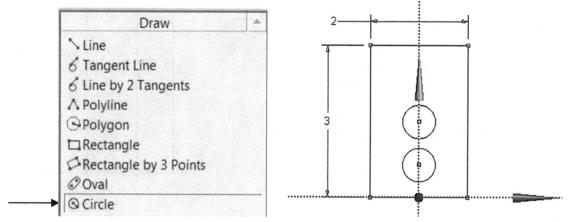

From Sketching Toolboxes, click the box called **Dimensions**. From the **Dimensions** panel, select **General**. Define 2 position dimensions: 1 and 2, and define a diameter dimension of 0.25 as shown. Users may use **Vertical** to define the 2 position dimensions.

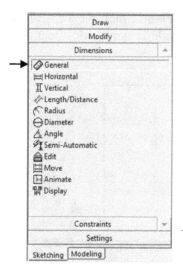

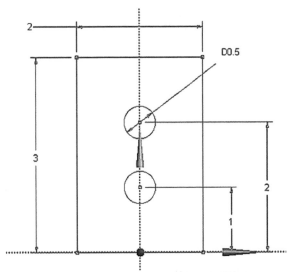

Upon completing the sketch, click the icon of **Extrude**. Click the created sketch (Sketch1) from **Tree Outline**, and click **Apply**. Specify 0.125 as its thickness. Click the **Generate** icon to obtain the 3D solid model, as shown.

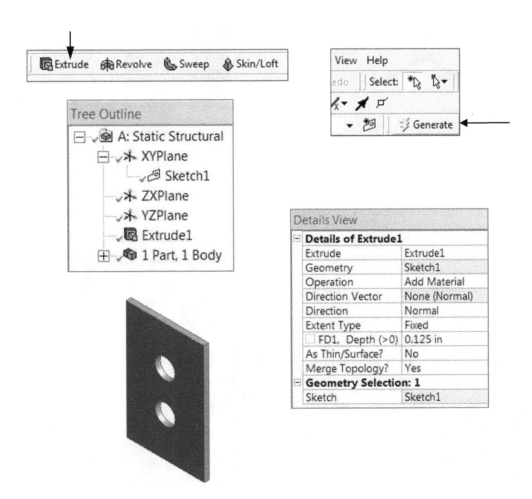

Click the icon of **New Plane**. In the Details of Plane4 window, specify **From Face** and pick the front surface of the created plate, as shown. In **Base Face**, click **Apply** to confirm the selection. Click **Generate** to finalize the creation of **Plane4**.

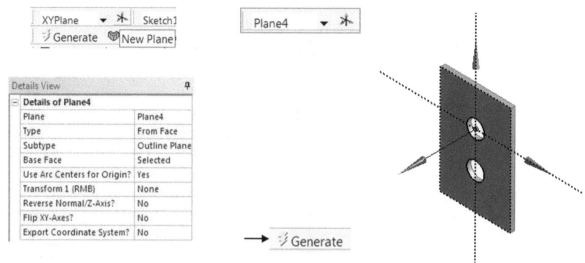

Click the **Z** orientation to orient Plane4 or this new sketching plane.

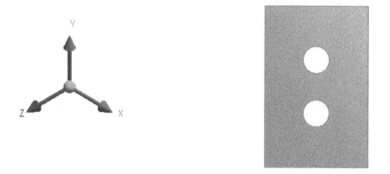

Click **New Sketch**. From **Draw**, select **Rectangle**. Sketch a rectangle as shown.

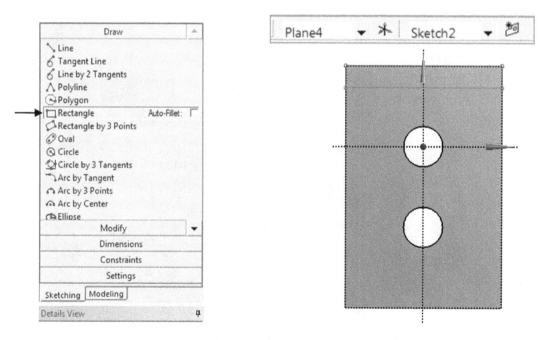

  Click **Dimensions** and from the **Dimensions** panel, select **General**. Pick the vertical line and make a left click to position the dimension. Specify 0.125 as its value. There is no need to define the dimension of 2 because of the alignment.

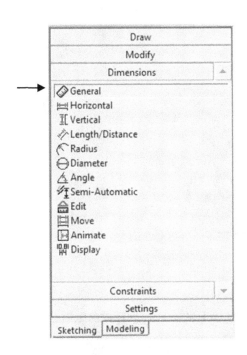

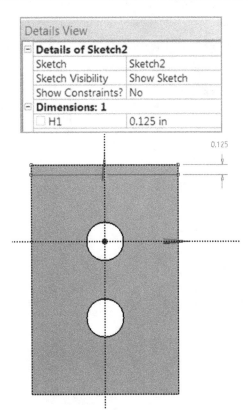

Upon completing the sketch, click the icon of **Extrude**. Click the created sketch (Sketch2) from the Tree Outline, and click **Apply**. Specify 2.875 as the distance (length). Click the icon of **Generate** to obtain the 3D solid model of the horizontal plate, as shown.

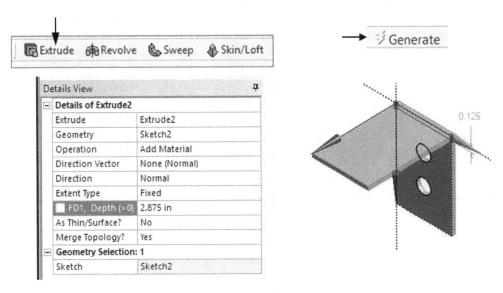

Now let us create the third feature: add a horizontal plate on the back side. The size dimension is 3 x 2.875 x 0.125.

Click the icon of **New Plane**. In the Details of Plane5 window, specify **From Face** and pick the back surface of the created plates. In **Base Face**, click **Apply** to confirm the selection. Click **Generate** to finalize the creation of **Plane5**.

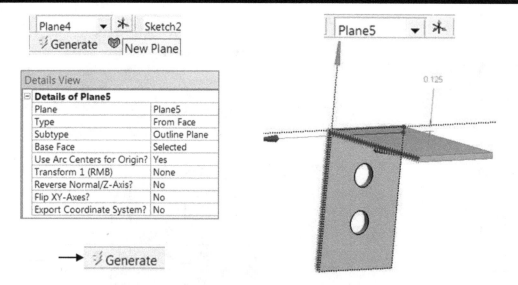

Click the icon of **Look At** to orient **Plane5** for easier sketching.

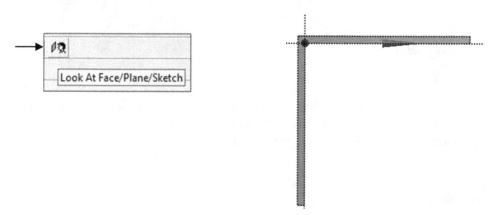

Click **New Sketch**. From **Draw**, select **Rectangle**. Sketch a rectangle as shown. There is no need to add dimensions because of the alignments. Users may add a dimension of 3 for each side to make sure that the two alignments are correctly made.

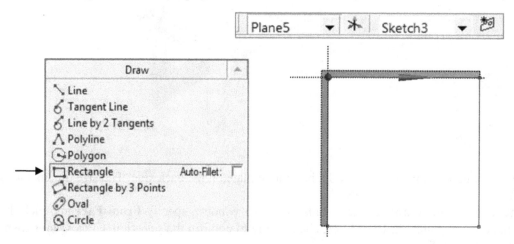

Upon completing the sketch, click the icon of **Extrude**. Click the created sketch (Sketch3) from the Tree Outline, and click **Apply**. Specify 0.125 as the height. In Direction, select **Reversed** for the extrusion direction. Click the icon of **Generate** to obtain the 3D solid model of the third plate, as shown.

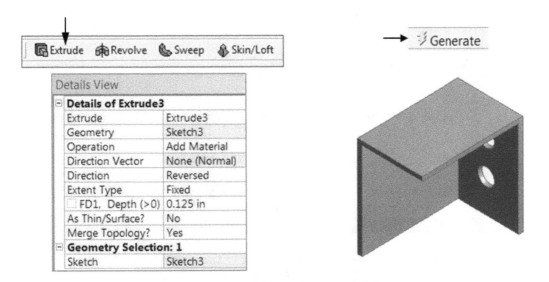

Upon completing the creation of this 3D solid model, we need to close Design Modeler. Click **File** > Close **DesignModeler**. Click **Save Project**. Specify **Welding Structure** as the file name.

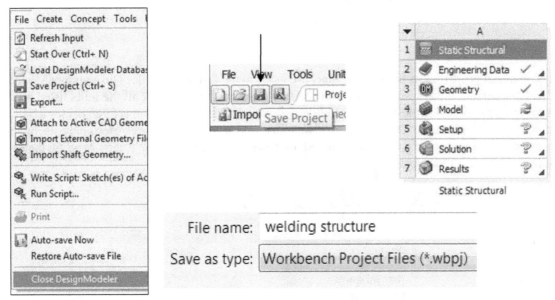

Step 4: Assign Material Properties and Perform FEA

In the Static Structural panel, double click **Engineering Data A2**. We enter the Engineering Data Mode. Click the icon of **Engineering Data Sources**. In Outline of Favorites, Structural Steel is listed and activated with the activated symbol listed in the column C location by system default.

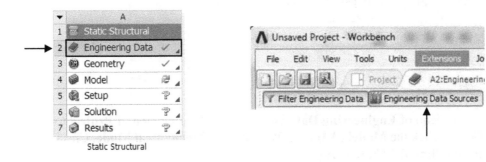

Engineering Data Sources

	A	B	C	D
1	Data Source	/	Location	Description
2	☆ Favorites			Quick access list and default items
3	📖 General Materials	☐	📖	General use material samples for use in various analyses.
4	📖 General Non-linear Materials	☐	📖	General use material samples for use in non-linear analyses.
5	📖 Explicit Materials	☐	📖	Material samples for use in an explicit analysis.

Outline of Favorites

	A	B	C	D	E	F
1	Contents of Favorites ⋧		Add	✅	ource	Description
2	⊟ Material					
3	🏷 Air	🔁		☐	🔗	
4	🏷 Structural Steel	🔁	📕	☐	🔗	Fatigue Data at zero mean stress comes from 1998 ASME BPV Code, Section 8, Div 2, Table 5-110.1

The material properties of **Structural Steel** are listed in the following table. The Young's Modules value is 2E11 Pascal and the Poisson's Ratio value is 0.3. Users may notice that all the units listed are SI units. It is important to note that the Young's Modules value in the inch unit system is 29e6 psi for a typical steel material. The Engineering Data Sources always uses SI units. Therefore, users may use a unit converter to enter the numerical value(s) in the SI unit system when creating a new type of material properties in Engineering Data Sources.

Properties of Outline Row 3: Structural Steel

	A	B	C
1	Property	Value	Unit
2	🔲 Density	7850	kg m^-3
3	⊞ 🔲 Isotropic Secant Coefficient of Thermal Expansion		
6	⊟ 🔲 Isotropic Elasticity		
7	Derive from	Youn...	
8	Young's Modulus	2E+11	Pa
9	Poisson's Ratio	0.3	
10	Bulk Modulus	1.6667E+1	Pa
11	Shear Modulus	7.6923E+1(	Pa
12	⊞ 🔲 Alternating Stress Mean Stress	▦ Tabular	
16	⊞ 🔲 Strain-Life Parameters		
24	🔲 Tensile Yield Strength	2.5E+08	Pa
25	🔲 Compressive Yield Strength	2.5E+08	Pa
26	🔲 Tensile Ultimate Strength	4.6E+08	Pa
27	🔲 Compressive Ultimate Strength	0	Pa

Afterwards, click the icon of **Engineering Data Sources** and click the icon of **Project** to go back to Project Schematic. Double-click the **Model (A4)** cell. We enter the **Mechanical Mode**. Check the unit system: **U.S. Customary (in, lbm, lbf, °F, s, V. A).**

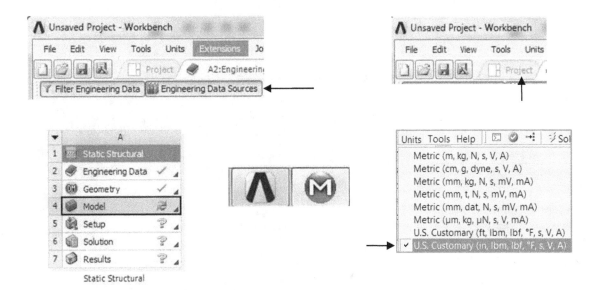

From the project tree, highlight **Mesh**. In the Details of "Mesh" window, expand Sizing, and set Resolution to 4. Right-click and select **Generate Mesh**.

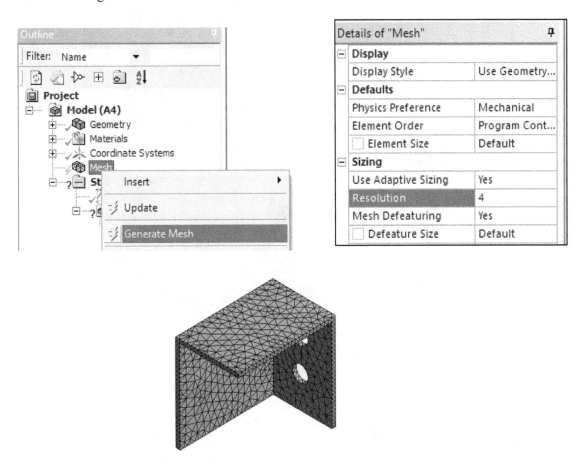

From the project tree, highlight **Static Structural (AS)** and right-click to pick **Insert**. Afterwards, select **Fixed Support**. From the screen, pick the two holes while holding down the Ctrl key, as shown. As a result, this 3D solid model is fixed to the ground through the two holes.

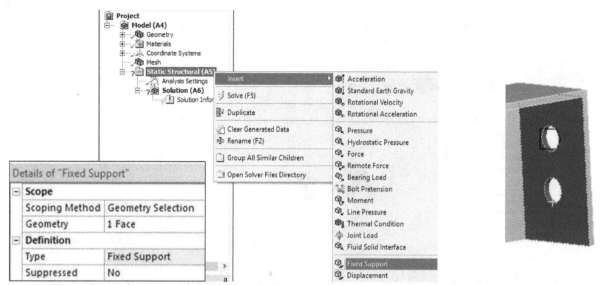

From the project tree, highlight **Static Structural (A5)** and right-click to pick **Insert**. Afterwards, select **Pressure**. From the screen, pick the top surface of the welding structure, as shown. Specify 10 psi as the magnitude of the pressure load.

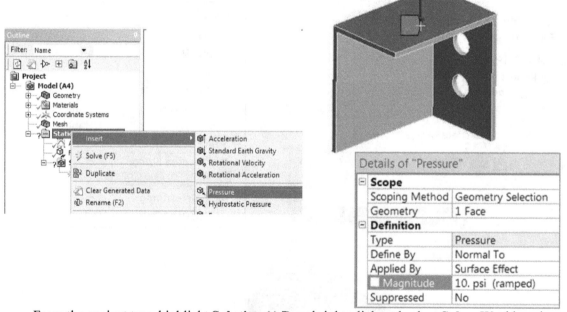

From the project tree, highlight **Solution (A5)** and right-click and select **Solve**. Workbench system is running FEA.

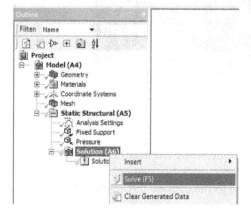

To plot the von Mises stress distribution, highlight **Solution (A6)** and pick **Insert**. Afterwards, highlight **Stress** and pick **Equivalent (von Mises).** Afterwards, highlight **Equivalent Stress** and right-click to select **Evaluate All Results**.

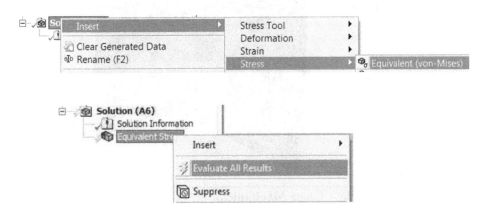

The maximum value is 9014.7 psi (equivalent to 62.2 MPa) located at the fixed end with the lower hole.

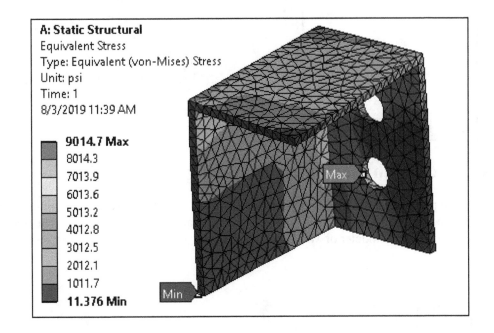

To plot the displacement distribution, highlight **Solution (A5)** and pick **Insert**. Highlight **Deformation** and pick **Total.** Afterwards, highlight **Total Deformation** and right-click to pick **Evaluate All Results**. The maximum deformation is 0.0075 inch (equivalent to 0.19 mm), located at the front corner, as shown.

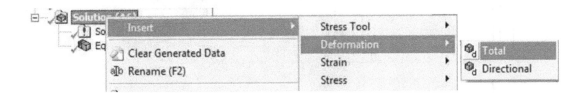

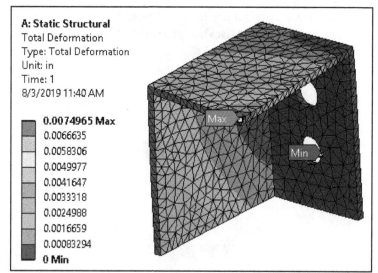

A: Static Structural
Total Deformation
Type: Total Deformation
Unit: in
Time: 1
8/3/2019 11:40 AM

0.0074965 Max
0.0066635
0.0058306
0.0049977
0.0041647
0.0033318
0.0024988
0.0016659
0.00083294
0 Min

From the top menu, click **File > Close Mechanical**. In this way, we return to the main page of Workbench, or Project Schematic. From the top menu, click **Save**.

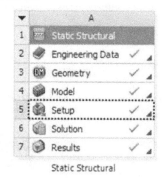

2.4   Case Study with Geometric Modeling

In this case study, we perform FEA for the object shown below. The material type is aluminum alloy. The bottom surface is fixed to the ground. A force is applied downward on the top surface. The magnitude is 1000 Newton. Use the Design Modeler of Workbench to create a 3D solid model and perform FEA.

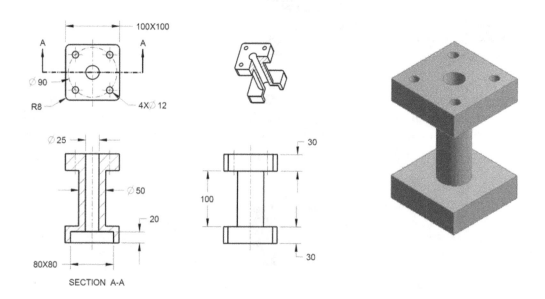

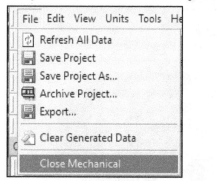

Step 1: Start Workbench.

To start a design process using ANSYS Workbench, select the icon, called "**Workbench 19**". This icon is displayed on the start menu. Click **ANSYS 19** > **Workbench 19**. Users may directly click Workbench 19 when the symbol of Workbench 19 is on display.

Workbench 19.2

Step 2: From **Toolbox**, highlight **Static Structural** and drag it to the **Project Schematic** area.

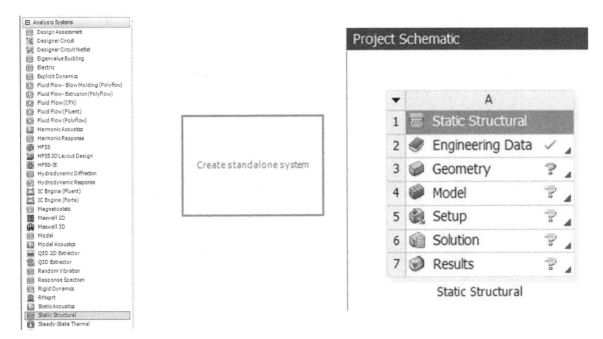

Step 3: Start Design Modeler and create a 2D sketch first.

Right-click the **Geometry** cell and pick **New DesignModeler Geometry** to start **Design Modeler**. Users may double click **Geometry** to directly enter **Design Modeler** if **New SpaceClaim Geometry** is not present. Click the **Units** drop down on the top menu, and select **Millimeter**.

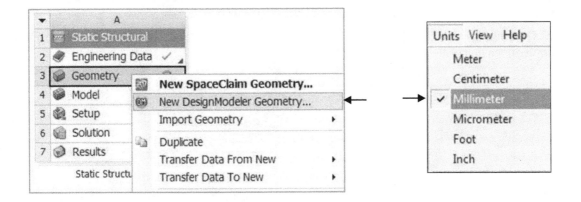

Select **ZXPlane** as the sketching plane, and click **Look At** to orient the sketching plane.

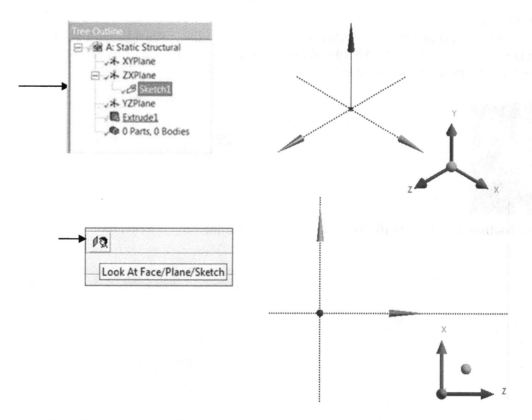

Click **Sketching**, and click **Draw**. In the Draw window, click **Circle**. Sketch a circle, as shown below.

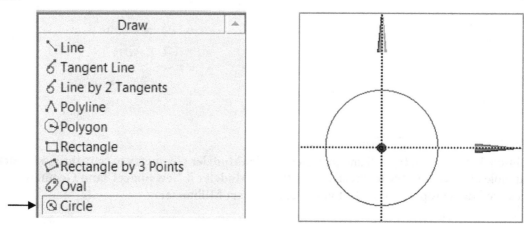

Click **Dimensions** and from the **Dimensions** panel, select **General**. Pick the circle and make a left click to position dimension. Specify 50 as its value.

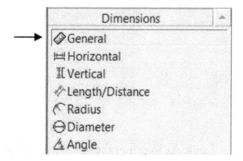

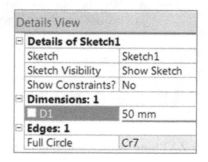

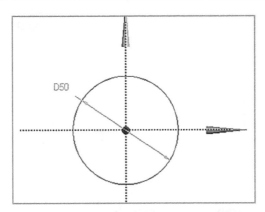

Upon completing the sketch, click the **Extrude** icon. Click the created sketch (Sketch1) from the Tree Outline, and click **Apply**. For Direction, select **Both-Symmetric** and specify 50 as the height (both = 100). Click the icon of **Generate** to obtain the 3D solid model of the cylindrical feature, as shown.

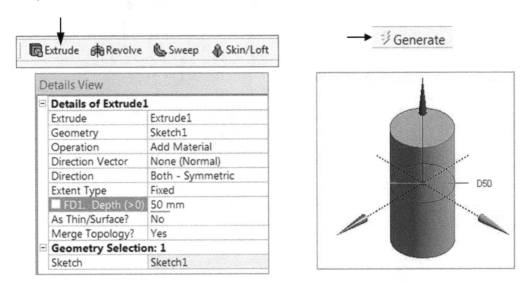

Now let us create the second feature: add a squared-block feature. The size dimension is 100 and the height value is 30 mm.

Click the icon of **New Plane**. In the Details of Plane4 window, specify **From Face** and pick the top surface of the created cylinder. In **Base Face**, click **Apply** to confirm the selection. Click **Generate** to finalize the creation of **Plane4**.

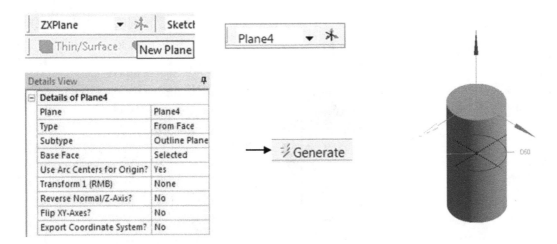

Click the icon of **Look At** to orient Plane4 or this new sketching plane.

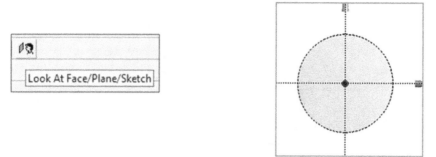

Click **New Sketch**. From **Draw**, select **Rectangle**. Sketch the rectangle as shown.

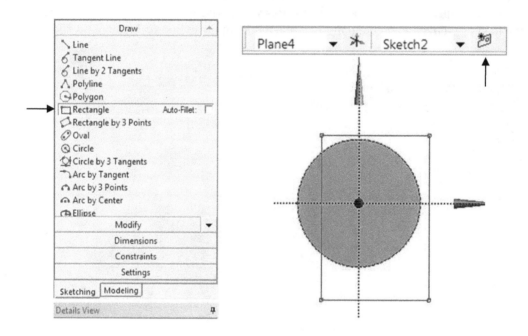

To add two symmetrical constraints, click **Constraints** and select **Symmetry**. Make a left click to pick the vertical axis as the line of symmetry, and pick the two vertical sides of the rectangle. On the screen, make a right-click and hold to pick Select new symmetry axis. Pick the horizontal axis as the line of symmetry, and pick the two horizontal sides of the rectangle.

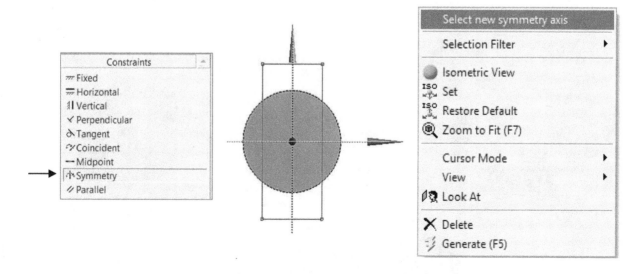

Click **Dimensions** and from the **Dimensions** panel, select **General**. Pick the horizontal line and make a left click to position the dimension. Specify 100 as its value and repeat this process. Pick the vertical line and make a left click to position the dimension. Specify 100 as its value.

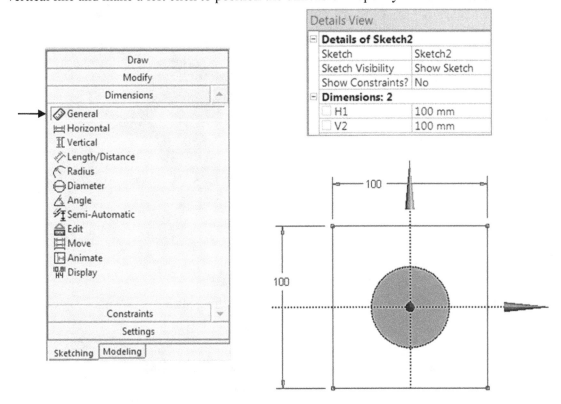

Upon completing the sketch, click the **Extrude** icon. Click the created sketch (Sketch2) from the Tree Outline, and click **Apply**. Specify 30 as the height. Click the icon of **Generate** to obtain the 3D solid model of the squared box feature, as shown.

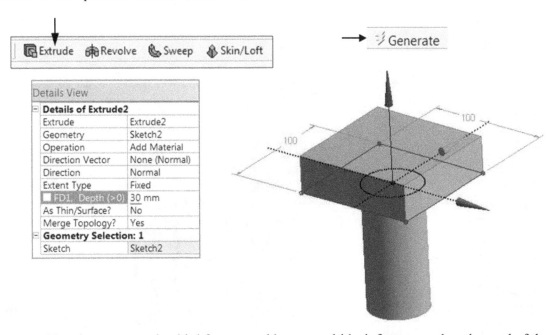

Now let us create the third feature: add a squared-block feature on the other end of the cylinder. The size dimension is 100 and the height value is 30 mm.

Click the icon of **New Plane**. In the Details of Plane5 window, specify **From Face** and pick the bottom surface of the created cylinder. In **Base Face**, click **Apply** to confirm the selection. Click **Generate** to finalize the creation of **Plane5**.

Click the icon of **Look At** to orient **Plane5** or this new sketching plane.

Click **New Sketch**. From **Draw**, select **Rectangle**. Sketch the rectangle as shown.

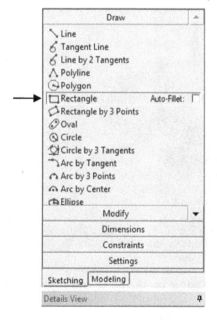

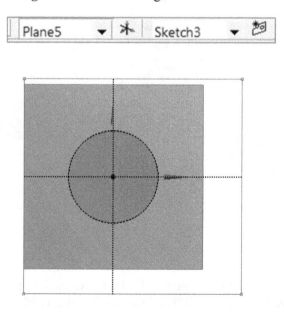

To add two symmetrical constraints, click **Constraints** and select **Symmetry**. Right-click and pick the vertical axis as the line of symmetry, and pick the two vertical sides of the rectangle. Right-click, pick the horizontal axis as the line of symmetry, and pick the two horizontal sides of the rectangle.

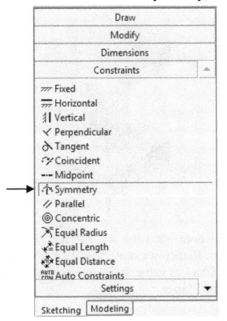

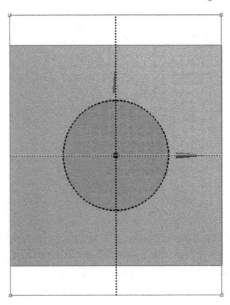

Click **Dimensions** and from the **Dimensions** panel, select **General**. Pick the horizontal line and make a left click to position the dimension. Specify 100 as its value. Pick the vertical line and make a left click to position the dimension. Specify 100 as its value.

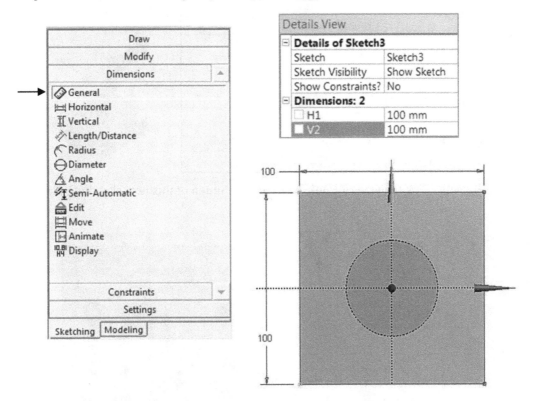

Upon completing the sketch, click the **Extrude** icon. Click the created sketch (Sketch3) from the Tree Outline, and click **Apply**. Specify 30 as the height. Click the icon of **Generate** to obtain the 3D solid model of the squared box feature, as shown.

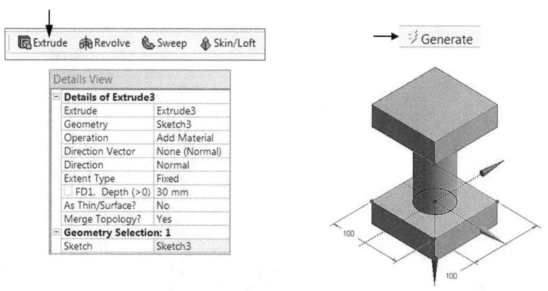

Now let us create the 4th feature: cut a through-all hole from the top surface. The diameter dimension is 25. Click the icon of **New Plane**. In the Details of Plane6 window, specify **From Plane** and pick ZX plane from the model tree. Select Offset and specify 80 as its value. In **Base Face**, click **Apply** to confirm the selection. Click **Generate** to finalize the creation of **Plane6**.

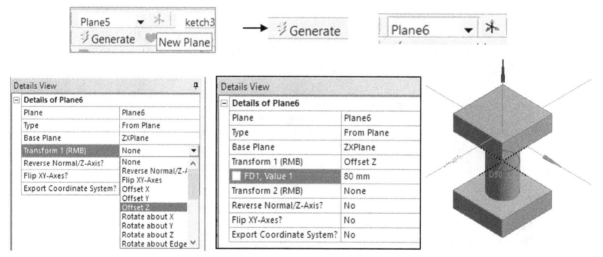

Click New Sketch. Click the icon of **Look At** to orient **Plane6** or this new sketching plane.

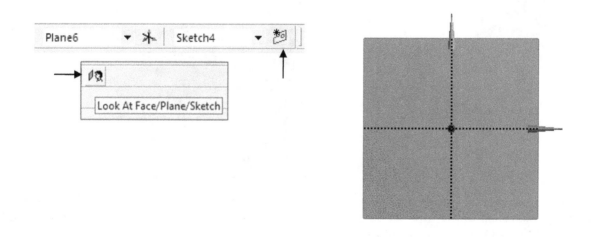

Click **Sketching**, and click **Draw**. In the Draw window, click **Circle**. Sketch a circle, as shown below. Click Dimensions and select **General**. Pick the circle and make a left click to position the diameter size dimension. Specify 25 as its value.

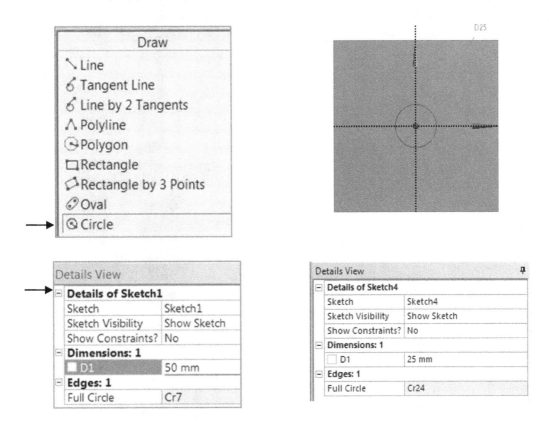

Upon completing the sketch, click the icon of **Extrude**. Click the created sketch (Sketch4) from the Tree Outline, and click **Apply**. Select **Cut Material, Reverse,** and **Through All** as the **Operation** and **Extent Type** respectively. Click the icon of **Generate** to obtain the 3D solid model of the hole going from the top surface to the bottom surface, as shown.

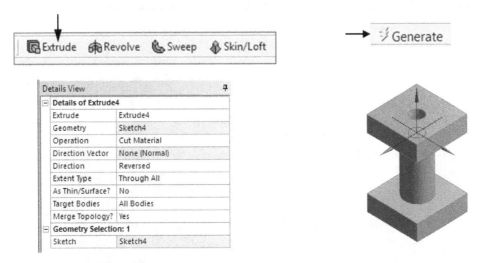

Now let us create the 5th feature: cut a through-all hole at the corner from the top surface. The diameter dimension is 12.

Click the icon of **New Plane**. In the Details of Plane7 window, specify **From Coordinates.** The coordinates are x = 0, y = 80 and z = 0. Set the normal vector [0, 1, 0]. Click **Generate** to finalize the creation of **Plane7**.

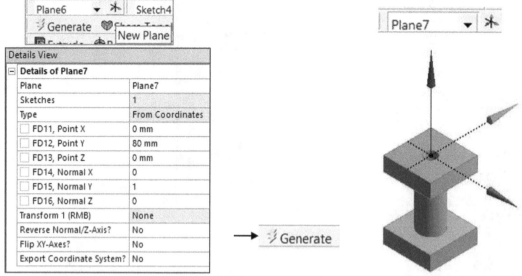

Click **New Sketch**. Click the icon of **Look At** to orient **Plane7** or this new sketching plane.

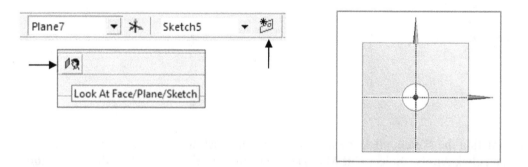

Click **Sketching**, and click **Draw**. In the Draw window, click **Circle**. Sketch a circle, as shown below. Click **Dimensions** and select **Diameter**. Pick the circle and make a left click to position the size dimension. Specify 12 as its value.

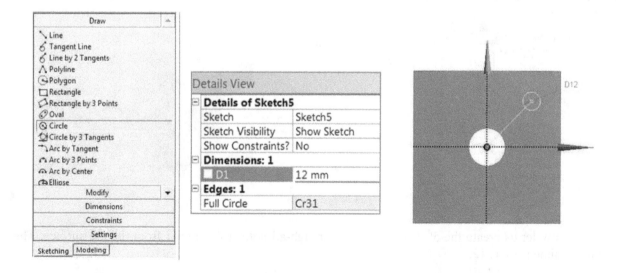

Now click **Line** to sketch a line connecting the origin and the center of the sketched circle, as shown. Specify 45 as the distance value, and specify 45 degrees as the angular dimension.

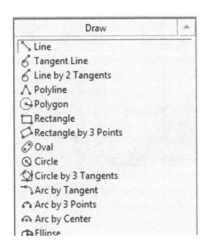

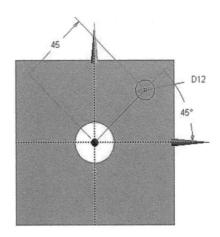

Upon completing the sketch and dimensioning, click the icon of **Extrude**. Click the created sketch (Sketch5) from **Tree Outline**, and click **Apply**. Specify **Cut Material**, **Reverse** and 30 as the depth of cut value. Click the icon of **Generate**, as shown.

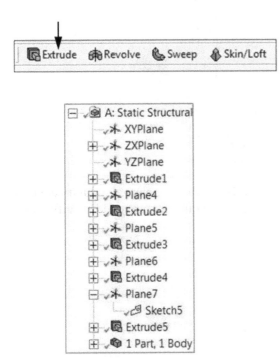

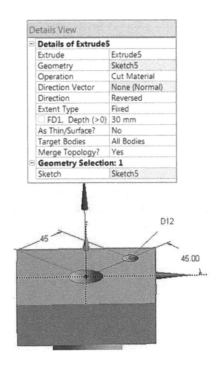

We use a pattern feature to create 3 more holes. Before patterning, we need a vertical axis of rotation. Select **YZPlane** as the sketching plane and click **New Sketch** to create Sketch6. Click **Look At**. Sketch a line along the Y axis or the horizontal axis, as shown. There is no need for a length dimension.

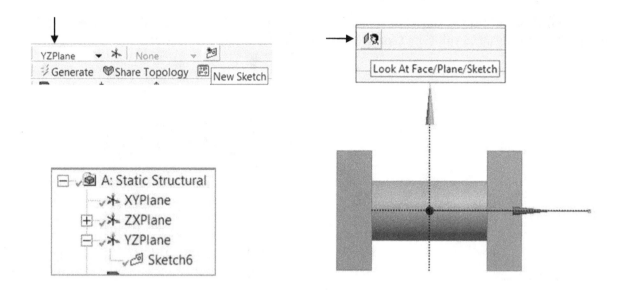

From the top menu, click the icon of **Selection Filter: Faces**. Select the cylindrical surface from the created hole.

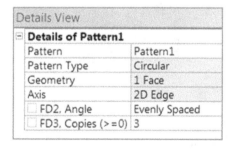

From the top menu, click **Create > Pattern**. Click **Apply** so that the selected cylindrical surface is selected for pattern. In Details View, select Circular, pick the vertical line as Axis, and specify 3 as the number of copies. Click **Generate**.

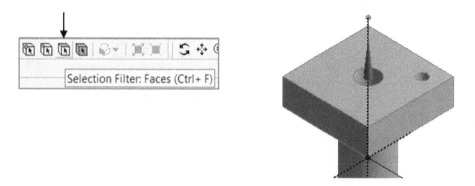

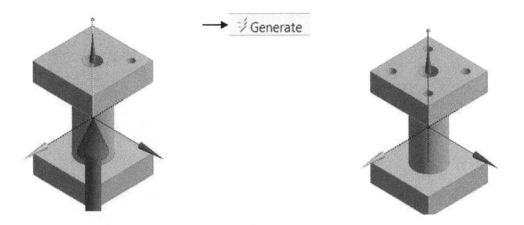

Now let us create the final feature. We cut a block volume from the bottom surface. The size dimensions are 80X80X20 where value of 20 serves as the value of the depth of cut.

Click the icon of **New Plane**. In the Details of Plane8 window, specify **From Coordinates.** The coordinates are x = 0, y = -80 and z = 0. Set the normal vector [0, 1, 0]. Click **Generate** to finalize the creation of **Plane8**.

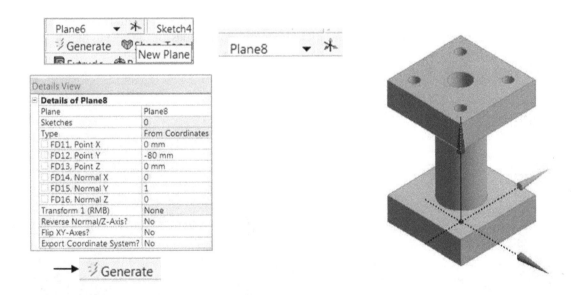

Click **New Sketch**. Click the icon of **Look At** to orient **Plane8** or this new sketching plane.

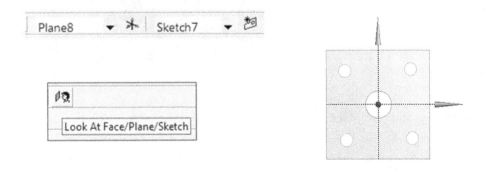

From **Draw**, select **Rectangle**. Sketch the rectangle as shown.

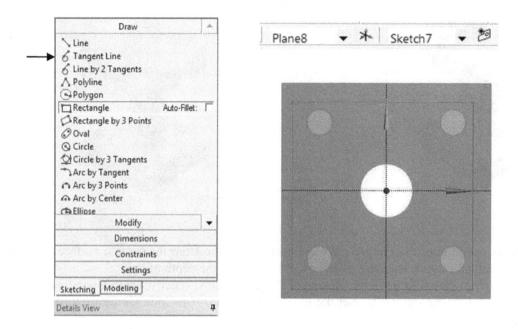

To add two symmetrical constraints, click **Constraints** and select **Symmetry**. Left-click to pick the vertical axis as the line of symmetry, and pick the two vertical sides of the rectangle. Right-click to pick Select new symmetry axis. Pick the horizontal axis as the line of symmetry, and pick the two horizontal sides of the rectangle.

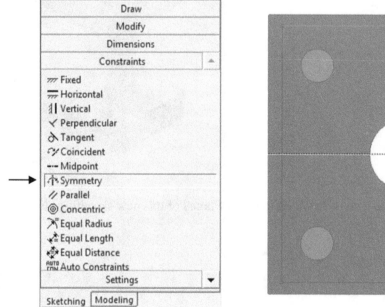

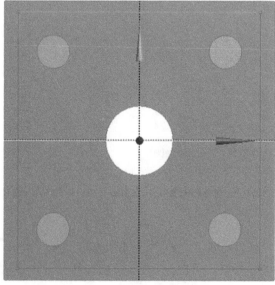

Click **Dimensions** and select **General**. Pick the horizontal line and make a left click to position the dimension. Specify 80 as its value. Pick the vertical line and make a left click to position the dimension. Specify 80 as its value.

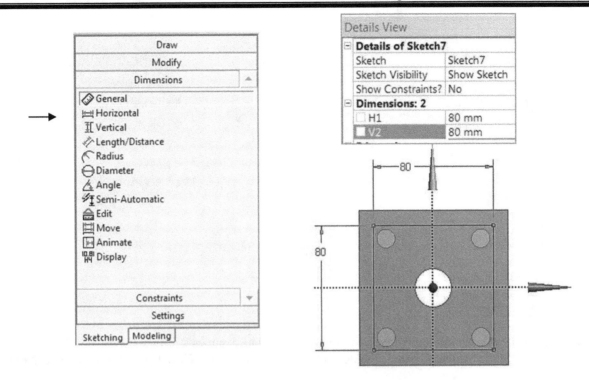

Upon completing the sketch and dimensioning, click the icon of **Extrude**. Click the created sketch (Sketch7) from **Tree Outline**, and click **Apply**. Specify **Cut Material**, and 20 as the depth of cut. Click the icon of **Generate**, as shown.

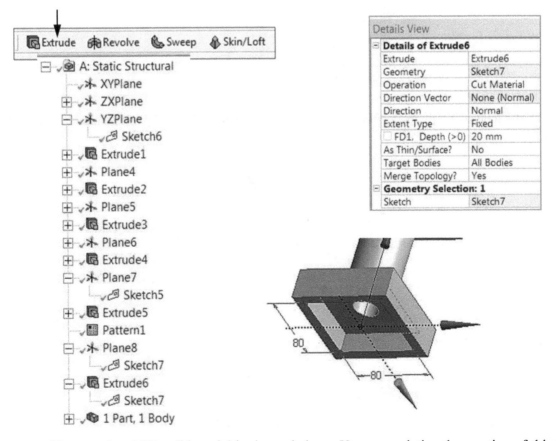

The completed 3D solid model is shown below. Upon completing the creation of this 3D solid model, we need to close Design Modeler. Click **File** > Close **DesignModeler**.

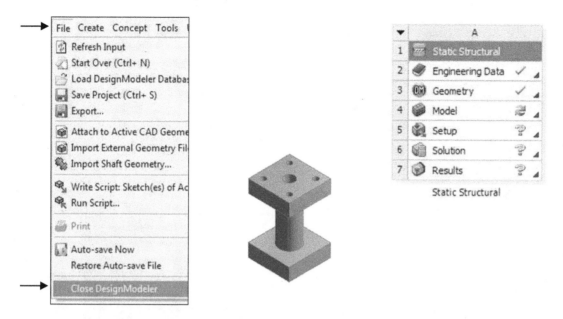

Step 5: Activate the Aluminum Alloy in the Engineering Data Sources

In the Static Structural panel, double click **Engineering Data**. We enter the Engineering Data Mode.

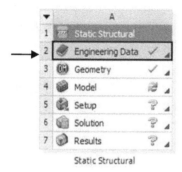

Static Structural

Click the icon of **Engineering Data Sources**. Click **General Materials**. We are able to locate Aluminum Alloy.

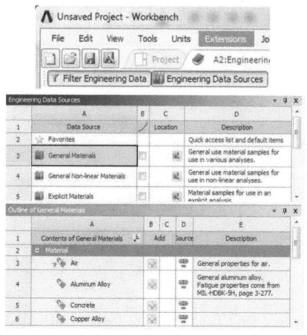

Click the plus symbol nearby Aluminum Alloy to activate it. Afterwards, click the icon of Engineering Data Sources and click the icon of Project to go back to Project Schematic.

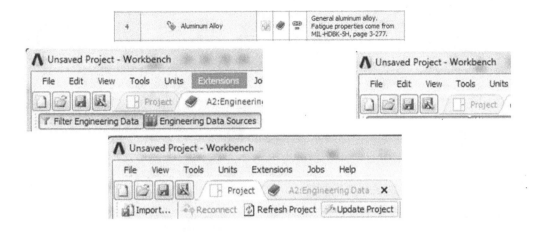

Step 6: Assign Aluminum Alloy to the 3D solid Model
  Double-click the icon of Model. We enter the Mechanical Mode. Check the unit system.

Now let us do the material assignment. From the Project tree, expand **Geometry**. Highlight **Solid**. In Details of Solid, click material assignment, pick **Aluminum Alloy** to replace Structural Steel, which is the default material assignment.

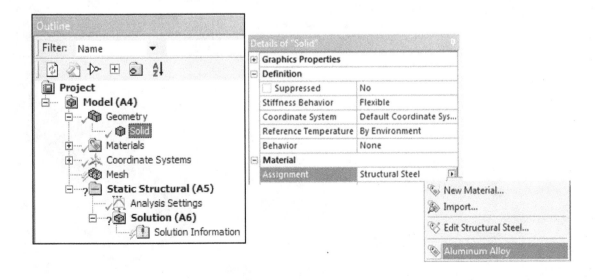

From the project model tree, highlight **Mesh.** Click Sizing and set Resolution to 4, and right-click to pick **Generate Mesh**.

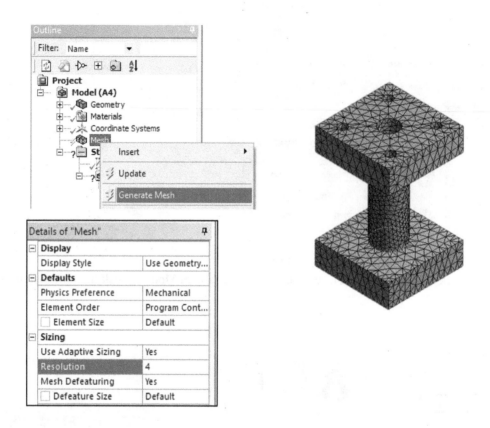

From the project tree, highlight **Static Structural (A5)** and right-click and select **Insert**. Afterwards, select **Fixed Support**. From the screen, pick the bottom surface of the lower rectangular plate, as shown. As a result, this 3D solid model is fixed to the ground through the bottom surface.

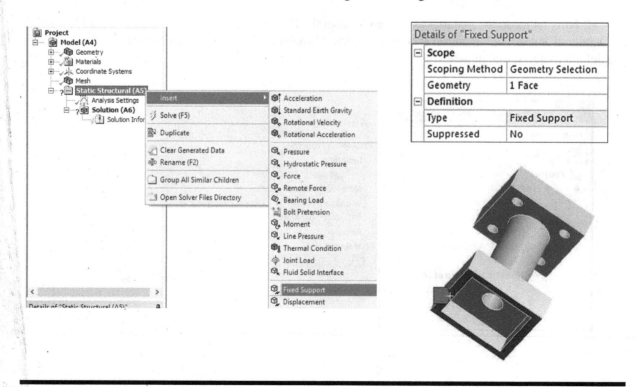

From the project tree, highlight **Static Structural (A5)** and right-click and select **Insert**. Afterwards, select **Force**. From the screen, pick the top surface of the rectangular block, as shown. Specify 1000 N as the magnitude of the force load.

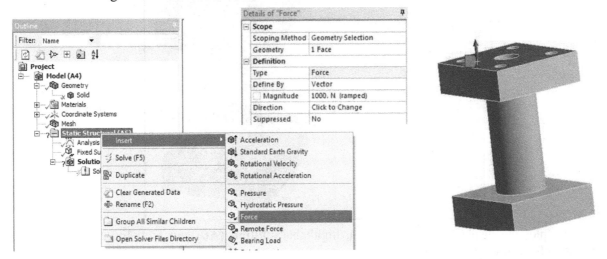

From the project tree, highlight **Solution (A6)** and right-click to pick **Solve**. Workbench system is running FEA.

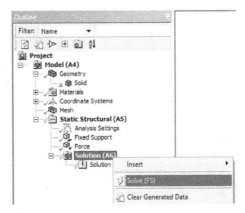

To plot the von Mises stress distribution, highlight **Solution (A6)** and pick **Insert**. Afterwards, highlight **Stress** and pick **Equivalent (von Mises)**. Highlight **Equivalent Stress** and right-click to select **Evaluate All Results**. The maximum value of von Mises stress is 2.58 MPa.

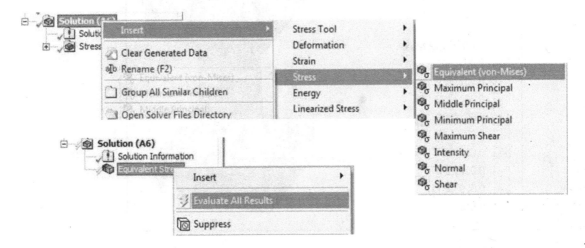

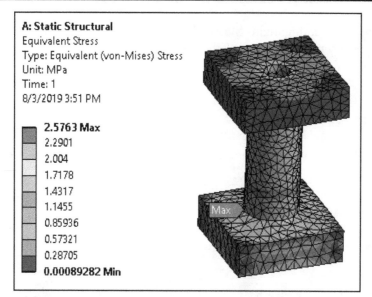

To plot the displacement distribution, highlight **Solution (A6)** and pick **Insert**. Afterwards, highlight **Deformation** and pick **Total.**

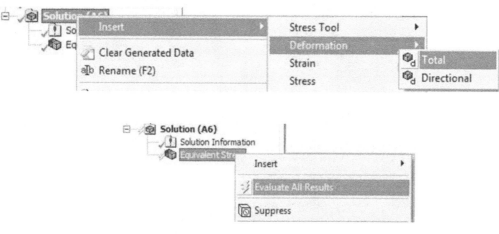

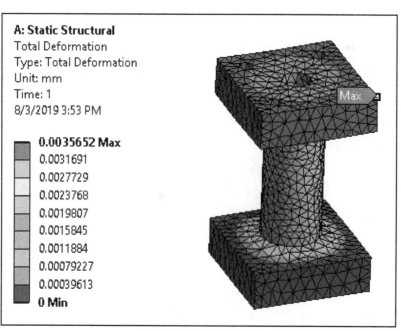

To create a section view to examine the stress distribution developed in the inner part of the 3D model, we use section views. To create a section view, first orient the model to the isometric view. For example, click the Z – axis.

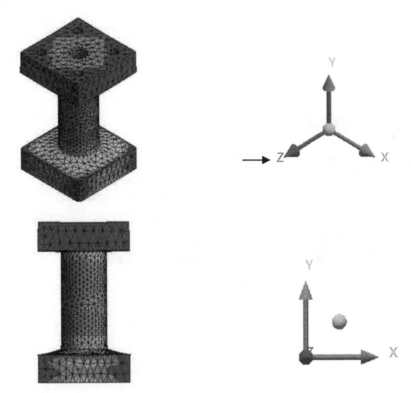

Afterwards, from the top menu, click the icon of **New Section Plane**. A new window called Section Planes appears. Press down the left button to sketch a vertical line on the display and release the left button. A half-section view is on display. Use the isometric view to display the section view showing the stress distribution on the inner part, as shown below.

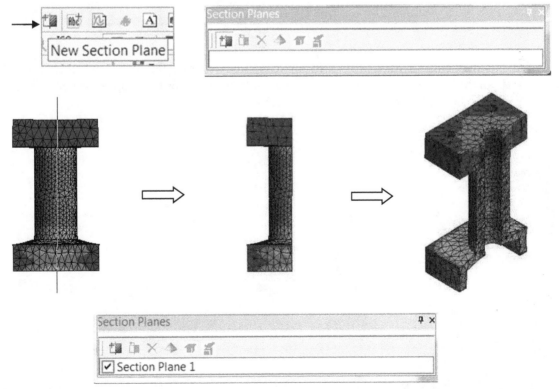

To recover the full view from the section view, just clear the box associated with Section Plane 1. Users may practice creating new section views by creating Section Plane 2, as illustrated below.

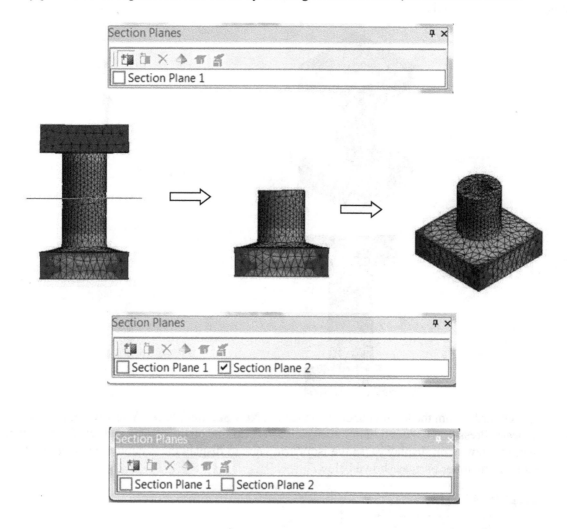

At this moment, we have completed this case study. From the top menu, click **File** > **Close Mechanical**. In this way, we return to the main page of Workbench or the Project Schematic page. Click Save. Specify Geometrical Modeling as its file name.

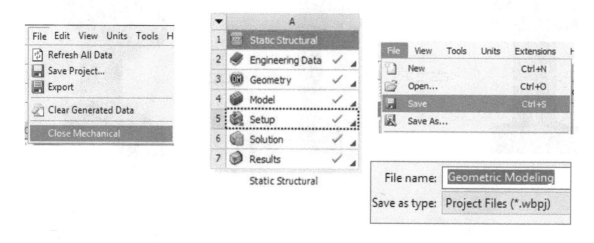

2.5 Modeling of a Tube when Subjected to Torsion

Torsion is the twisting of an object due to an applied torque. The following figure illustrates how the free end of the shaft tends to turn about its longitudinal axis while the other end is held fast.

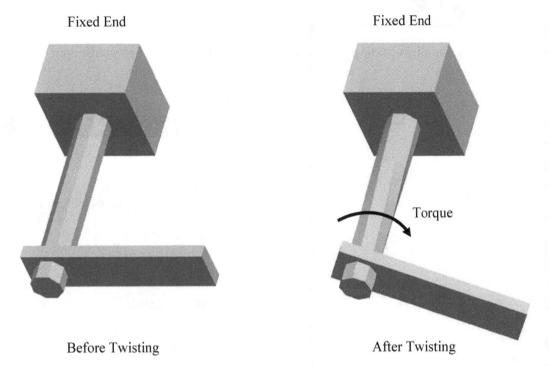

Fixed End   Fixed End

Torque

Before Twisting   After Twisting

In this case study, we create a #D solid model representing the tube component and use it to perform FEA, as shown below. The left end of the tube is fixed to the ground. The material type is structural steel. A torque acts on the free end with a magnitude equal to 10 N-M. Perform FEA to obtain the angle of twist θ and the shear strainγ.

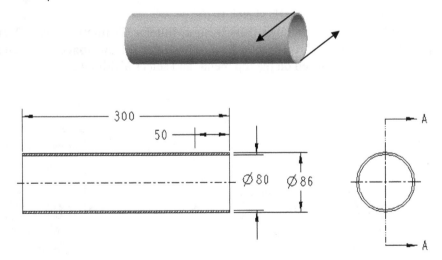

Step 1: Start Workbench.

To start a design process using ANSYS Workbench, select the icon, called "**Workbench 19**". This icon is displayed on the start menu. Click **ANSYS 19 > Workbench 19**. Users may directly click Workbench 19 when the symbol of Workbench 19 is on display.

  Workbench 19.2

Step 2: From **Toolbox**, highlight **Static Structural** and drag it to the **Project Schematic** area.

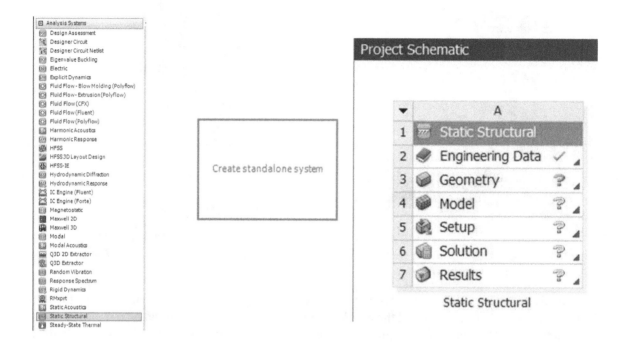

Step 3: Start Design Modeler and create a 2D sketch first.

Right-click the **Geometry** cell and pick **New DesignModeler Geometry** to start **Design Modeler**. Users may double click **Geometry** to directly enter **Design Modeler** if New **SpaceClaim Geometry** is not present. Click the **Units** drop down listed on the top menu, and select **Millimeter**.

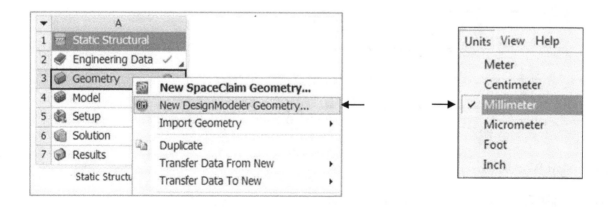

Select **XYPlane** as the sketching plane. Click **New Sketch**.

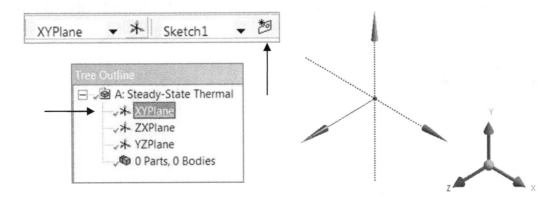

To orient the sketching plane, from the top menu, click the icon of **Look At** so that the sketching plane is parallel to the screen and ready for preparing a sketch.

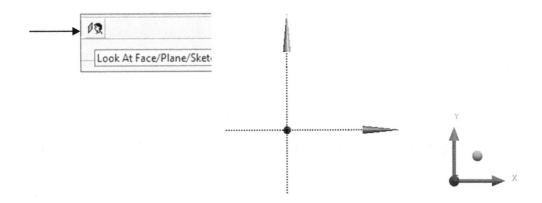

Click **Sketching**, and click **Draw**. In the **Draw** window, click **Rectangle**. Sketch a rectangle, as shown below.

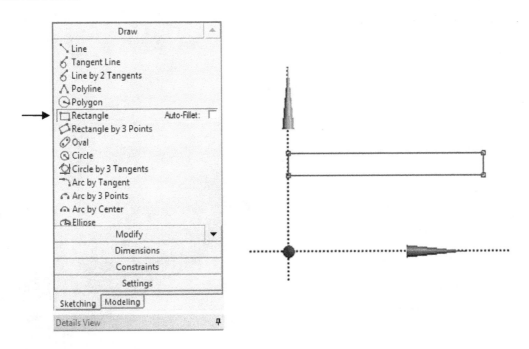

Click **Dimensions** and from the **Dimensions** panel, select **Vertical** to define 2 radius values from the origin of 43 and 40. Select **General** to define the length value of 300, as shown.

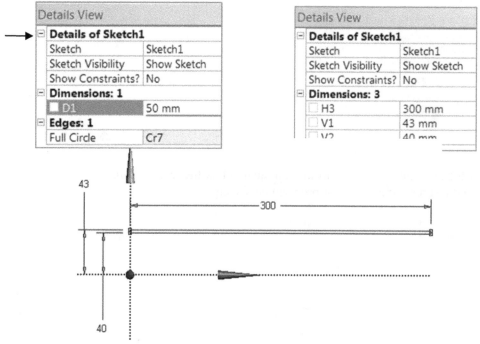

Upon completing the sketch, click the icon of **Revolve**. Click the created sketch (Sketch1) from the Tree Outline, and click **Apply.** Click the icon of **Generate** to create the cylindrical feature, as shown.

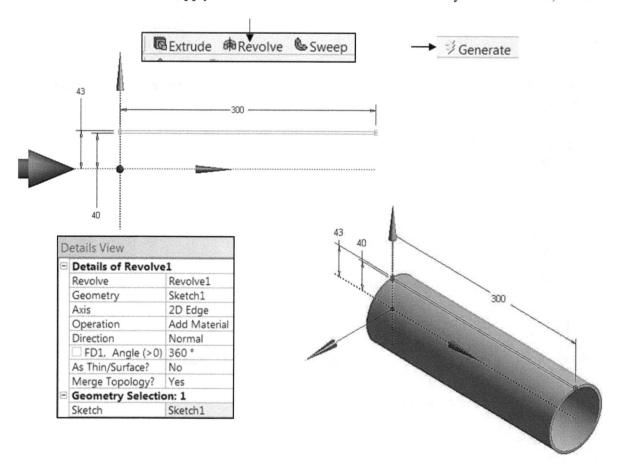

Upon completing the creation of this 3D solid model, we need to close Design Modeler. Click **File** > Close **DesignModeler**.

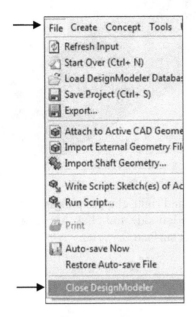

Static Structural

Step 4: Accept the Default Material Assignment of Structural Steel and Perform FEA

Double-click the icon of **Model**. We enter the Mechanical Mode. Check the unit system.

Static Structural

Now let us do the material assignment. From the Project tree, expand **Geometry**. Highlight **Solid**. In Details of Solid, In Assignment, **Structural Steel** is listed. Structural Steel is the default material assignment.

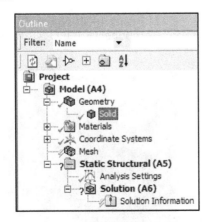

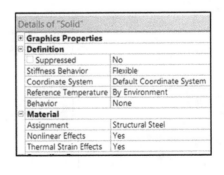

From the project model tree, highlight **Mesh.** In Sizing, set Resolution to 4. Right-click to pick **Generate Mesh**.

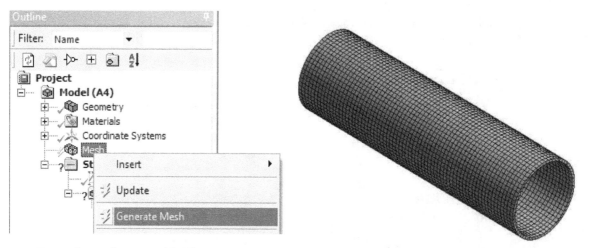

From the project tree, highlight **Static Structural (A5)** and right-click to pick **Insert**. Afterwards, select **Fixed Support**. From the screen, pick the surface on the left side and click Apply, as shown. As a result, this 3D solid model is fixed to the ground through the left surface.

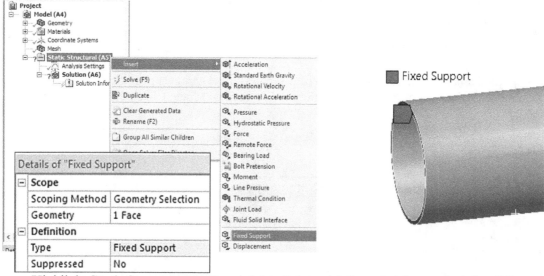

Highlight **Static Structural (A5)** and right-click to pick **Insert**. Afterwards, select **Moment**. From the screen, pick the surface on the right-side of the tube, as shown. Specify 10000 N-mm as the magnitude.

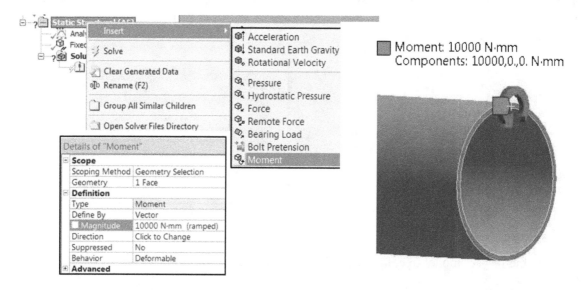

From the project tree, highlight **Solution (A6)** and right-click to pick **Solve**. Workbench system is running FEA.

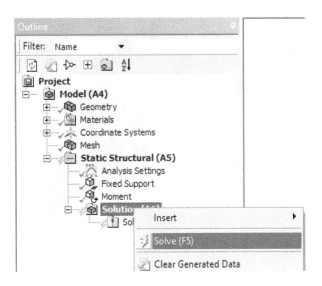

To plot the shear stress distribution, highlight **Solution (A6)** and pick **Insert**. Afterwards, highlight **Stress** and pick **Maximum Shear.** Highlight **Maximum Shear** and right-click to select **Evaluate All Results**. The maximum value of Maximum Shear Stress is 0.32 MPa located on the cylindrical surface with the radius value equal to 43 mm.

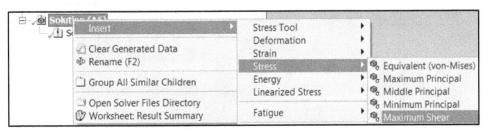

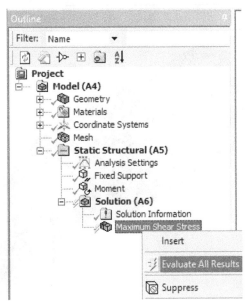

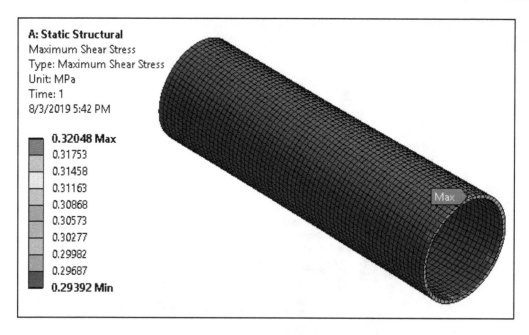

To plot the displacement distribution, highlight **Solution (A6)** and pick **Insert**. Afterwards, highlight **Deformation** and pick **Total**. The maximum deformation is 0.00124 mm.

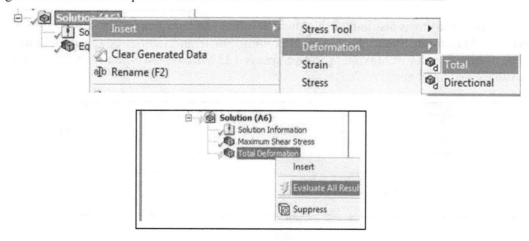

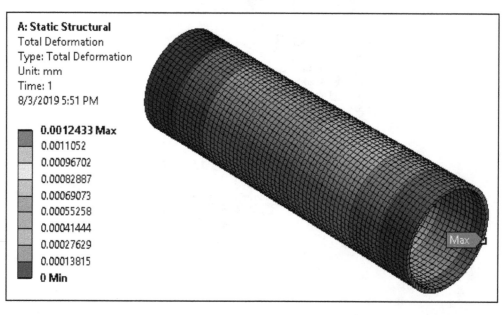

To plot the strain distribution, highlight **Solution (A6)** and pick **Insert**. Afterwards, highlight **Strain** and pick **Shear**. Click Evaluation All Results. The maximum shear strain is 0.00000415.

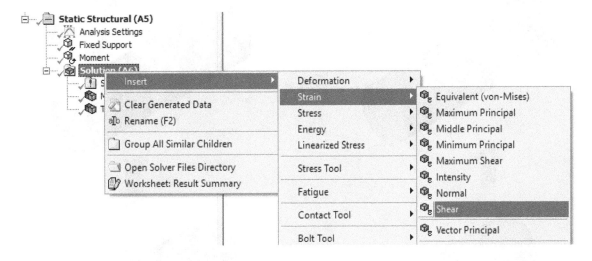

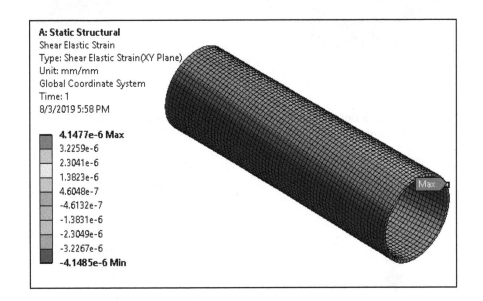

Users may create an Animation to observe the twisting action. In the Graph window, click the icon of Play to observe. Users may adjust the number of Frames and the time duration to control the speed of animation.

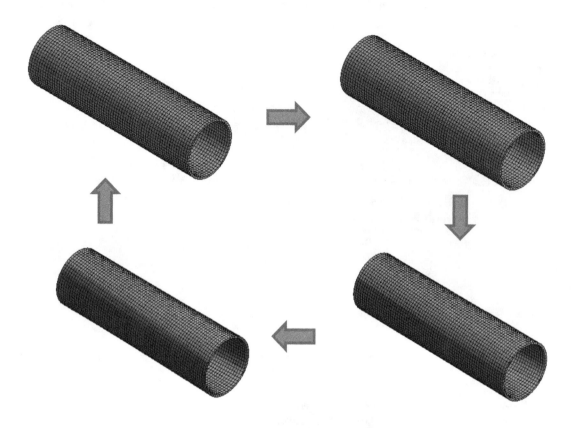

From the top menu, click **File** > **Close Mechanical**. In this way, we return to the main page of Workbench, or the Project Schematic page.

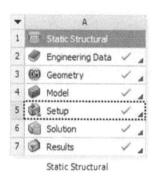

From the top menu, click **Save Project**. Specify torsion as the file name and click **Save**. Type torsion as the file name.

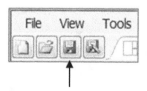

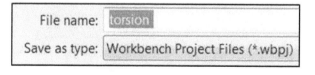

From the top menu, click **File** > **Exit**.

Verification of the FEA results.

(1).   Verify that the value of the maximum shear stress obtained from FEA is close to the value calculated by the analytical formula listed below:

$$\tau_{max} = \frac{Torque * R_{max}}{J} = \frac{10*(0.043)}{(1.349x10^6)} = 0.31875(MPa)$$

The calculation of moment of inertia J is given by

$$J = \frac{\pi}{2}[(43)^4 - (40)^4] = 1.349x10^6 \ (mm^4) = 1.349x10^{-6}(m^4)$$

The FEA result is 0.32048 (MPa). Therefore, these two values are well matched.

(2).   Verify that the value of the angle of twist evaluated from FEA results is close to the value calculated by the analytical formula listed below:

$$\theta = \frac{Torque * L}{G * J} = \frac{(10)(0.3)}{(78x10^9)(1.349x10^{-6})} = 0.028511x10^{-3} \ (radian)$$

G = 78 GPa (shear modulus of rigidity or shear modulus).

The FEA result gives a ratio of 0.0012433/43 or 0.000028914 radian. Therefore, these two values are also well matched.

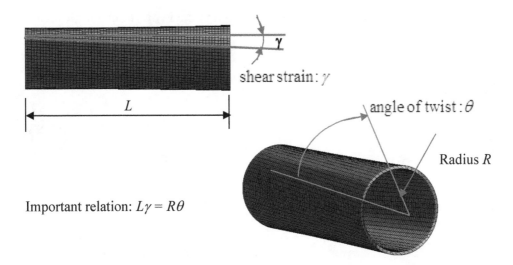

Important relation: $L\gamma = R\theta$

2.6   FEA of an Allen Wrench with Surface Regions

In this case study, we perform FEA for the wrench shown below. The material type is steel. The Young's modules is given by $2x10^{11}$ (Pascal) and the Poisson's ration is given by 0.30. The surface at the right side is fixed to the ground. A pressure load is acting on the 2 surface regions. Its magnitude is 0.2 MPa, as shown. Make two plots: von Mises stress distribution and displacement distribution under the load and constraint conditions. Use the Design Modeler of Workbench to create a 3D solid model and perform FEA.

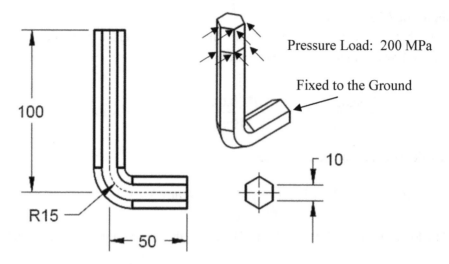

Step 1: Start Workbench.

To start a design process using ANSYS Workbench, select the icon, called "**Workbench 19**". This icon is displayed on the start menu. Click **ANSYS 19** > **Workbench 19**. Users may directly click Workbench 19 when the symbol of Workbench 19 is on display.

Step 2: From **Toolbox**, highlight **Static Structural** and drag it to the **Project Schematic** area.

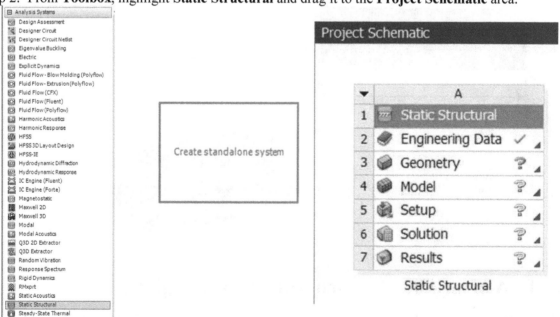

Step 3: Start Design Modeler and create a 2D sketch first.

Right-click the **Geometry** cell and pick **New DesignModeler Geometry** to start **Design Modeler**. Users may double click **Geometry** to directly enter **Design Modeler** if New **SpaceClaim Geometry** is not present. Click the **Units** drop down listed on the top menu, and select **Millimeter**.

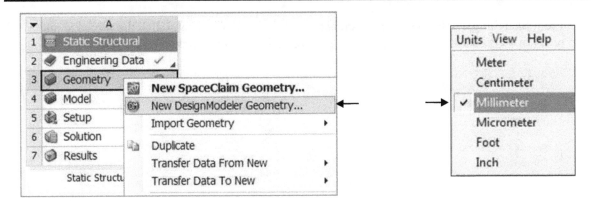

In **Tree Outline**, select **XYPlane** as the sketching plane. Click the icon of **Look At** to orient the selected sketching plane.

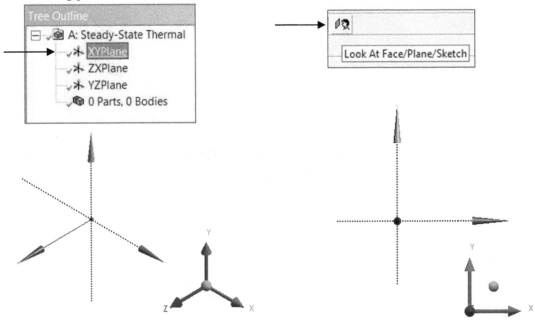

At the bottom of **Tree Outline**, click the icon of **Sketching**. The Sketching Toolboxes window appears. Click the box called **Draw**. Select **Polygon** and pick 6 for hexagon. To sketch a hexagon, click the origin, move the cursor along the horizontal axis, and make the second left click, as shown below.

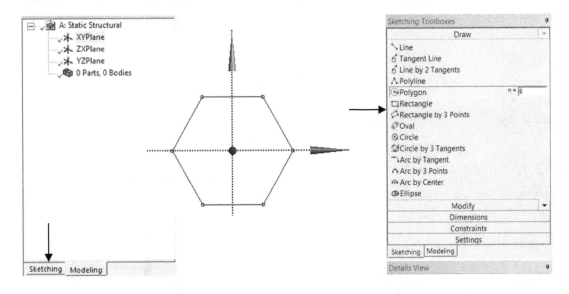

From Sketching Toolboxes, click **Dimensions**. And select **General**. Click the top edge and a left click to position the size dimension. Specify 10. This value is displayed in the **Details View** window.

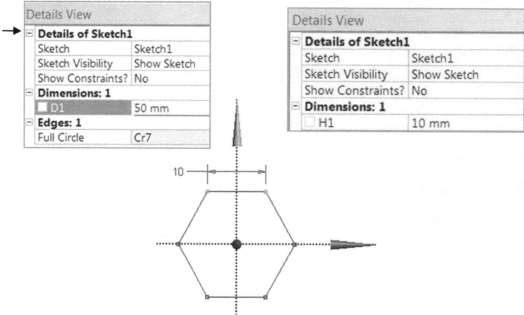

Upon completing this sketch (profile), click **Modeling** for creating the second sketch (path). In the **Tree Outline**, select **YZPlane** as the sketching plane.

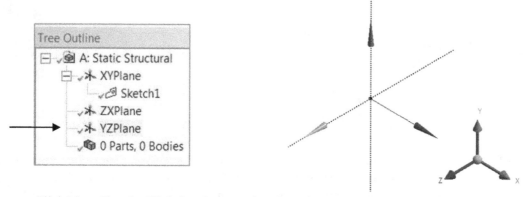

Click New Sketch. Click **Look At** so that the selected sketching plane is oriented to be parallel to the screen and ready for making a sketch.

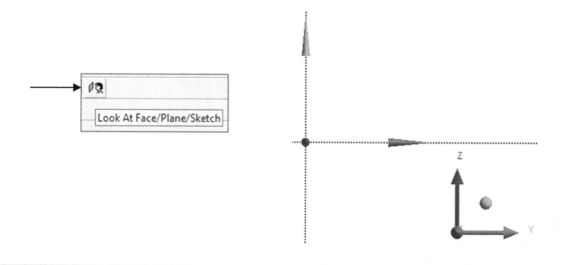

At the bottom of **Tree Outline**, click **Sketching**. Afterwards, click **Draw**. Select **Line**. Sketch 2 lines, as shown.

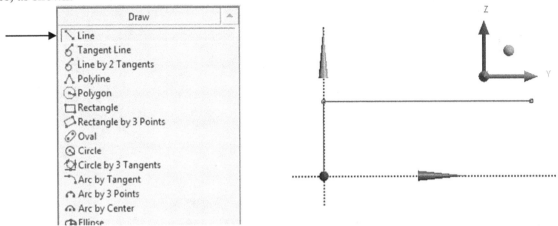

From Sketching Toolboxes, click **Dimensions**, and select **General**. Specify 50 and 100 as the 2 dimensions. These 2 values are displayed in the **Details View** window.

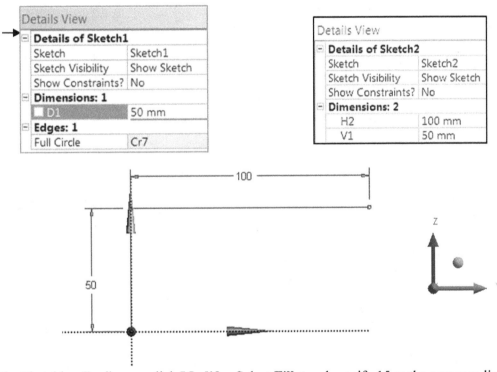

In the Sketching Toolboxes, click **Modify**. Select **Fillet** and specify 15 as the corner radius value. On the screen, left click the 2 lines. The arc on the corner is created with the radius value equal to 15, as shown.

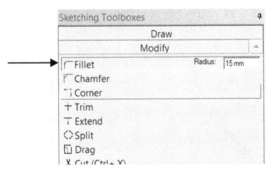

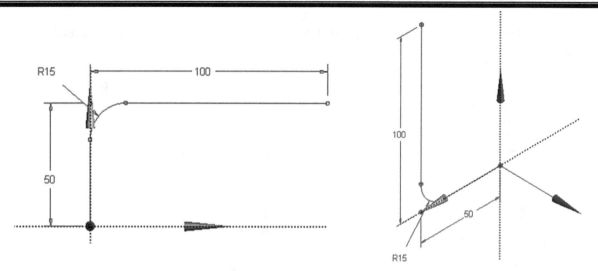

Upon completing this sketch (path), click **Modeling.** Click **Sweep**. Click **Sketch1** (profile) first and click **Apply.** Afterwards, click **Sketch2** (path) and **Apply.** Click **Generate** to obtain the 3D solid model of the wrench, as shown.

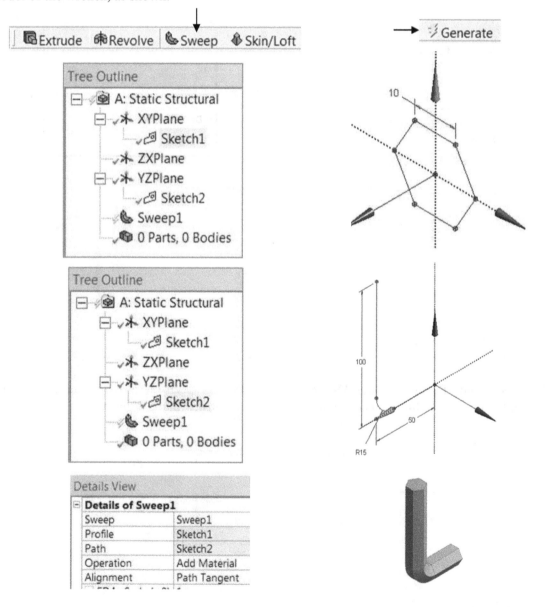

Step 4: Create 2 rectangular surface regions through the use of imprinting 2 rectangular sketches to accommodate the pressure load acting on portions of the side surfaces of the wrench.

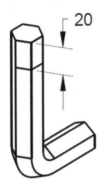

20

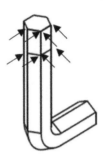

Pressure Load: 200 MPa

Click the icon of **New Plane**. In Details View, select From Face in the Type field. In the Base Plane field, left click to show Apply and pick the side surface of the wrench model. Click **Apply**. Click **Generate**. Plane4 is created.

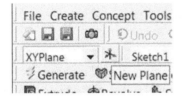

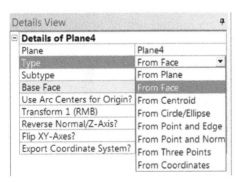

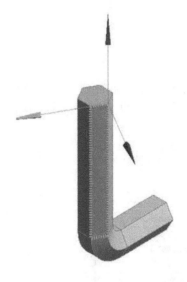

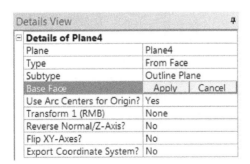

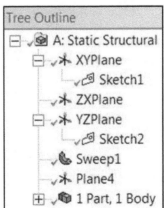

Now click the icon of **New Sketch**. Click Sketching and select Rectangle to sketch a corner rectangle, as shown.

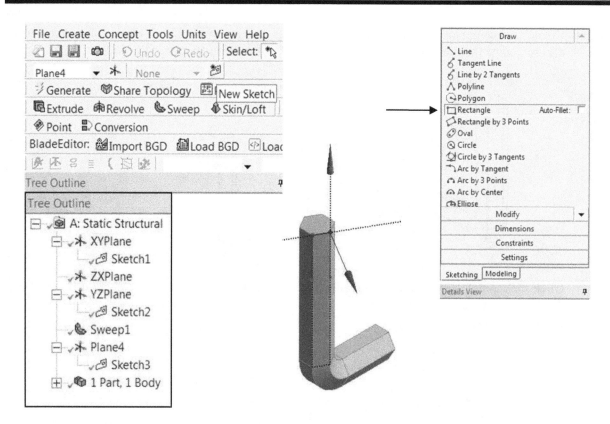

From Sketching Toolboxes, click **Dimensions**. Select **General**. Click the edge and a left click to position the size dimension. Specify 20.

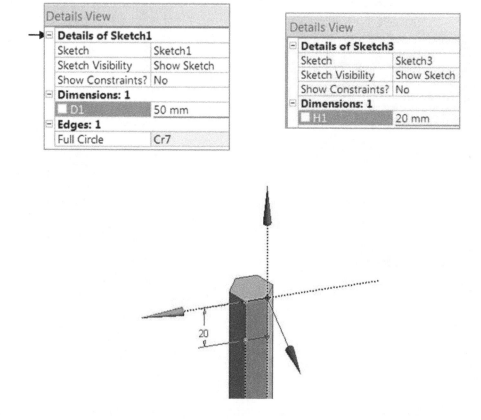

Click **Modeling** > **Extrude**. Select **Imprint Faces** from the **Operation** field. Click **Generate**.

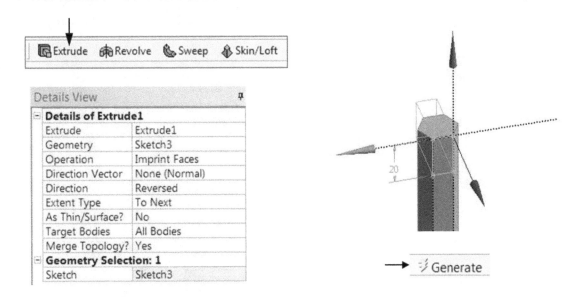

Repeat this process to create a second surface region. Click the icon of **New Plane**. In Details View, select **From Face** in the Type field. In the Base Plane field, left click to show Apply and pick the side surface of the wrench model. Click **Apply**. Click **Generate**. Plane5 is created.

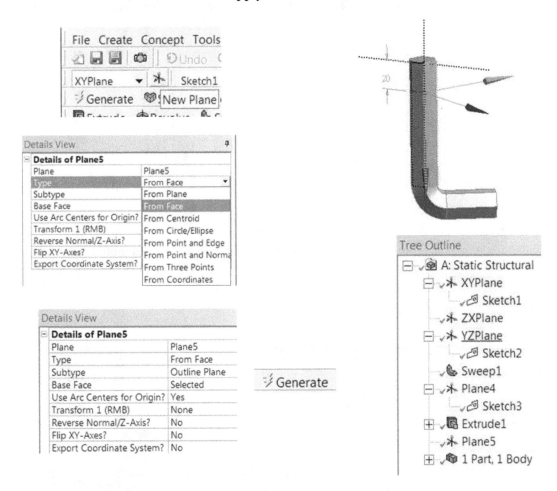

Now click the icon of **New Sketch**. Click Sketching and select **Rectangle** to sketch a corner rectangle, as shown. There is no need to add a dimension because the sketched rectangle is aligned with the rectangle sketched previously.

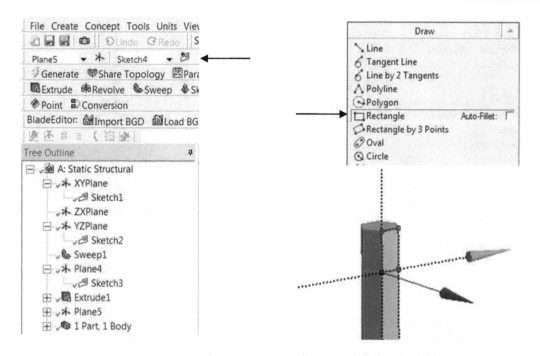

Click **Modeling** > **Extrude**. Select **Imprint Faces** from the **Operation** field. Click **Generate**.

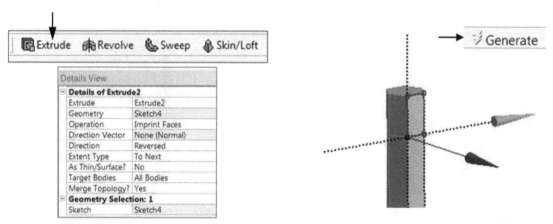

Upon completing the process of adding 2 imprinted surfaces to the wrench model, we need to close Design Modeler. Click **File** > Close **Design Modeler**, and go back to **Project Schematic**. Change the name to **wrench**.

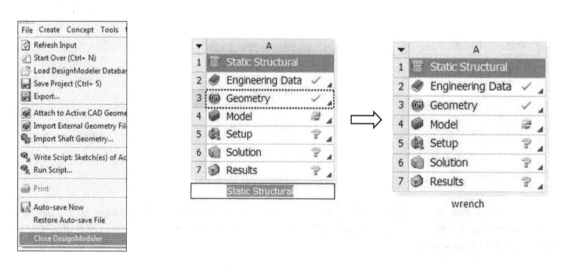

Click **Save** and specify wrench as the name of the file. Click **Save**.

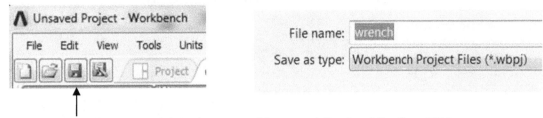

Step 5: Accept the Default Material Assignment of Structural Steel and Perform FEA

Double-click the icon of **Model**. We enter the **Mechanical Mode**. Check the unit system, and pick Metric (mm, t, N, s, mV, mA).

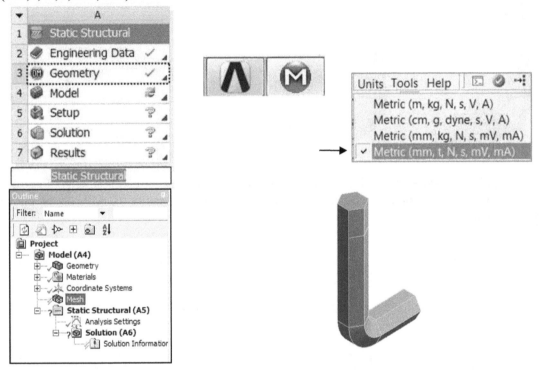

First highlight **Mesh**. In Sizing, set Resolution to 4. Right-click to pick **Generate Mesh**. The mesh is generated.

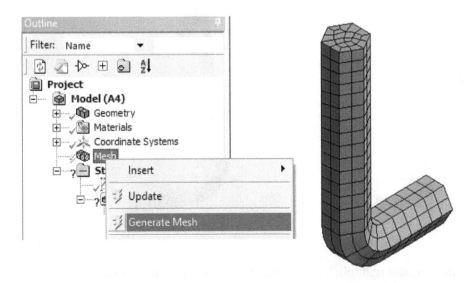

Highlight **Static Structural (A5)** and right-click to pick **Insert**. Afterwards, select **Fixed Support**. From the screen, pick the surface on the right side, as shown. Click **Apply**.

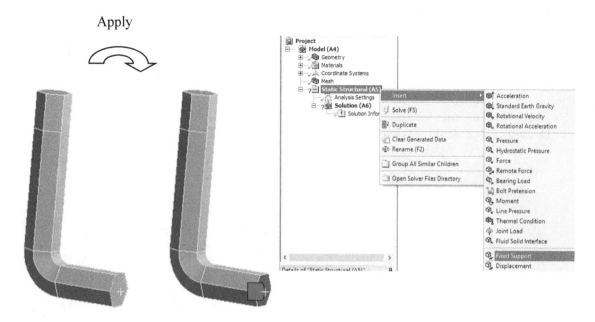

Apply

Highlight **Static Structural (A5)** and right-click to pick **Insert**. Afterwards, select **Pressure**. From the screen, pick the 2 imprinted surfaces, as shown. Click **Apply**. Specify 0.2 MPa as the magnitude of the pressure load.

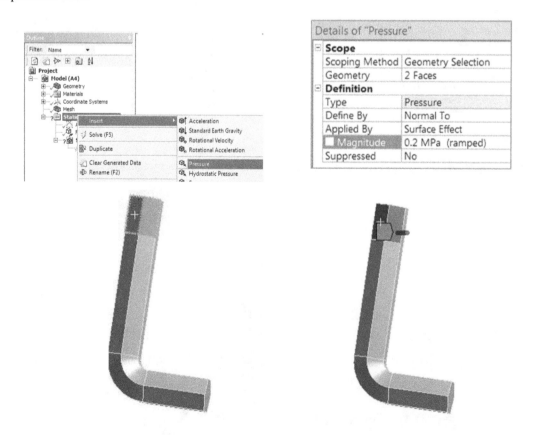

Highlight Solution (**A6**) and right-click to pick **Insert > Deformation > Total.**

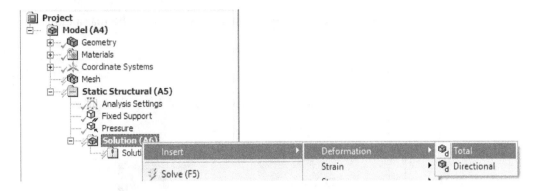

Highlight Solution (**A6**) and right-click to pick **Insert >Stress> Equivalent (von Mises).**

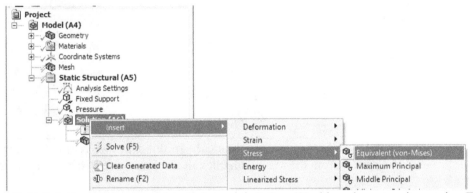

Highlight Solution (**A6**) and right-click to pick **Solve**. Workbench system is running FEA.

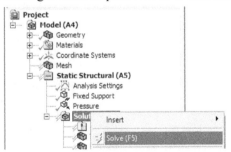

To plot the deformation, right click **Evaluate All Results**. Highlight **Total Deformation**. The maximum value is 0.055 mm.

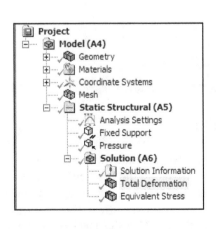

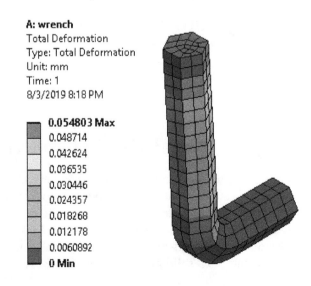

To plot the von Mises Stress distribution, highlight **Equivalent Stress**. The maximum value is 14.2 MPa.

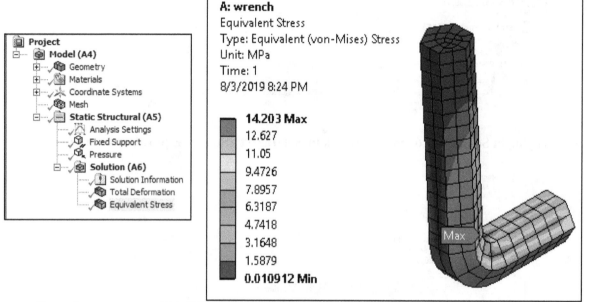

From the top menu, click **File > Close Mechanical**. In this way, we return to the main page of Workbench, or Project Schematic. From the top menu, click the icon of **Save Project.** From the top menu, click **File > Exit**.

2.7   Design and Analysis of a Connecting Rod

In this case study, we first design a connecting rod. Afterwards, we evaluate the rod by using FEA. The material type is structural steel. The Young's modules is 2×10^{11} Pascal and the Poisson's ratio is 0.30. The cylindrical surface at the left side is fixed to the ground. A bearing load is acting on the cylindrical surface at the right side. Its magnitude is 800 N. Make two plots: von Mises stress distribution and displacement distribution under the bearing load and constraint condition. Use the Design Modeler of Workbench to create a 3D solid model and perform FEA.

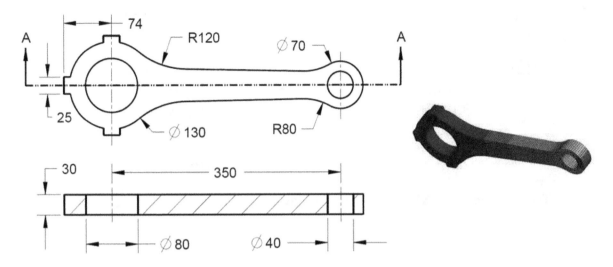

Step 1: Open Workbench.
From the start menu, click **ANSYS 19 > Workbench 19**. Users may directly click **Workbench 19** when the symbol of **Workbench 19** is on display.

Step 2: From **Toolbox**, highlight **Static Structural** and drag it to the **Project Schematic** area.

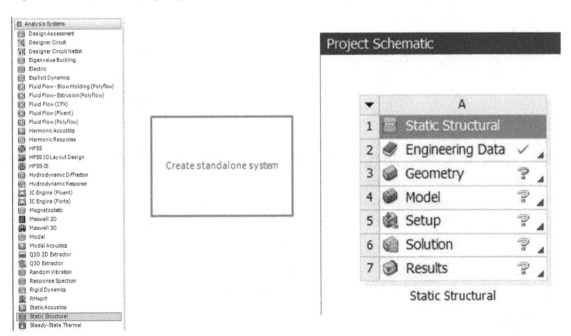

Step 3: Start Design Modeler and create a 2D sketch first.
Right-click the **Geometry** cell and pick **New DesignModeler Geometry** to start **Design Modeler**. Users may double click **Geometry** to directly enter **Design Modeler** if **New SpaceClaim Geometry** is not present. Click the icon of **Units** listed on the top menu, and select **Millimeter**.

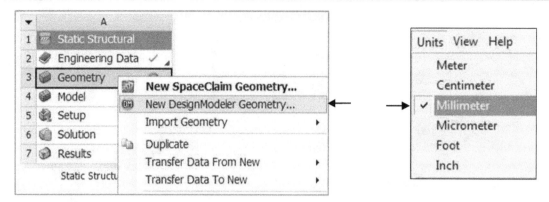

In **Tree Outline**, select **XYPlane** as the sketching plane.

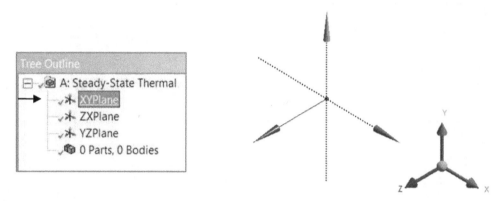

To orient the sketching plane, from the top menu, click the icon of **Look At** so that the selected sketching plane is oriented to be parallel to the screen and ready for preparing a sketch.

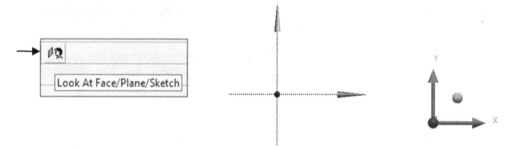

Click **Sketching**, and click **Draw**. In the Draw window, click Circle. Sketch a circle, as shown below.

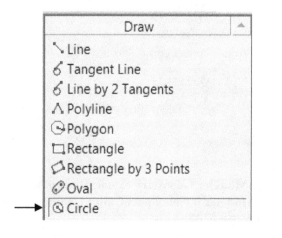

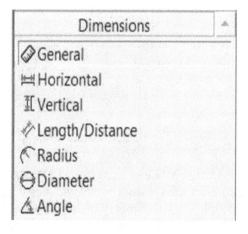

Click **Dimensions** and from the **Dimensions** panel, select **General**. Pick the circle and make a left click to position the diameter dimension. Specify 80 as its value.

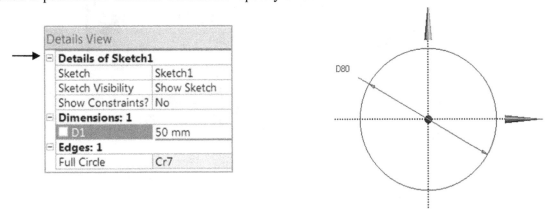

Sketch a second circle with its center on the horizontal axis, as shown.

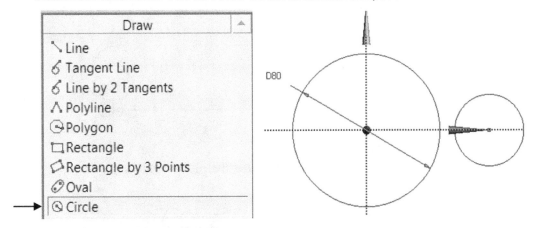

Select **General**. Pick the circle and make a left click to position its diameter dimension. Specify 40 as its value. Also specify 350 as the distance value between the 2 centers.

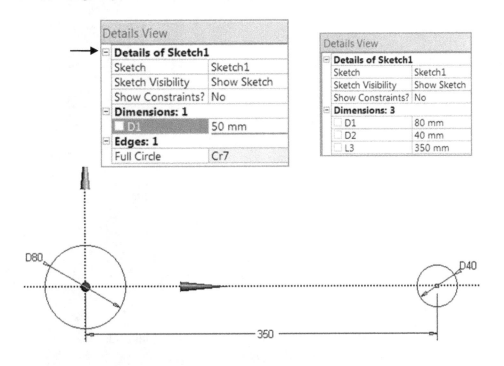

Sketch 2 more circles, as shown. Specify 130 and 70 as their diameter values.

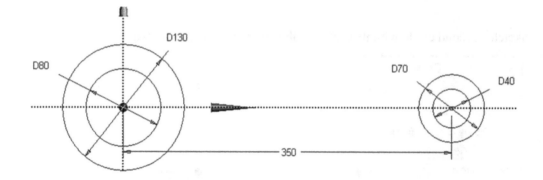

Sketch 2 vertical lines, as shown. Specify 50 and 40 as their length values.

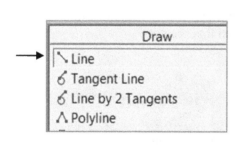

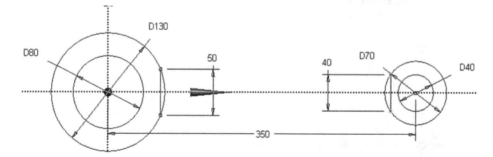

Sketch 2 lines connecting the 2 lines just sketched, as shown. Make sure that the symbol of P is shown when sketching these 2 lines. No dimensions are needed.

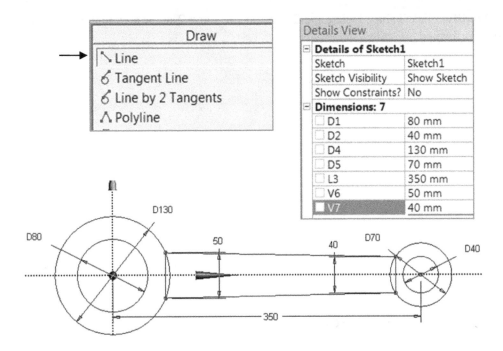

In Modify, click Fillet. Specify 120 as the radius value. On the screen display, click the circle with diameter value equal to 130 and the upper connecting line. An arc is created, as shown. Repeat this process: click the circle with diameter value equal to 130 and the lower connecting line. An arc is created, as shown.

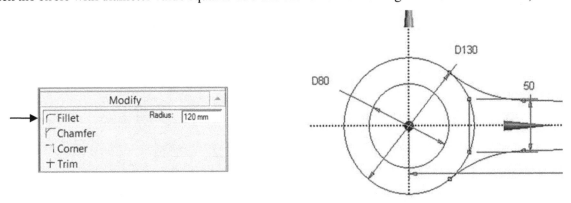

In Modify, click Fillet. Specify 80 as the radius value. On the screen display, click the circle with diameter value equal to 70 and the upper connecting line. An arc is created, as shown. Repeat this process: click the circle with diameter value equal to 70 and the lower connecting line. An arc is created, as shown.

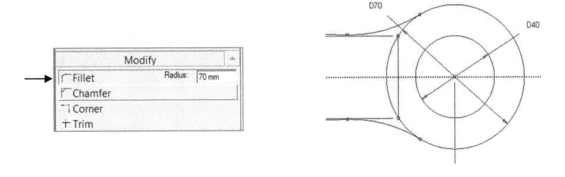

In **Modify**, click **Trim**. Delete the 2 vertical lines and 2 inner arcs that are not needed so that a closed sketch is formed, as shown below.

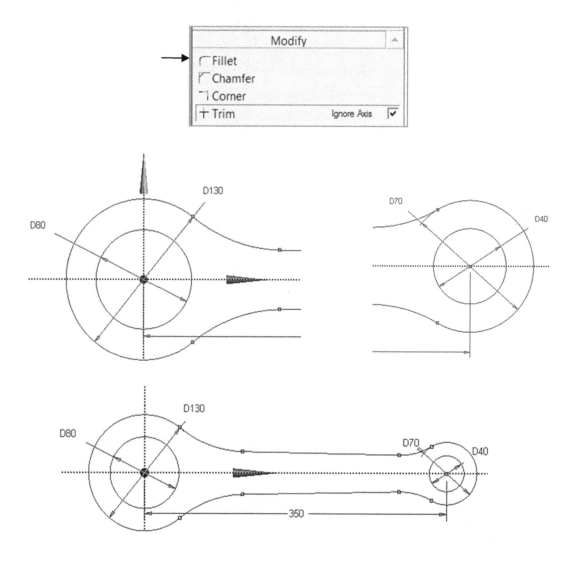

Now let us add three (3) rectangular steps. In Draw, click **Rectangle** and sketch 3 small rectangles on the circle with the diameter value equal to 130, as shown.

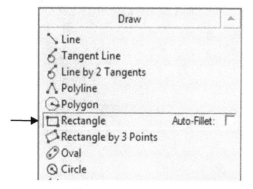

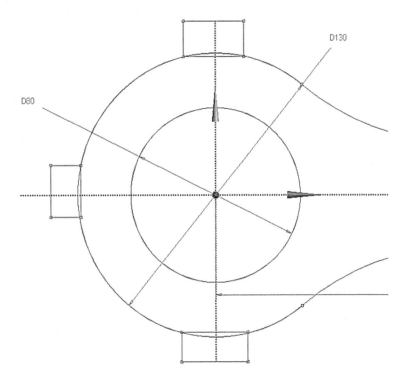

Click Dimensions. Specify 25 and 74 as the 2 size dimensions for each of the 3 sketched steps, as shown.

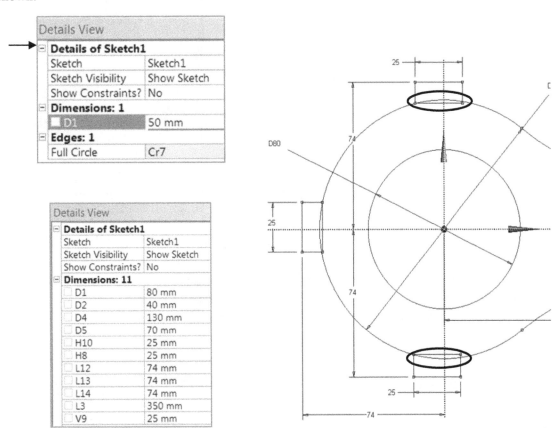

Note that some of the corners may not lie on the circle. As a result, the symmetric nature is not present. Under such circumstance(s), click Coincident from Constraint. Pick the corner point and the arc to enforce the corner point on the arc, as shown below.

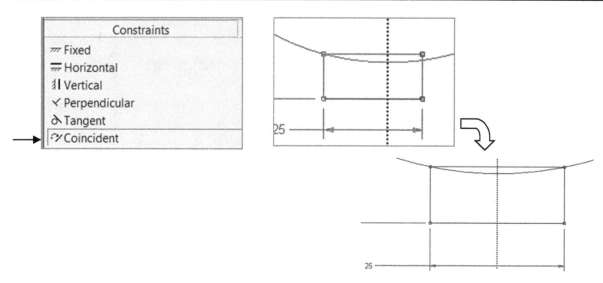

In Modify, click **Trim**. Delete the 2 vertical lines and 2 inner arcs that are not needed so that a closed sketch is formed, as shown below.

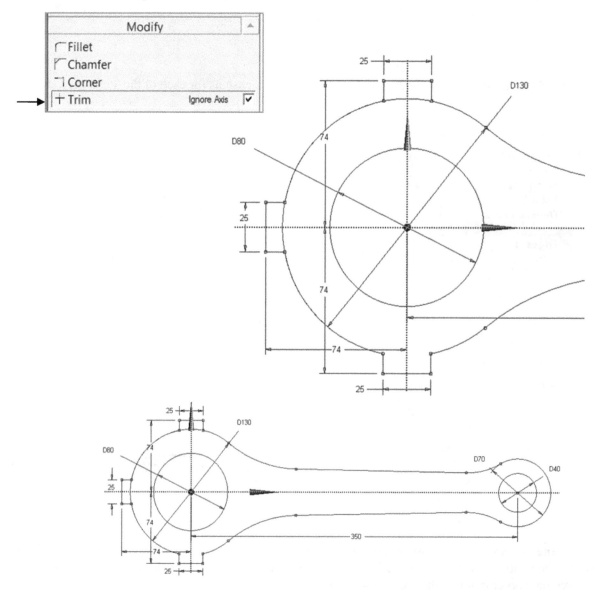

Upon completing the sketch, click the icon of **Extrude**. Click the created sketch or sketch 1 from the Tree Outline, and click **Apply**. Specify 30 as the thickness value. Click the icon of **Generate** to obtain the 3D solid model of the connecting rod feature, as shown.

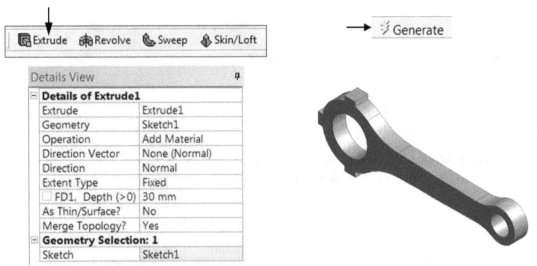

Now let us close **Design Modeler**. Click **File** > Close **DesignModeler**. We go back to **Project Schematic**. Click **Save** and specify connecting rod design as the name of the file. Click **Save**.

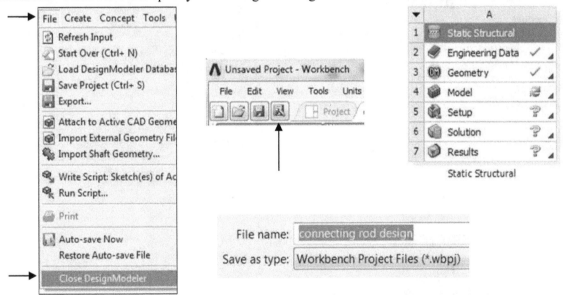

Step 4: Accept the Default Material Assignment of Structural Steel and Perform FEA

Double-click the icon of **Model**. We enter the **Mechanical Mode**. Check the unit system, and pick Metric (mm, t, N, s, mV, mA).

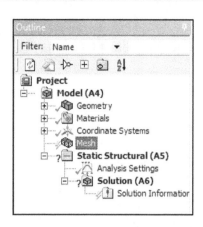

First highlight **Mesh.** To control the mesh size, in Details of "Mesh", expand Sizing. Set Resolution to 6. Now right-click **Mesh** to pick **Generate Mesh**. The mesh is generated.

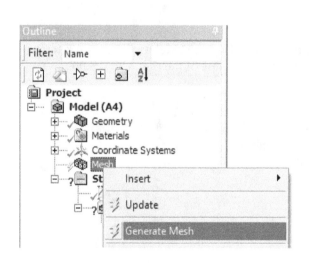

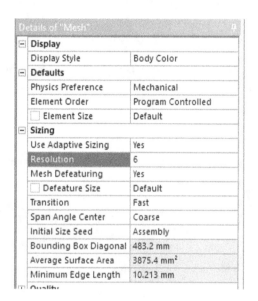

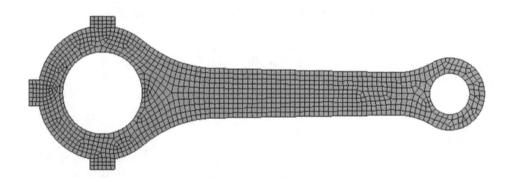

Highlight **Static Structural (A5)** and right-click to pick **Insert**. Afterwards, select Fixed Support. From the screen, pick the cylindrical surface from the hole on the left side, as shown. Click **Apply**.

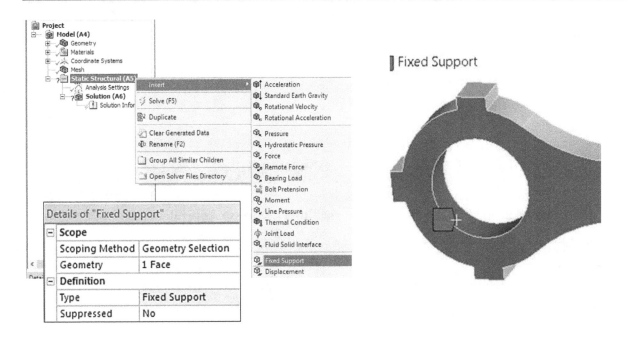

Highlight **Static Structural (A5)** and right-click to pick **Insert**. Afterwards, select **Bearing Load**. Select the cylindrical surface of the hole. Click **Apply**. In Details of "Bearing Load", in Coordinate System, select Global Coordinate System. Specify 1200 N in X Component and keep 0 as Y and Z Components.

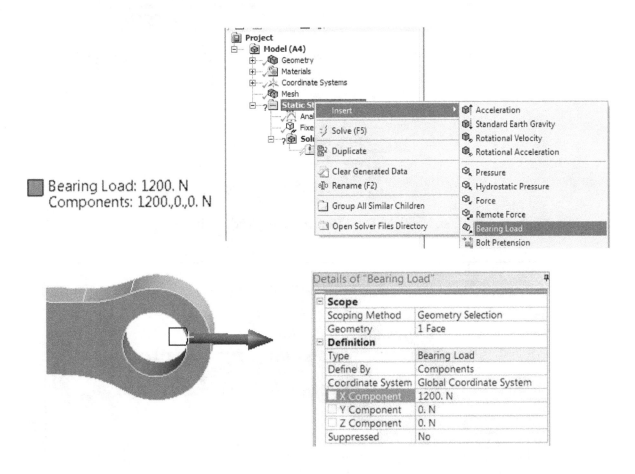

Highlight Solution (**A6**) and right-click to pick **Insert > Deformation > Total.**

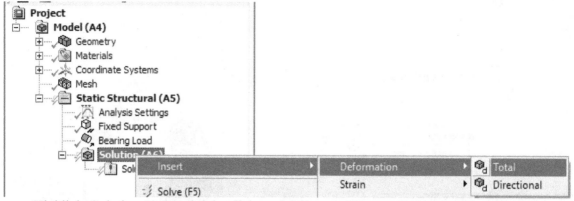

Highlight Solution (**A6**) and right-click to pick **Insert >Stress> Equivalent (von Mises).**

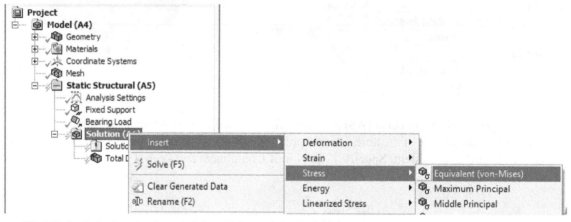

Highlight Solution (**A6**) and right-click to pick **Solve**. Workbench system is running FEA.

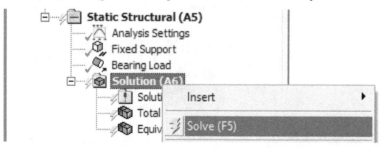

To plot the deformation, highlight **Total Deformation**. The maximum value is 0.0028 mm.

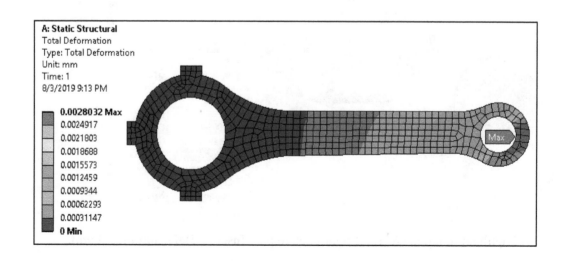

To plot the von Mises Stress distribution, highlight **Equivalent Stress**. The maximum value is 6.35 MPa.

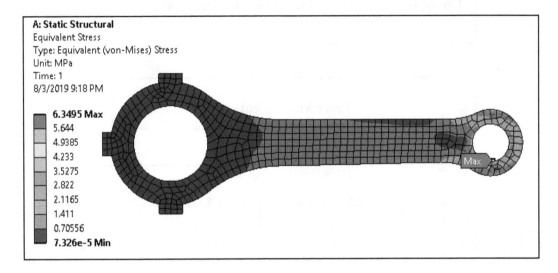

From the top menu, click **File > Close Mechanical**. In this way, we return to the main page of Workbench, or Project Schematic. From the top menu, click the icon of **Save Project.** From the top menu, click **File > Exit**.

2.8   Summary

Chapter 2 presents the general procedure to preform FEA in the ANSYS Workbench. When a user double-click or drag an analysis system from the Toolbox onto the Project Schematic, an analysis system appears. As an example, the Static Structural analysis system has the 7 components/cells, starting from Engineering Data, Geometry, Model, Setup, Solution, and Results. The ANSYS Workbench provides such a template for users to carry out a finite element analysis in a systematic approach. A double-click on one cell will lead to an action. Click a cell in an orderly fashion will lead to a complete and successful FEA run. The six case studies have provided the users with 6 opportunities to practice such a systematic approach to run FEA. The following flowchart summarizes and depicts such an information flow to carry out an FEA run.

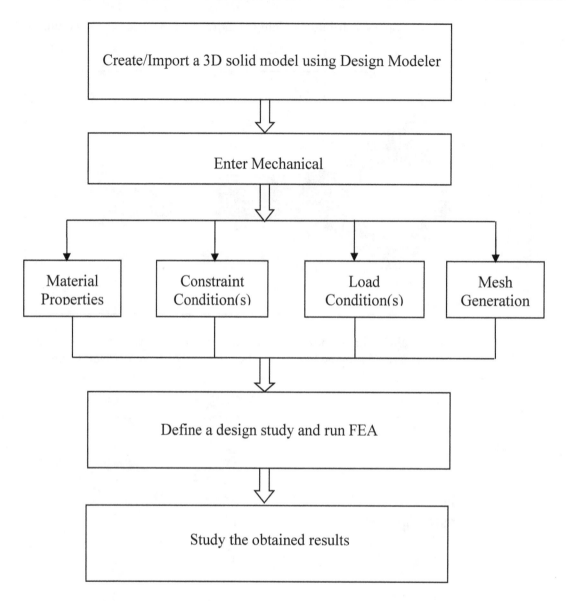

Chapter 2 also focuses on different operations needed for creating geometric models. Users should be familiar with Extrude, Revolve and Sweep. These three (3) operations are used most frequently in geometry creation. Users should also have learned the importance of setting a unit system, which best fits his/her design need. As a conclusion, users have already learned that his/her own hands-on practice with the commands to create geometric models is the most efficient way to gain the needed skills of using the ANSYS Design Modeler.

2.9   References

1.   ANSYS Parametric Design Language Guide, Release 15.0, Nov. 2013, ANSTS, Inc.
2.   ANSYS Workbench User's Guide, Release 15.0, Nov. 2013, ANSYS, Inc.
3.   X. L. Chen and Y. J. Liu, Finite Element Modeling and Simulation with ANSYS Workbench, 1st edition, Barnes & Noble, January 2004.
4.   X. S. Ding and G. L. Lin, ANSYS Workbench 14.5 Case Studies (in Chinese), Tsinghua University Publisher, Feb. 2014.
5.   G. L. Lin, ANSYS Workbench 15.0 Case Studies (in Chinese), Tsinghua University Publisher, October 2014.

6. K. L. Lawrence. ANSYS Workbench Tutorial Release 13, SDC Publications, 2011.
7. K. L. Lawrence. ANSYS Workbench Tutorial Release 14, SDC Publications, 2012
8. Huei-Huang Lee, Finite Element Simulations with ANSYS Workbench 14, Theory, Applications, Case Studies, SDC Publications, 2012.
9. Huei-Huang Lee, Finite Element Simulations with ANSYS Workbench 16, Theory, Applications, Case Studies, SDC Publications, 2015.
10. E. H. Dill, The Finite Element Method for Mechanics of Solids with ANSYS Applications, Barnes & Noble, September 2011.
11. Jack Zecher, ANSYS Workbench Software Tutorial with Multimedia CD Release 12, Barnes & Noble, 2009.
12. G. M. Zhang, Engineering Design and Creo Parametric 3.0, College House Enterprises, LLC, 2014.

2.10 Exercises

1.   The engineering drawing of a beam structure is shown below. The unit is millimeter. Assume that a uniformly distributed load, acts on the entire top surface. The magnitude of this load is 1000 Newton and downward. The material type is Structural Steel. It is assumed that the left side of the beam is fixed, or the beam is a cantilever beam. The reader is asked to create a 3D solid model of this beam in the ANSYS DesignModeler. Perform an FEA analysis to determine two values: the total deformation and von Mises stress. Pay a special attention to the location of the maximum von Mises stress.

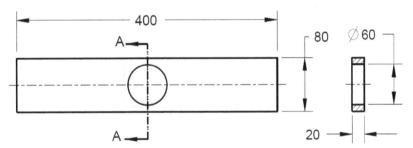

Afterwards, the user is asked to use Duplicate to create a new analysis system. Double-click the Geometry cell to modify the hole diameter from the current 60 mm to a new diameter value of 62 mm, as shown below. Perform an FEA analysis to determine two values: the total deformation and von Mises stress. Pay a special attention to the location of the maximum von Mises stress. Use Probe to identify the coordinates of this location.

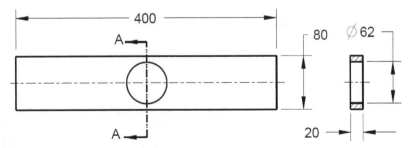

For the purpose of comparison, fill in the numerical values with their corresponding units in the following table. Also compare the locations of the maximum von Mises stresses obtained from the 2 FEA runs. The user should notice that this location has been changed from Case 1, indicating the importance of the design of such a cantilever beam in determining the location of the maximum von Mises stress.

Hole Diameter (mm)	Max Total Deformaton (mm)	Max von Mises Stress (MPa)	Location (fixed end or middle)
Case 1: 60			
Case 2: 62			
Case 3: 64			

2. The engineering drawing of a beam structure is shown below. The unit is millimeter. Assume that a uniformly distributed load, acts on the part of the top surface. The size of this area is 60x10 mm². The magnitude of this load is 1000 Newton and downwards. The material type is Structural Steel. It is assumed that the left side of the beam is fixed, or the beam is a cantilever beam. The reader is asked to create a 3D solid model of this beam in the ANSYS DesignModeler. Make a surface region of 60x10 mm² is also created using Printed Face. Perform an FEA analysis to determine two values: the total deformation and von Mises stress. Pay a special attention to the location of the maximum von Mises stress. Use Probe to identify the coordinates of this location.

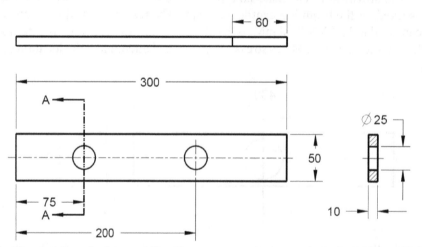

Afterwards, the user is asked to use Duplicate to create a new analysis system. Double-click the Geometry cell to add a new surface region: 40x10 mm². The load magnitude is 500 N and downwards. Perform an FEA analysis to determine two values: the total deformation and von Mises stress. Pay a special attention to the location of the maximum von Mises stress.

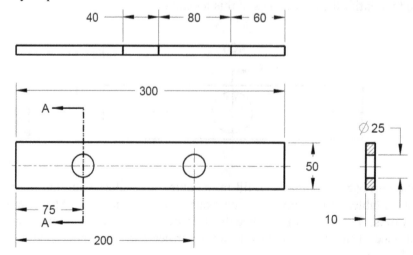

For the purpose of comparison, fill in the numerical values in the following table. Also compare the locations of the maximum von Mises stresses obtained from the 2 FEA run. The user should notice that this location of the maximum von Mises stress. Use Probe to identify the coordinate of this location.

Surface Region	Max Total Deformation	Max von Mises Stress
Surface Region: 60x10 mm^2		
2 Surface Regions: 60x10 mm^2 and 40x10 mm^2		

3. The engineering drawing of a beam structure is shown below. The unit is millimeter. Assume that a bearing load acts on the cylindrical surface of the hole. The magnitude of this load is 500 Newton and downwards (Y component only). The material type is Structural Steel. It is assumed that the left side of the beam is fixed, or the beam is a cantilever beam. The reader is asked to create a 3D solid model of this beam in the ANSYS DesignModeler. Perform an FEA analysis to determine two values: the total deformation and von Mises stress. Pay a special attention to the location of the maximum von Mises stress.

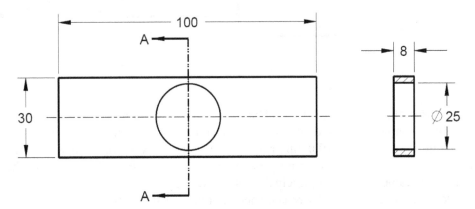

Afterwards, the user is asked to use Duplicate to create a new analysis system. Double-click the Geometry cell to modify 100 mm to 200 mm, 30 mm to 45 mm, 8 mm to 12 mm and diameter value of 25 mm to diameter value of 30 mm. In addition, add a new fixed support constraint to the surface on the right side (this beam is supported at both ends). Perform an FEA analysis to determine two values: the total deformation and von Mises stress. Pay a special attention to the location of the maximum von Mises stress. Use Probe to identify the coordinates of this locations.

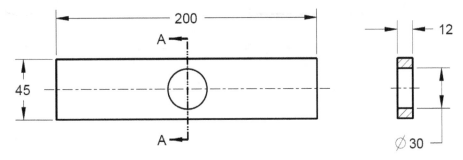

For the purpose of comparison, fill in the numerical values in the following table. Also compare the locations of the maximum von Mises stresses obtained from the 2 FEA run. The user should notice that this location of the maximum von Mises stress. Use Probe to identify the coordinates of this location.

Displacement Constraint(s)	Max Total Deformation	Max von Mises Stress
Fixed Support on the Left Side		
Fixed Supports on Both Left and Right Sides		

4. The engineering drawing of a shaft component is shown below. The unit is millimeter. Assume that a tensile force acts on the circular surface on the right end. The magnitude of this load is 1000 Newton. The material type is Structural Steel. It is assumed that the left side of the shaft component is fixed. The reader is asked to create a 3D solid model of this shaft component in the ANSYS DesignModeler. Perform an FEA analysis to determine two values: the total deformation and maximum principal stress. Pay a special attention to the location of the maximum value of the maximum principal stress. Its location is on the edge or the intersection of the two cylindrical features. Also use Probe to determine the stress developed at the right end, which is called Reference Stress. Its value should be closely equal to 0.5 MPa.

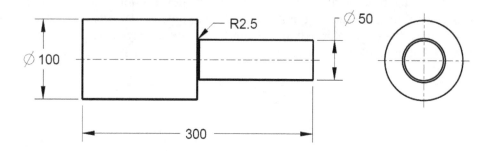

Afterwards, the user is asked to use Duplicate to create 3 more new analysis systems. For each of the 3 new analysis system, double-click the Geometry cell to modify the radius value from 2.5 mm to 5.0 mm, 7.5 mm, and 10.0 mm, respectively. Just click Solve to update the information on the maximum value of the maximum principal stress. Pay a special attention to the location of the maximum value of the first principal stress. Use Probe to identify the coordinates of this location for each of those 3 cases. There is no need to run FEA three more times.

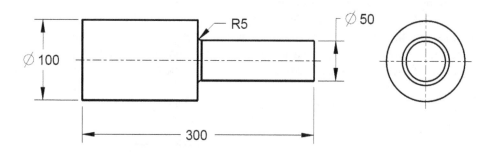

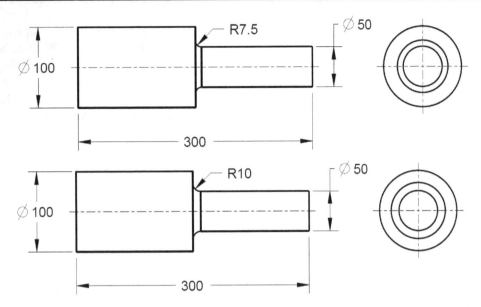

For the purpose of comparison, fill in the numerical values in the following table and calculate the ration of max value of maximum principal stress and reference stress for each of the 4 cases. The user is asked to construct a curve: Maximum Value of the Maximum Principal Stress vs Value of the Corner Radius. Such a curve is also called Stress Concentration Curve in the engineering design community.

Corner Radius	R2.5	R5.0	R7.5	R10.0
Max Value of Maximum Principal Stress (MPa)				
Reference Stress (MPa)				
Ratio of Max Value of Max Principal Stree and Reference Stress				

5.   The engineering drawing of a lock strip is shown below. The unit is inch. You are asked to use sweep operation to create a 3D solid model in the ANSYS Workbench. The path is a horizontal line connected to 2 semi-circles. The 2 semi-circles are extended at their end. The profile is a square with the side dimension equal to 0.50 inch. Assume that a uniformly distributed pressure load, acts on a surface area. The surface area is 1.50x0.4 inch^2. The magnitude of this pressure load is 1.0 MPa. The material type is Structural Steel. It is assumed that the two opening surfaces are subjected to the fixed support constraint. Perform an FEA analysis to determine two values: the total deformation and von Mises stress. Pay special attention to the locations of the maximum von Mises stress, and the maximum total deformation.

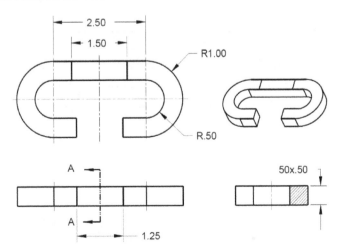

6. The engineering drawing of a conic-shaped support is shown below. The unit is inch. You are asked to use the revolve operation to create a 3D solid model in the ANSYS Workbench. There is a pressure load acts on the top surface. Assume the magnitude of this pressure load is 5.0 MPa. The material type is Structural Steel. It is assumed that the bottom surface is subjected to the fixed support constraint. Perform an FEA analysis to determine two values: the total deformation and von Mises stress. Pay special attention to the locations of the maximum von Mises stress, and the maximum total deformation.

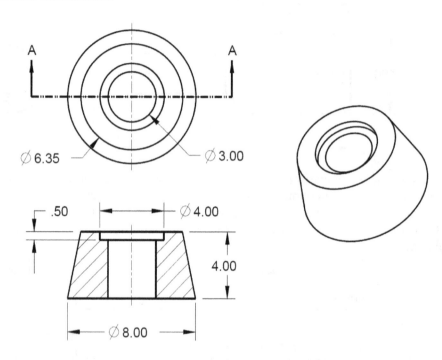

7. The engineering drawing of a beam is shown below. The unit is mm. Create a 3D solid model in ANSYS Workbench. There is a pressure load of 5 MPa acting on the top surface. The material is steel. The bottom surface is fixed to the ground. Create the model and perform an FEA to determine the total deformation and von Mises stress. Pay special attention to the locations of the maximum von Mises stress, and the maximum total deformation.

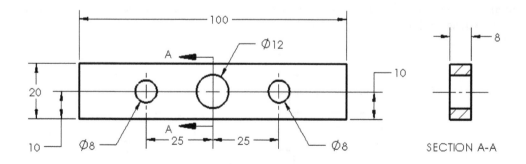

CHAPTER 3

FEA with Assemblies

3.1 Introduction

In Chapter 2, the discussion focuses on FEA with components. Components are important because a product is an assembly of components. Without components, there would be no products. Our focus of this chapter is FEA with assemblies. Section 3.2 is a case study to show how to assemble 2 components to create an assembly model. Because we have two components, we may need to specify 2 material types and each material type has to be assigned to a specific component. Section 3.2 also introduces the concept of contact configuration. At the contact surfaces, the 2 components are bonded together without any penetration and sliding when subjected to a force action. Section 3.2 also illustrates the procedure to display the stress/deformation for the entire assembly and for each of the individual components as well, resembling the approach of free-body diagrams used in engineering mechanics. Section 3.3 deals with an assembly consisting of 2 thin layers. The difference in the thermal expansion coefficients induces the internal stress, leading to the system deformation. Section 3.4 presents a case study where an assembly has three layers, and each layer has a designated material type. When subjected to a temperature load, the entire assembly system is deflected with induced internal stress. Section 3.5 presents the design of a gearbox. In this case study, we present the feature-based modeling approach signifying the capabilities of the DesignModeler in creating geometry.

3.2 FEA with a Support and Pin Assembly

The following drawing shows an assembly consisting of a support and a pin. The material type of support is structural steel and the material type of pin is aluminum alloy. The bottom surface is fixed to the ground. The loading condition is a force acting on the surface of the pin. The magnitude is 1000 Newton. Use the DesignModeler of Workbench to create a 3D solid model of this assembly and perform FEA.

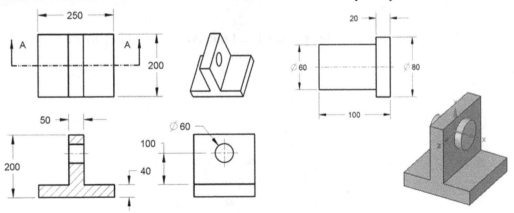

Step 1: Open Workbench. From the start menu, click **ANSYS 19 > Workbench 19**. Users may directly click Workbench 19 when the symbol is displayed on screen.

Workbench 19.2

Step 2: From **Toolbox**, highlight **Static Structural** and drag it to the main screen.

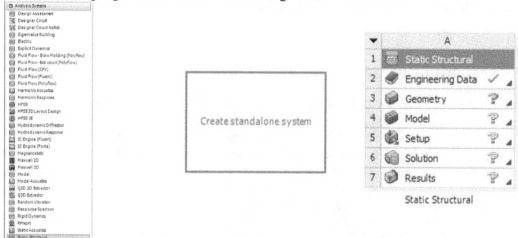

Step 3: Double click Geometry to start Design Modeler.

Right-click the **Geometry** cell and pick **New DesignModeler Geometry** to start **Design Modeler**. Users may double click **Geometry** to directly enter **Design Modeler** if **New SpaceClaim Geometry** is not present. In the DesignModeler window, select the unit system. In this case, select Millimeter.

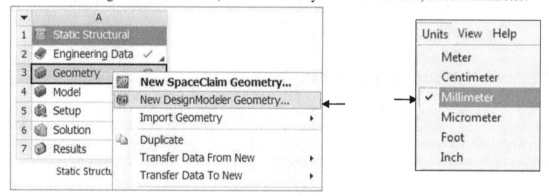

Select **YZPlane** as the sketching plane. Click **Look At** to orient the selected sketching plane.

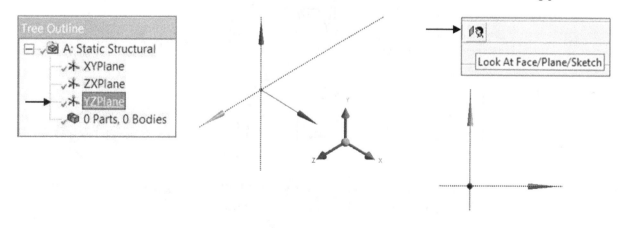

Click **Sketching** > **Draw**. In the **Draw** window, click **Circle**. Sketch a circle.

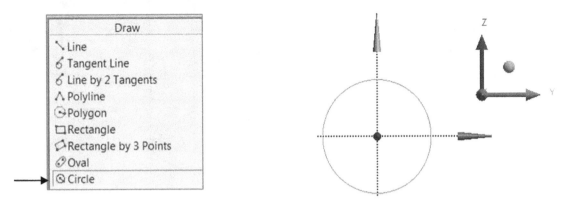

Click **Dimensions** and select **General**. Pick the circle and make a left click to position the size dimension. Specify 60 as its value.

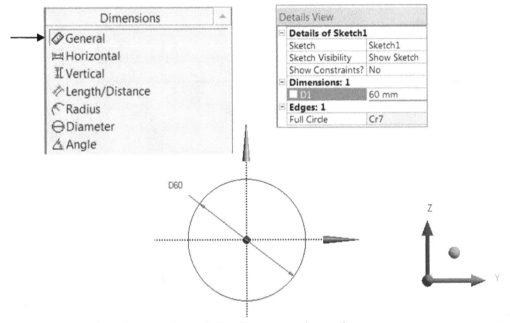

From **Draw**, select **Rectangle**, and sketch a rectangle, as shown.

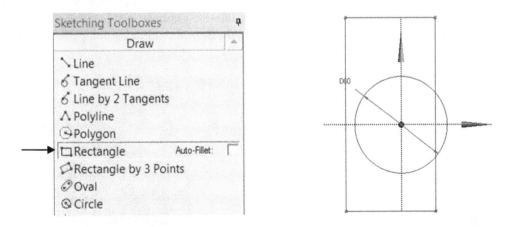

From **Constraints**, select **Symmetry**. Add a symmetry constraint so that the sketched rectangle is symmetric about the horizontal or Y-axis.

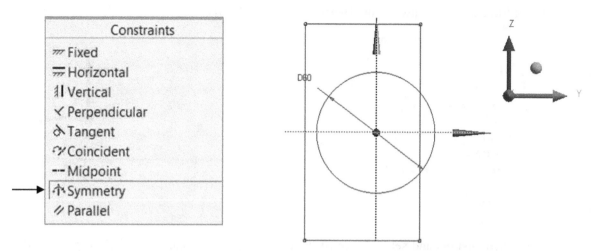

Click Dimensions > General. Add three dimensions: 200. 100 and 60, as shown.

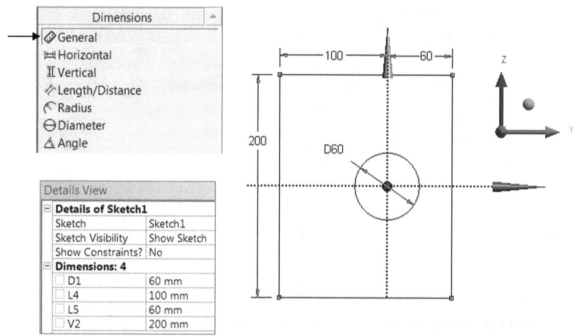

Upon completing the sketch, click the icon of **Extrude**. Click the created sketch or sketch 1 from the Tree Outline, and click **Apply**. For Direction, select **Both-Symmetric** and specify 20 as the thickness value (both = 40). Click the icon of **Generate** to obtain the 3D solid model of the plate feature, as shown.

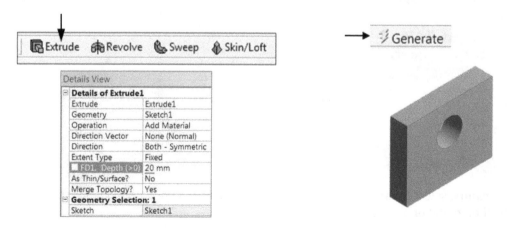

Now let us create the second feature: From Tree Outline, highlight the **XY** Plane (selecting it as the sketching plane) and click **New Sketch**. Click **Look At**. From **Draw** to select **Rectangle**. Sketch a rectangle as shown.

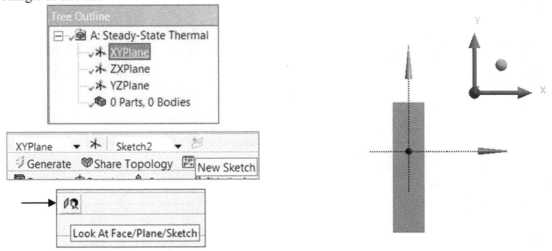

From **Draw** to select **Rectangle**. Sketch a rectangle as shown.

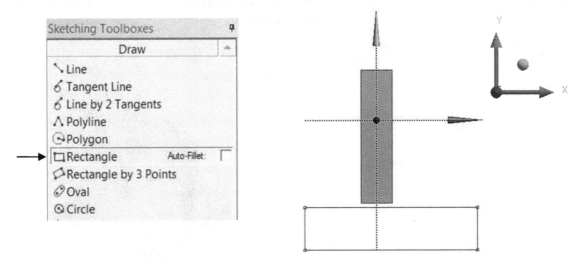

From Constraints, select Symmetry. Add a symmetric constraint so that the sketched rectangle is symmetry about the vertical axis, as shown.

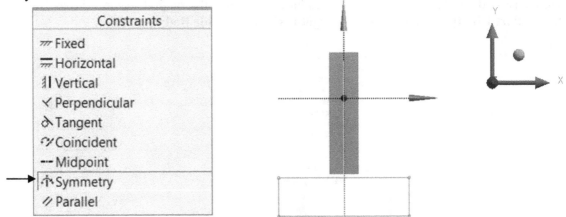

From **Dimensions,** select **General**. Specify a position dimension of 100 first. Afterwards, add 2 size dimensions of 40 and 200.

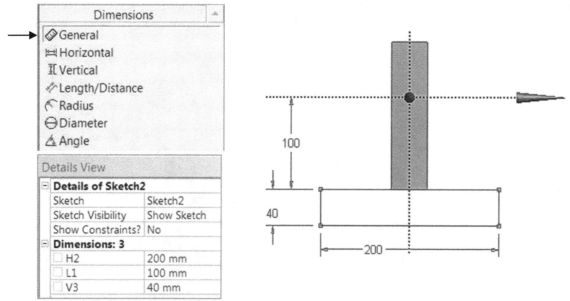

Upon completing the sketch, click the icon of **Extrude**. Click the created sketch or sketch 2 from Tree Outline, and click **Apply**. For Direction, select **Both-Symmetric** and specify 100 as the extrusion value (both = 200). Click **Generate** to create the plane geometry.

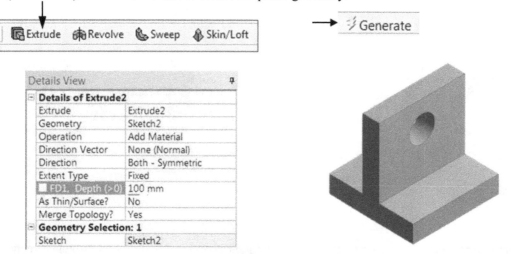

Up to this moment, we have completed the creation of a 3D solid model for Support. Now let us create the second component – pin. First let us hide the created the support feature. From Tree Outline, expand **1 Part 1 Body**. Highlight **Solid** and right-click to pick **Hide Body**.

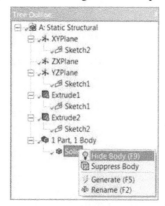

Now select **XY Plane**, click **New Sketch** and click **Look At** to create **Sketch3**.

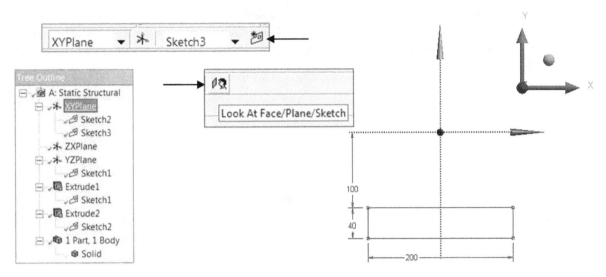

From **Draw**, select **Line**. Along the X-axis, sketch the following with 6 lines. Pay attention to using the symbol P to connect the 2 neighboring lines.

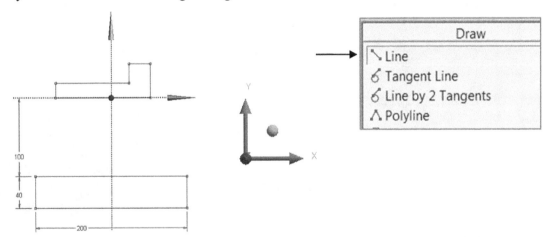

Click **Dimensions** and select **Horizontal**. Click the vertical Y-axis and the vertical line on the left side. Specify 60 as its value. Afterwards, specify 80 and 100. Click **Vertical** to specify 30 and 60, as shown.

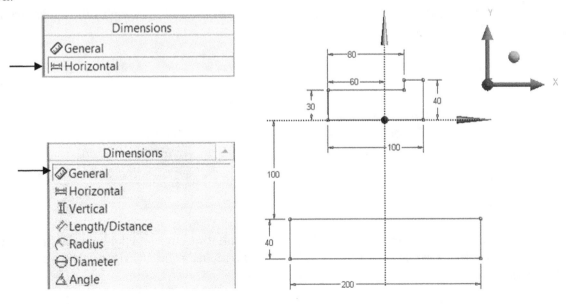

Click Modeling, and click Revolve. From Tree Outline, select Sketch3, and click Apply. In Axis and select the X-axis. In Operation, select Add Frozen. Click **Generate**. The pin component is created. The Add Frozen operation add the created feature as a new separate part, Therefore, in Tree Outline, **2 Parts, 2 Bodies** are listed.

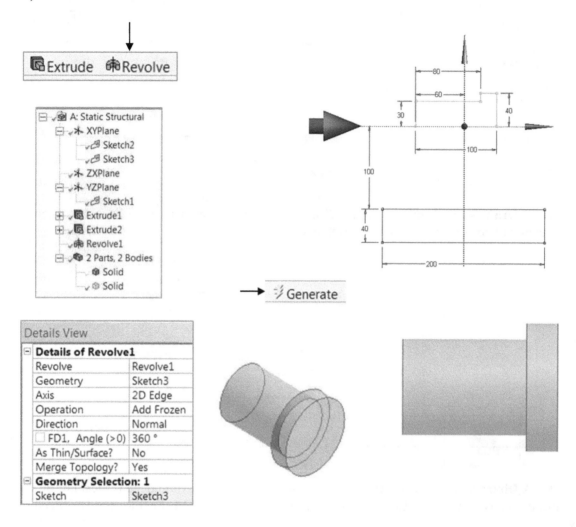

Highlight the support solid and right-click to pick **Show Body**. The assembly of support and pin is completed and shown below.

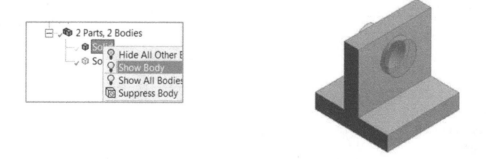

Let us rename the 2 created solids to Support and Pin, respectively. Highlight the first Solid and right-click to pick Rename and type Support. Highlight the second solid and right-click to pick Rename and type Pin.

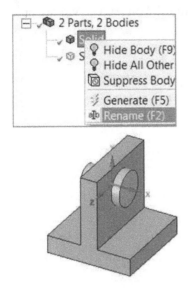

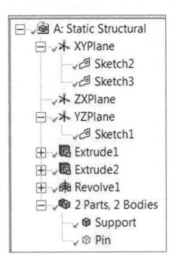

Upon completing the creation of this 3D solid model for the assembly, we need to close Design Modeler. Click **File** > Close **DesignModeler**. Go back to Project Schematic. Click Save Project and specify Support-Pin Assembly as the file name, and click **Save**.

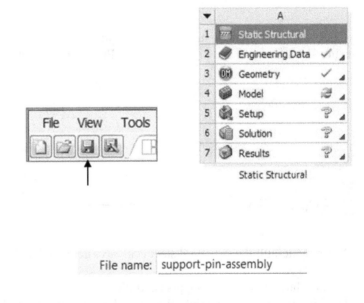

Step 5: Activate the Aluminum Alloy and Define the Contact Region

We need two types of material: Structural Steel for Support and Aluminum Alloy for Pin. In the Static Structural panel, double click Engineering Data. We enter the Engineering Data Mode.

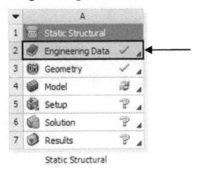

Click the icon of **Engineering Data Sources**. Click **General Materials**. We are able to locate Aluminum Alloy.

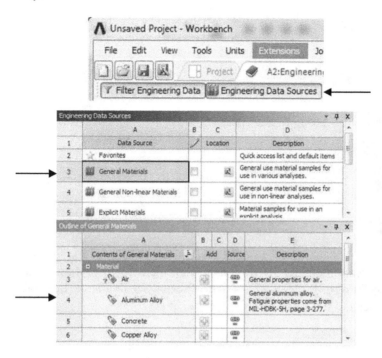

Click the plus symbol nearby Aluminum Alloy to activate it. Afterwards, click the icon of **Engineering Data Sources** and click the icon of Project to go back to **Project Schematic**.

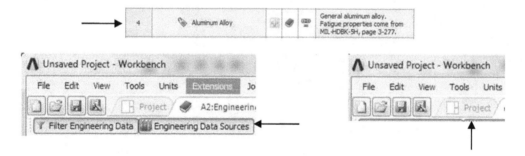

Note: by default the structural steel is always activated and assigned.

Step 6: Assign Material Types to Support and Pin.

Double-click the cell: Model. We enter the Mechanical Mode. Check the unit system: Metric (mm, t, N, s, mV, mA).

Let us do the material assignment. From the Project tree, expand Geometry. Highlight the listed Support. In Details of "Support", Structural Steel is already assigned to Support.

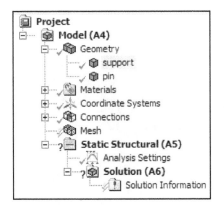

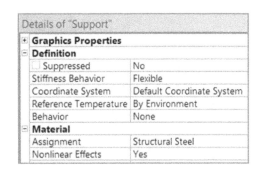

Highlight the listed Pin. In Details of "Pin", click material assignment, pick Aluminum Alloy to replace Structural Steel, which is the default material assignment.

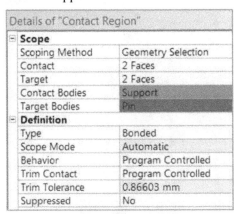

Expand Connections > Contacts > Contact Region. The contact surfaces between Support and Pin are colored immediately. In Details of "Contact Region", there are 2 pairs of faces are in contract. The first pair of faces are the 2 cylindrical surfaces of Support and Pin. The second pair of surfaces are the flat surfaces of Support and Pin.

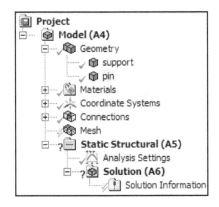

On the right side of the screen, Support is defined as Contact Body and Pin is defined as Target Body. The type of Contact is defined as Bonded. This is the default configuration for contact region, meaning no sliding or separation between faces or edges is allowed.

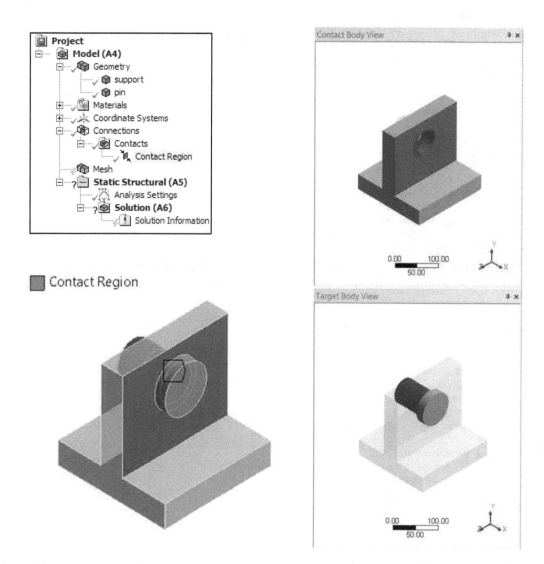

Step 6: Perform FEA

From the project model tree, highlight **Mesh.** In Details of "Mesh", expand **Sizing**, Set Resolution to 4. Highlight Mesh, and right-click to pick **Generate Mesh**.

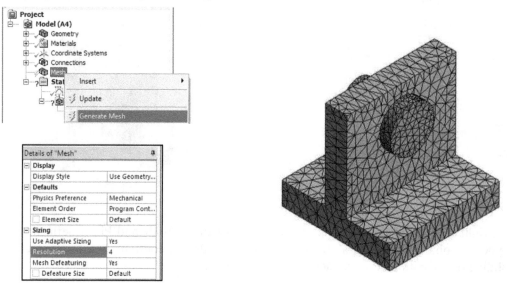

From the project tree, highlight **Static Structural (A5)** and right-click to pick **Insert**. Afterwards, select **Fixed Support**. From the screen, pick the bottom surface of Support, as shown. As a result, this 3D solid assembly model is fixed to the ground through the bottom surface.

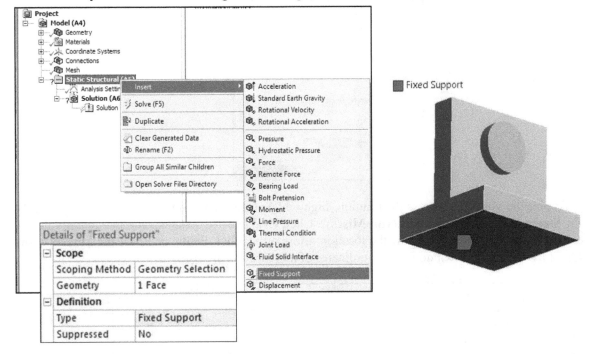

From the project tree, highlight **Static Structural (A5)** and right-click to pick **Insert**. Afterwards, select **Force**. From the screen display, pick the surface on the left side of the pin, as shown. Specify -1000 N as the magnitude of the force load. The direction of force is towards the left side.

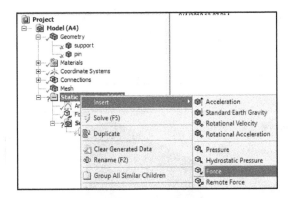

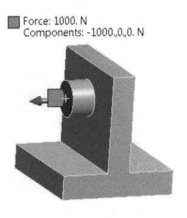

From the project tree, highlight **Solution (A6)** and right-click to pick **Solve**. Workbench system is running FEA.

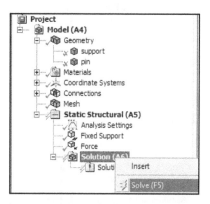

To plot the von Mises stress distribution, highlight **Solution (AS)** and pick **Insert**. Afterwards, highlight **Stress** and pick **Equivalent (von Mises).** Click Evaluate All Results. The maximum value of von Mises stress is 1.76 MPa. To know the location information, click **Probe** and pick the Max location. In the Graphics Annotations window, the coordinates of this location are listed.

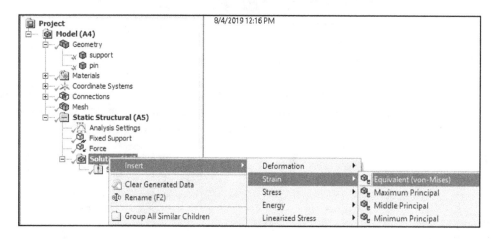

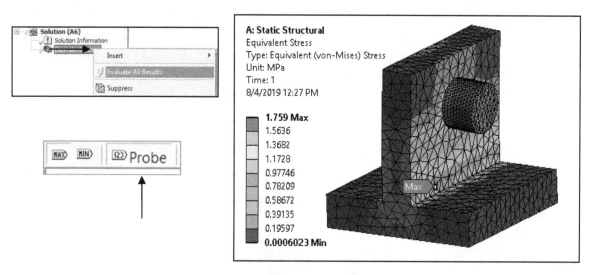

Graphics Annotations						
Type	Value	Note	Unit	Location X	Location Y	Location Z
Result	1.7191		MPa	-20.286438	-91.728647	-47.312189

To plot the displacement distribution, highlight **Solution (A6)** and pick **Insert**. Afterwards, highlight **Deformation** and pick **Total.** Click Evaluate All Results. The maximum total deformation is 0.004 mm.

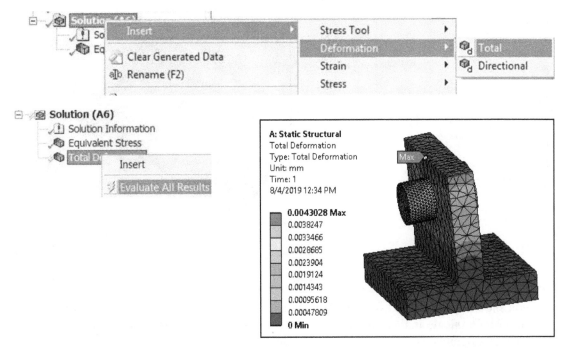

To display an individual component, say the support component only, go to Project tree, highlight Support and right-click to pick Hide All Other Bodies.

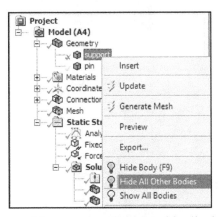

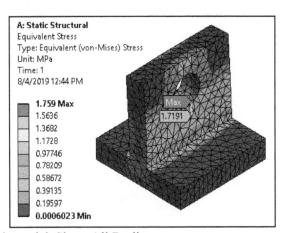

Now recover the assembly display, right-click to pick Show All Bodies.

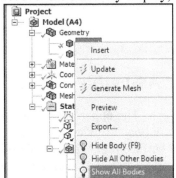

To display the pin component only, go to Project tree, highlight Pin and right-click to pick Hide All Other Bodies. It is important to display the maximum value of von Mises stress with the pin component.

Click Probe. At the Max symbol location, make a left click to display the value of von Mises stress. In the Graphics Annotations window located at the bottom of the computer screen, the value of von Mises stress and the coordinates of this location are also listed, x = -24.99, y = 5.99 and z = -29.26.

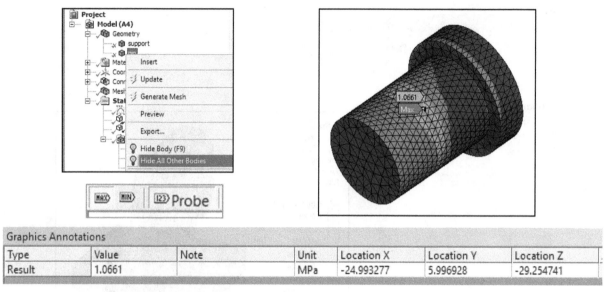

Graphics Annotations						
Type	Value	Note	Unit	Location X	Location Y	Location Z
Result	1.0661		MPa	-24.993277	5.996928	-29.254741

Now you should return to the assembly display by clicking Show All Bodies. From the top menu, click **File** > **Close Mechanical**. In this way, we return to the main page of Workbench. Click Save Project. Specify Support-Pin Assembly as its name, and click Save.

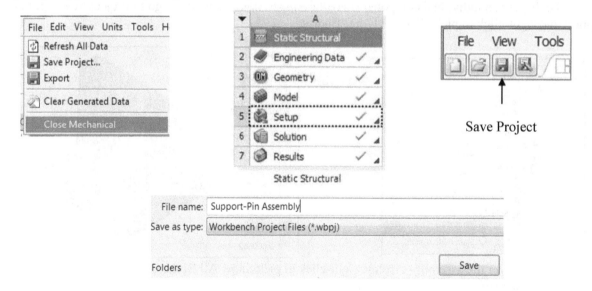

3.3   FEA with a Steel and Copper Layer Assembly

The following drawing shows an assembly consists of two layers. The upper layer material is structural steel and the lower layer material is copper alloy. Both layers are bonded. The right end has a constraint condition: Fixed Support. A pressure load acts on a rectangular surface region at the left end, which is set free. The pressure magnitude is 0.05 MPa. The rectangular area is 5x2 mm^2 with a position dimension equal to 1.5 mm. The assembly is also subjected to temperature load. The environment temperature is 10°C, and the temperature of the two layers of the assembly is 30°C.

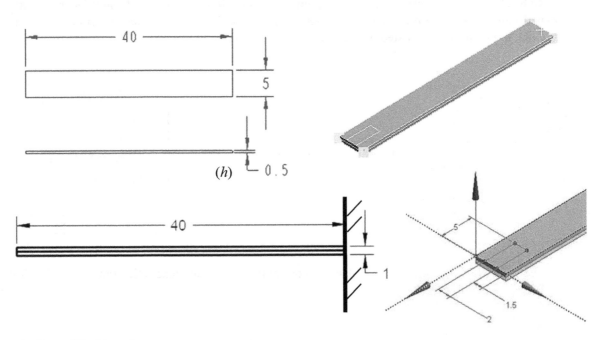

(h)

Step 1: Open Workbench.

From the start menu, click **ANSYS 19** > **Workbench 19**. Users may directly click Workbench 19 when the Workbench symbol is on display.

Workbench 19.2

Step 2: From **Toolbox**, highlight **Static Structural** and drag it to the main screen.

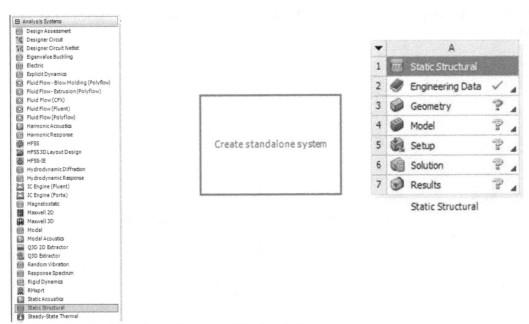

Step 3: Start Design Modeler and create a 2D sketch first.

Right-click the **Geometry** cell and pick **New DesignModeler Geometry** to start **Design Modeler**. Users may double click **Geometry** to directly enter **Design Modeler** if New **SpaceClaim Geometry** is not present. Click the icon of **Units** listed on the top menu, and select **Millimeter**.

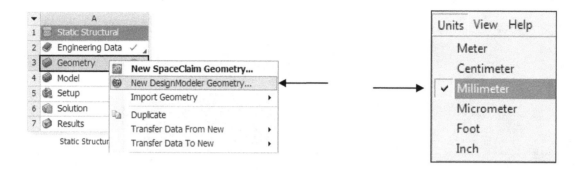

Select **ZXPlane** as the sketching plane. To orient the sketching plane, click the icon of **Look At.**

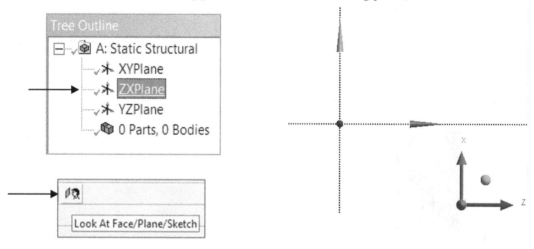

Click **Sketching**, and click **Draw**. In the **Draw** window, click **Rectangle**. Sketch a rectangle, as shown below.

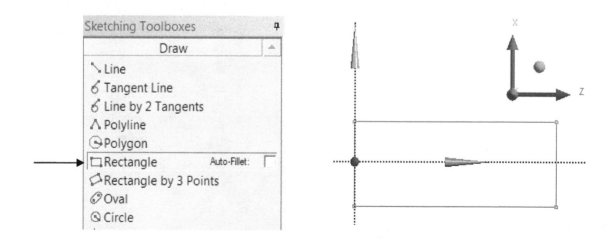

Let us add a symmetric constraint. Click the Z-axis first and click the two horizontal sides afterward.

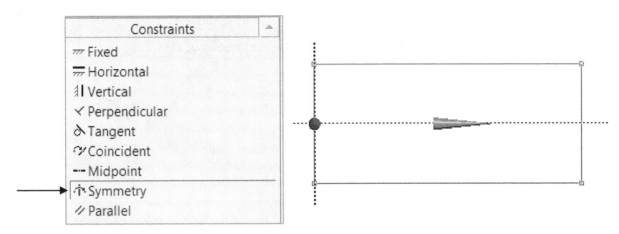

Click **Dimensions** and from the **Dimensions** panel, select **General**. Specify 40 and 5 as the 2 size dimensions of the sketched rectangle.

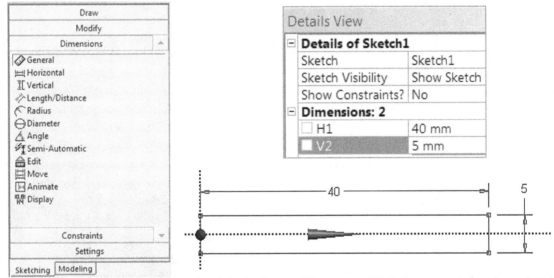

Upon completing the sketch, click the icon of **Extrude**. Click the created sketch or sketch 1 from the Tree Outline, and click **Apply**. For Direction, select **Both-Symmetric** and specify 0.5. Note the total thickness is 1 mm. Click the icon of **Generate** to obtain the 3D solid model of the plate feature.

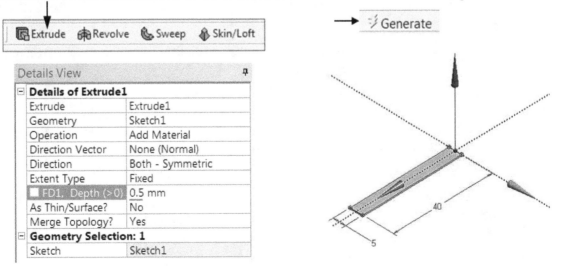

Now let us create a surface region on the top surface of the plate feature. Highlight ZX plane in Tree Outline. Click the icon of New Plane. In Details View, select From Face in Subtype. Pick the top surface, as shown. Click Apply. Click Generate to complete the creation of Plane4.

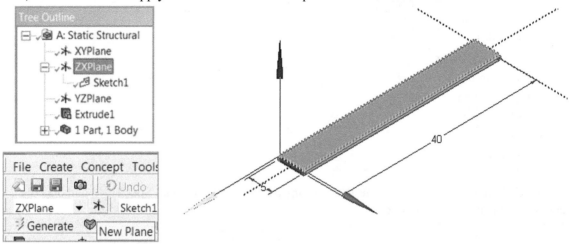

Now use Plane4 as the sketching plane. Click the icon of **New Sketch**. From Sketching, select Draw > Rectangle. Sketch a rectangle, as shown.

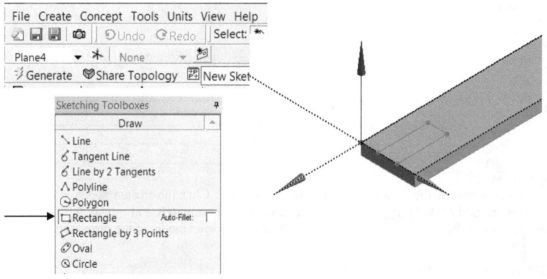

Click Dimensions, select General. Specify 5 and 2 as the size dimensions of the sketched rectangle. Also specify 1.5 as the position dimension, as shown.

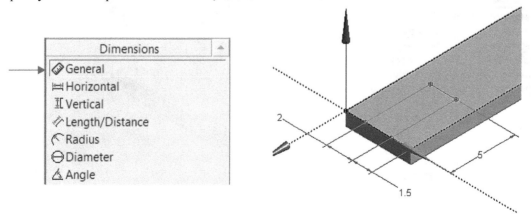

Upon completing the sketch, click the icon of **Extrude**. Click the created sketch or sketch 2 from the Tree Outline, and click **Apply**. Select Imprint Faces in Operation. Click the icon of **Generate** to create the surface region, as shown.

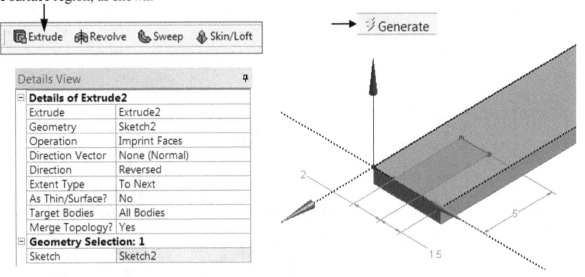

Now let us examining the information listed in Tree Outline. Extrude1 represents a plate of 40 x 5 x 1 mm. Extrude2 represents the surface region for defining the pressure load. The item 1 Part, 1 Body is listed, indicating this model is a component model. This model is not an assembly model.

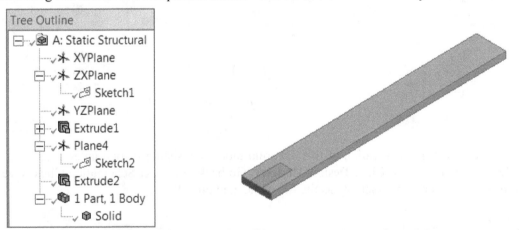

To split the plate feature into 2 separate layers, of which the thickness is 0.50 mm, we perform a **Slice** operation. From the top menu, click the icon of **Slice**. Highlight ZXPlane listed in Tree Outline. Click **Apply**. Click **Generate**. One Solid is split into 2 Solids listed in Tree Outline.

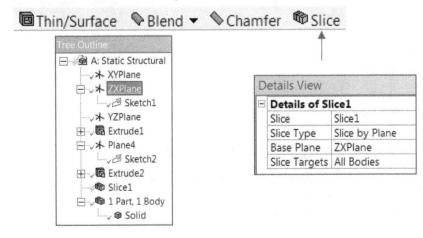

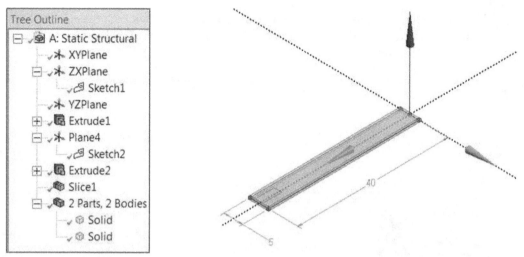

Let us rename the first Solid as Upper Steel and the second Solid as Lower Copper by right-clicking to pick **Rename**.

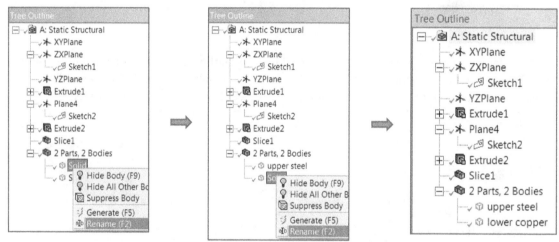

Upon completing the creation of this 3D solid model consisting of two layers, we need to close Design Modeler. Click **File** > Close **Design Modeler**. Go back to Project Schematic. Click Save Project and specify steel-copper layer assembly as the file name, and click Save.

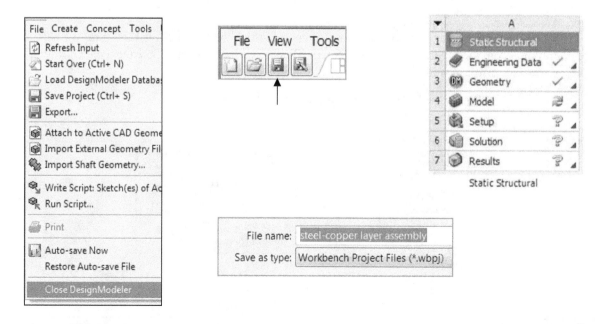

Step 5: Activate the Copper Alloy Material Properties and Define the Contact Region

We need two types of materials: Structural Steel for upper steel layer and copper alloy for lower copper layer. In the Static Structural panel, double click Engineering Data. We enter the Engineering Data Mode.

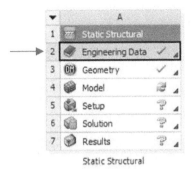

Static Structural

Click the icon of **Engineering Data Sources**. Click **General Materials**. We are able to locate Copper Alloy.

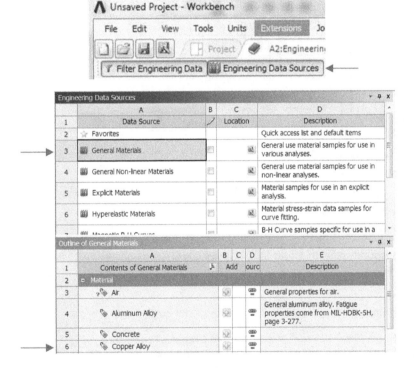

Click the plus symbol nearby Aluminum Alloy to activate it. Afterwards, click the icon of Engineering Data Sources and click the icon of Project to go back to Project Schematic.

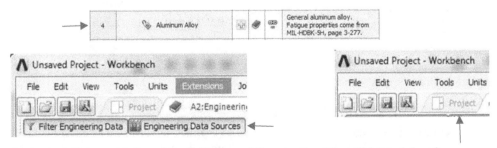

Step 6: Assign Structural Streel and Copper Alloy to the 3D solid Models of upper and lower layers, respectively. Double-click the icon of Model. We enter the Mechanical Mode. Check the unit system: Metric (mm, t, N, s, mV, mA).

Now let us do the material assignment. From the Project tree, expand Geometry. Highlight the listed lower copper. In Details of "Lower Copper", click **Assignment**, pick Copper Alloy.

Note Structural Steel is already assigned to upper steel by the default material assignment.

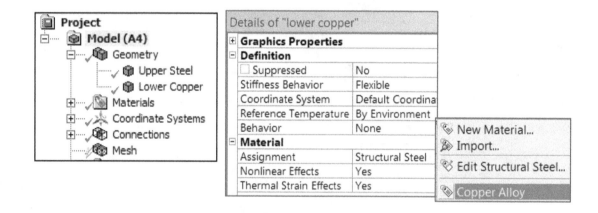

Highlight the listed upper steel. In Details of "Upper Steel", Structural Steel is already assigned to steel layer by default.

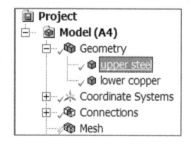

Expand Connections > Contacts > Contact Region. The contact surfaces between upper steel layer and lower copper layer colored immediately. In Details of "Contact Region", there is a pair of faces are in contract. The pair of faces are the bottom surface of the upper steel layer and the top surface of the lower copper layer.

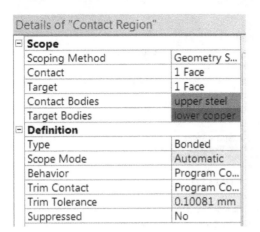

Details of "Contact Region"

Scope	
Scoping Method	Geometry S...
Contact	1 Face
Target	1 Face
Contact Bodies	upper steel
Target Bodies	lower copper
Definition	
Type	Bonded
Scope Mode	Automatic
Behavior	Program Co...
Trim Contact	Program Co...
Trim Tolerance	0.10081 mm
Suppressed	No

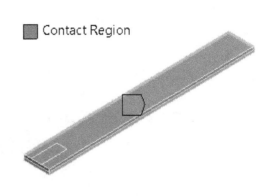

On the right side of the screen, the upper steel layer is defined as Contact Body and the lower copper layer is defined as Target Body. The type of Contact is defined as Bonded. This is the default configuration for contact region, meaning no sliding or separation between faces or edges.

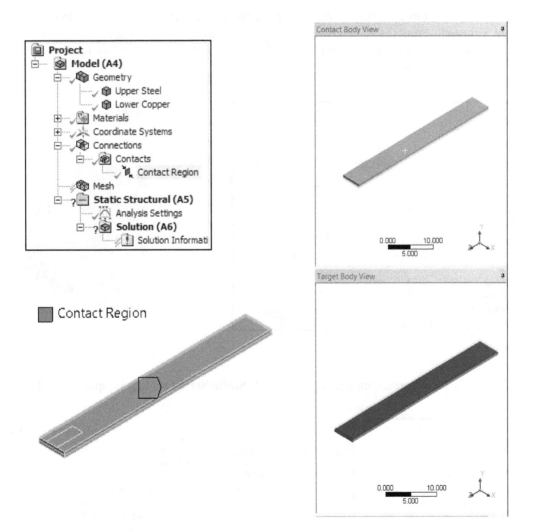

Step 6: Perform FEA

From the project model tree, highlight **Mesh.** Expand **Sizing**, and set Resolution to 6. Highlight **Mesh**, and right-click to pick **Generate Mesh**.

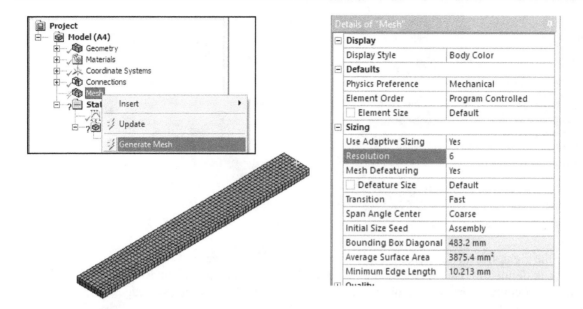

From the project tree, highlight **Static Structural (A5)** and right-click to pick **Insert**. Afterwards, select **Fixed Support**. From the screen, while holding down the Ctrl key, pick the 2 surfaces on the right-sides of the two layers, as shown. As a result, this 3D solid assembly model is fixed to the ground through the two surfaces on the right side.

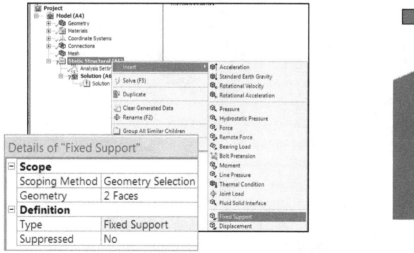

To specify the environment temperature, click **Static Structural (A5),** specify 10 °C in Details of "Static Structural".

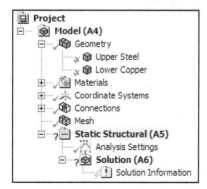

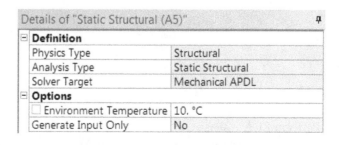

To define the thermal conditions, right-click on Static Structural (A5) to pick Thermal Condition. While holding down the **Ctrl** key, pick the 2 layers. And specify 30 °C as the temperature setting.

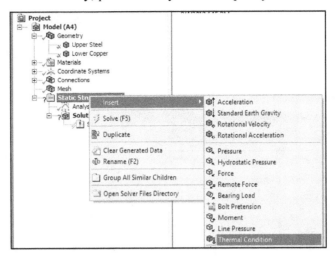

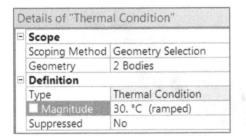

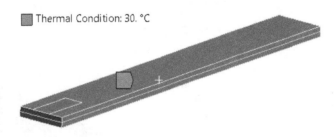

From the project tree, highlight **Solution (A6)** and right-click to pick **Solve**. Workbench system is running FEA.

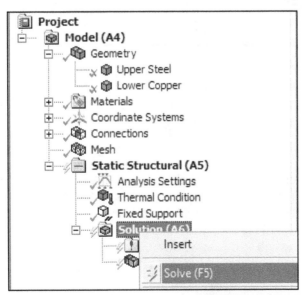

To plot the von Mises stress distribution, highlight **Solution (A6)** and pick **Insert**. Afterwards, highlight **Stress** and pick **Equivalent (von Mises).** Click **Evaluate All Results.** The maximum value of von Mises stress is 152 MPa. Click the icon of **Probe** to identify the coordinates of the maximum value location. Those coordinates are shown in the Graphics Annotations window.

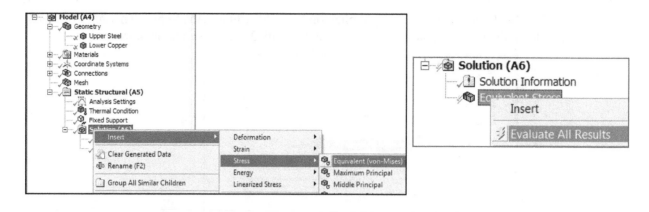

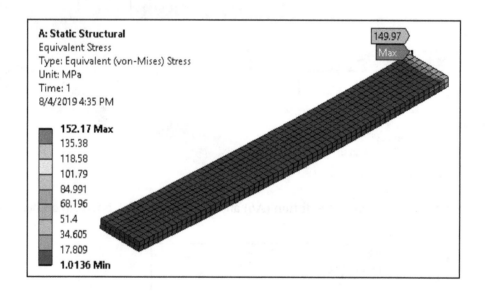

Graphics Annotations						
Type	Value	Note	Unit	Location X	Location Y	Location Z
Result	149.97		MPa	-2.488770	0.488610	0.000000

To plot the displacement distribution, highlight **Solution (A6)** and pick **Insert**. Afterwards, highlight **Deformation** and pick **Total.** Click **Evaluate All Results.** The maximum value of total deformation is 0.147 mm. Click the icon of **Probe** to identify the coordinates of the maximum value location. Those coordinates are shown in the Graphics Annotations window.

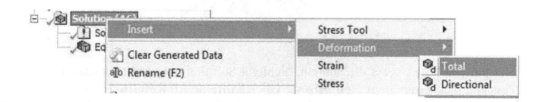

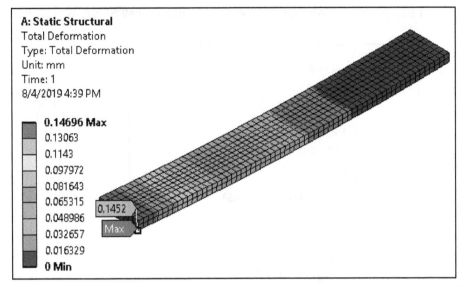

A: Static Structural
Total Deformation
Type: Total Deformation
Unit: mm
Time: 1
8/4/2019 4:39 PM

0.14696 Max
0.13063
0.1143
0.097972
0.081643
0.065315
0.048986
0.032657
0.016329
0 Min

0.1452
Max

Graphics Annotations						
Type	Value	Note	Unit	Location X	Location Y	Location Z
Result	0.1452		mm	2.190550	1.487895	39.982627

Now let us add the second load: pressure load. Highlight Static Structural (A5) and right-click to pick Insert > Pressure. Pick the created surface region and click Apply. Specify 0.05 MPa as the magnitude of pressure.

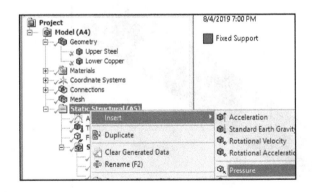

Details of "Pressure"

Scope	
Scoping Method	Geometry Selection
Geometry	1 Face
Definition	
Type	Pressure
Define By	Normal To
Applied By	Surface Effect
☐ Magnitude	5.e-002 MPa (ramped)
Suppressed	No

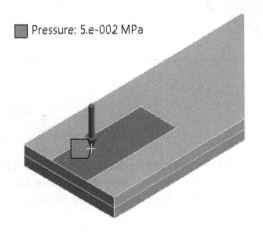

Pressure: 5.e-002 MPa

From the project tree, highlight **Solution (A6)** and right-click to pick **Solve**. Workbench system is running FEA.

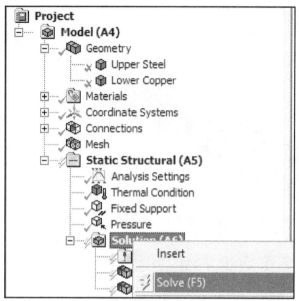

To plot the von Mises stress distribution, highlight Equivalent Stress. The maximum value of von Mises stress is 144.68 MPa. Click the icon of **Probe** to identify the coordinates of the maximum value location. Those coordinates are shown in the Graphics Annotations window.

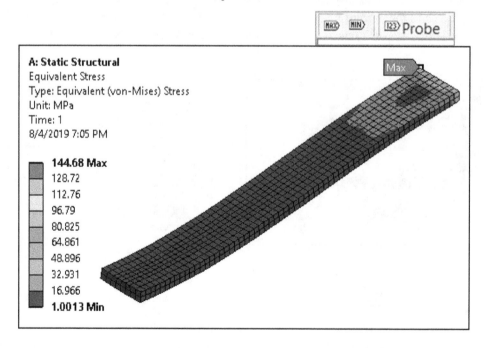

Graphics Annotations

Type	Value	Note	Unit	Location X	Location Y	Location Z
Result	138.62		MPa	-2.501542	0.481405	0.019647

To plot the displacement distribution, highlight Total Deformation. The maximum value of total deformation is 0.020 mm. Click the icon of **Probe** to identify the coordinates of the maximum value location. Those coordinates are shown in the Graphics Annotations window

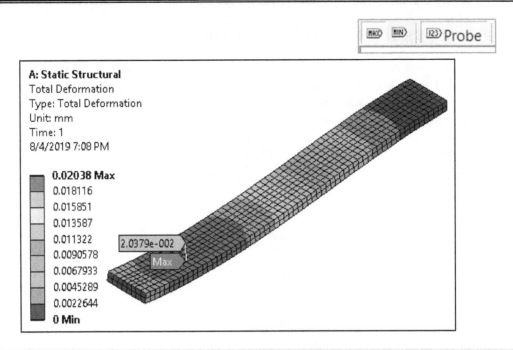

Graphics Annotations						
Type	Value	Note	Unit	Location X	Location Y	Location Z
Result	2.0379e-002		mm	-0.018353	-2.504468	33.187750

We list the obtained results in the following table for the purpose of making comparison. It is obvious that the combined thermal and pressure loads reduce the degree of deformation and the magnitude of maximum von Mises stress. More important is the location of the maximum value of total deformation has been moved from the free end to a new location with much less deformation.

Load Conditions	Max Total Deformaton (mm)	Max von Mises Stress (MPa)
Thermal Load only	0.147	152.17
Thermal Load and Pressure Load	0.020	144.68

From the top menu, click **File > Close Mechanical**. In this way, we return to the main page of Workbench. Click the icon of Save Project.

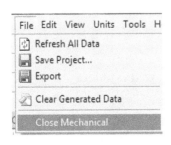

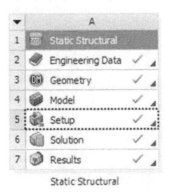

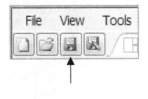

3.4 FEA with a Dental Layer Assembly

In this exercise, we work with a layer structure used in industry. The structure consists of 3 layers: root (composite), cement and glass. The load condition is a temperature variation. The reference temperature is 20°C and the model temperature is 40°C. Investigate the stress development and the pattern of deformation when subjected the temperature variation.

For the boundary condition, we assume that the dentin root structure is fixed to the ground through a local contact. The surface region is a circle. The diameter is 5 mm.

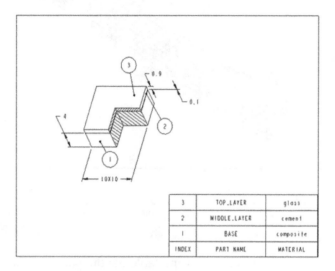

INDEX	PART NAME	MATERIAL
3	TOP.LAYER	glass
2	MIDDLE.LAYER	cement
1	BASE	composite

The geometrical shape is made of three rectangular blocks. The 3 thickness values are 4, 0.1, and 0.9, respectively. The properties of the three types of material are listed below:

	glass	cement	root
coef. of thermal expansion (/C)	5.00E-06	6.00E-05	3.00E-05
Young's modulus (N/mm^2)	8000	6000	15000
Poisson's ratio	0.27	0.33	0.33

Step 1: Open Workbench.
From the start menu, click **ANSYS 18 > Workbench 18**. Users may directly click Workbench 18 when the symbol is on display.

Workbench 19.2

Step 2: From **Toolbox**, highlight **Static Structural** and drag it to the main screen.

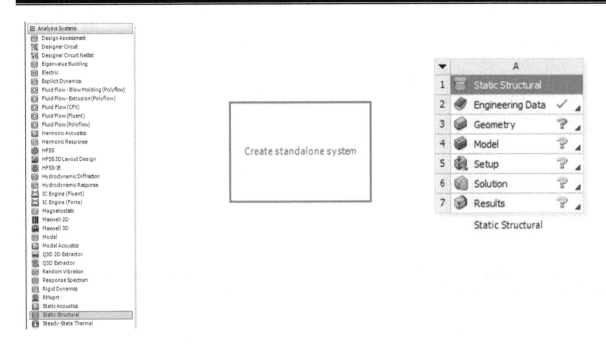

Step 3: Double click Geometry to start Design Modeler and create a 2D sketch first.

Right-click the **Geometry** cell and pick **New DesignModeler Geometry** to start **Design Modeler**. Users may double click **Geometry** to directly enter **Design Modeler** if New **SpaceClaim Geometry** is not present. Click the icon of **Units** listed on the top menu, and select **Millimeter**.

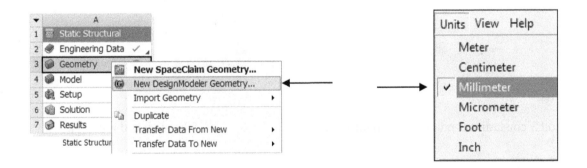

Select **ZXPlane** as the sketching plane.

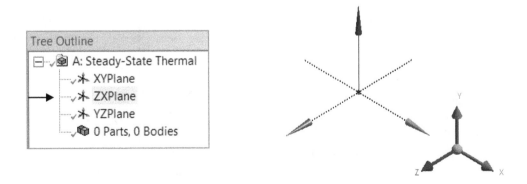

To orient the sketching plane, from the top menu, click the icon of **Look At** so that the sketching plane is parallel to the screen and ready for preparing a sketch.

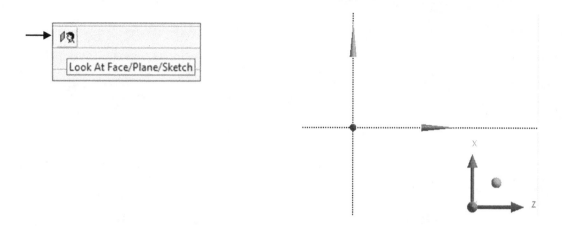

Click **Sketching**, and click **Draw**. Click **Polyline**. Sketch 4 lines shown below. At the end point, right-click and pick **Closed End**.

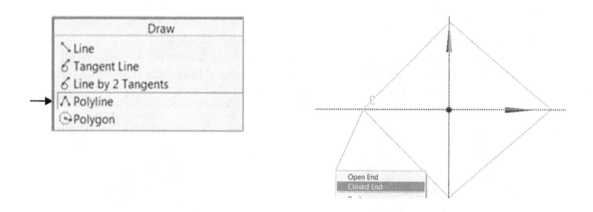

Add 3 constraint conditions: **Perpendicular**. Click the 4 sides of the sketched rectangle.

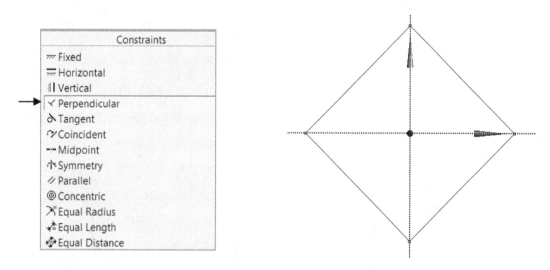

Click **Dimensions** and from the **Dimensions** panel, select **General**. Pick one of the four lines, and make a left click to position the size dimension. Specify 10 as its value.

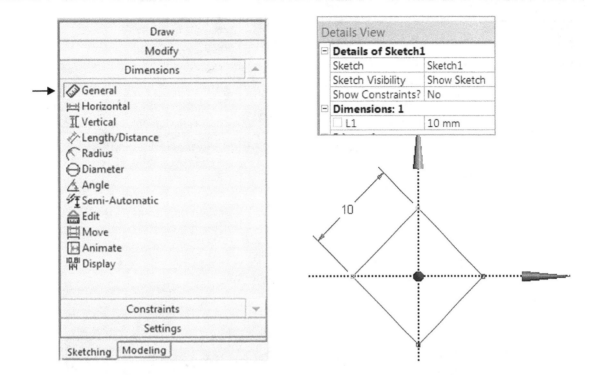

Upon completing the sketch, click the icon of **Extrude**. Click the created sketch or sketch 1 from the Tree Outline, and click **Apply**. For Operation, select **Add Material** and specify 5 as the height value. Click the icon of **Generate** to obtain the 3D solid model of the block feature, as shown.

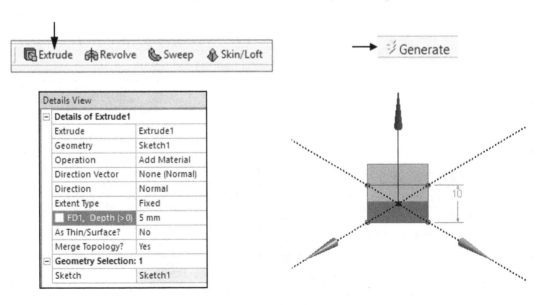

Now let us create the second feature, which is a datum plane. Highlight **ZX** Plane and click **New Plane**. In Type, select **From Plane.** Select Offset Z and specify 4. Click **Generate**. Plane4 is created.

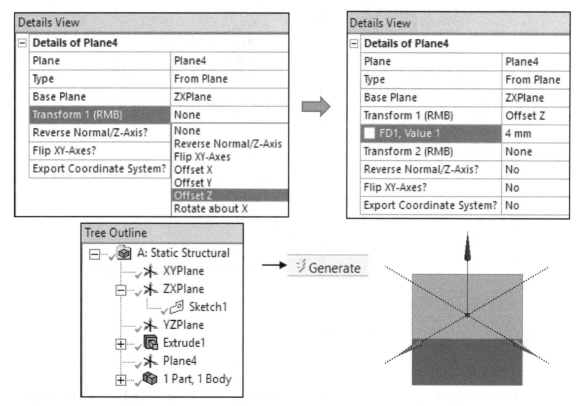

Now let us create the third feature, which is also a datum plane. Highlight **ZX** Plane and click **New Plane**. In Type, select **From Plane**. Select Offset Z and specify 4.1. Click **Generate**. Plane5 is created.

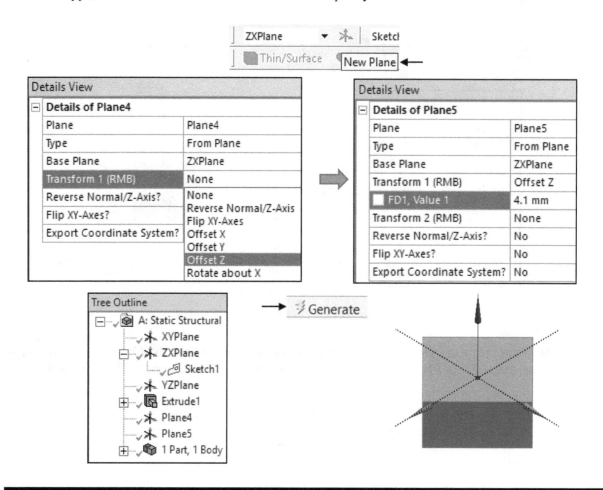

To create a surface region for defining a "fixed to the ground" constraint with a circular shape at the bottom surface of Root, select ZXPlane, and click **New Sketch.** Click **Look At** to orient the sketching plane.

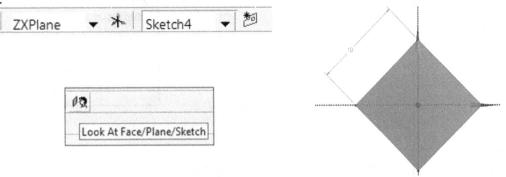

Click **Sketching**, and click **Draw**. Click **Circle**. Sketch a circle and specify 5 as its diameter value.

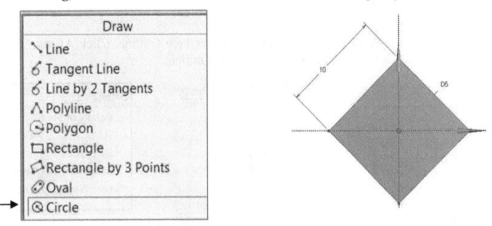

Upon completing the sketch, click the icon of **Extrude**. Click the created sketch or sketch 2 from the Tree Outline, and click **Apply**. For Operation, select **Imprint Faces**. Click the icon of **Generate.**

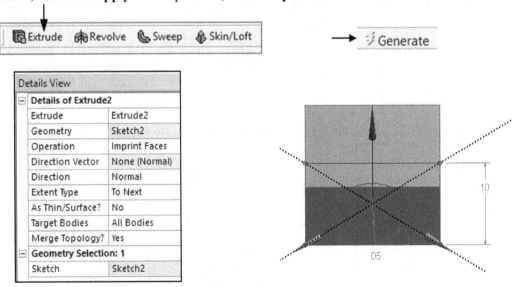

To split the block feature into 3 small block features, of which the thickness values are 4.0 mm, 0.10 mm and 0.90 mm, we perform two **Slice** operations. From the top menu, click the icon of **Slice**. Highlight Plane4 listed in Tree Outline. Click **Apply**. Click **Generate**. One Solid is split into 2 Solids listed in Tree Outline.

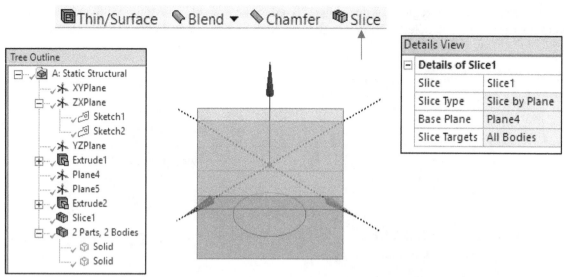

Click the icon of **Slice**. Highlight Plane5 listed in Tree Outline. Click **Apply**. Click **Generate**. One of the two Solids is split into 2 Solids listed in Tree Outline.

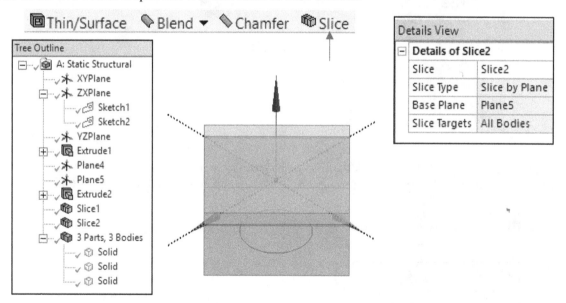

Let us rename the first Solid as Glass, the second Solid as Root, and the third Solid as Cement.

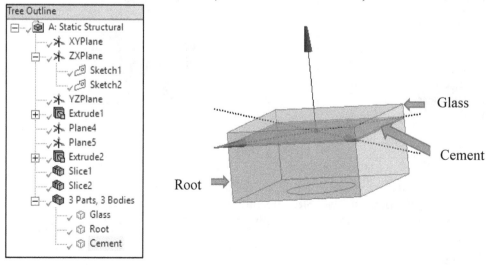

Upon completing the process of creating the geometry, click **File** > Close **Design Modeler**, and go back to **Project Schematic**. Specify "Dental Layer Structure Model" as the file name. Click **Save**.

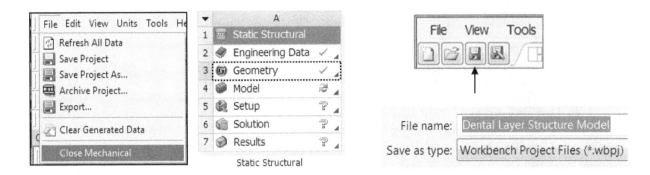

Step 4: Define Three (3) New Types of Material for the 3 Layers in the Engineering Data Sources
In the Steady-State Thermal panel, double click **Engineering Data**. We enter the Engineering Data Mode.

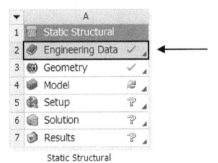

Static Structural

Click the icon of **Engineering Data Sources**. There is box called **Click here to add a new material**. Click this box and type Root as the name of new material type.

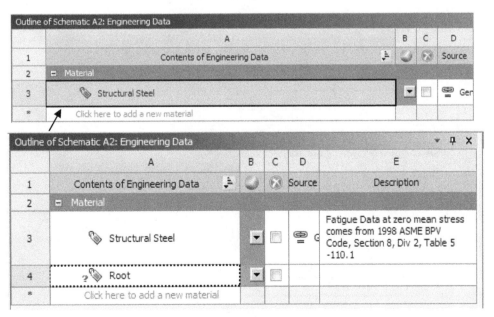

In Tool Box, expand Physical Properties. Double-click Isotropic Secant Coefficient of Thermal Expansion. The Properties window appears. The software system is waiting for the user to enter the value of thermal expansion coefficient. Specify 3e-05. Also enter 20 as the reference temperature.

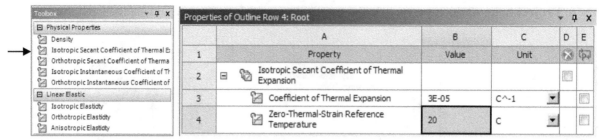

In Tool Box, expand Linear Elasticity. Double-click Isotropic Elasticity. The Properties window appears. The software system is waiting for the user to enter the value of Elasticity (Young's modulus). Specify 1.5e10 as the Young's modulus value. Also enter 0.33 as the Poisson's Ratio value. Note that other material properties, such as shear modulus equal to 1.4706E+10 Pa, and bulk modulus equal to 5.6391E+09 Pa, have default values that are calculated from the entered Young's modulus value and the Poisson's ratio value.

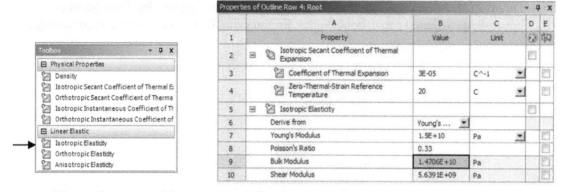

We need to repeat this process to define the material properties for the cement layer. Click the box below Root, and type Cement as the name of new material type.

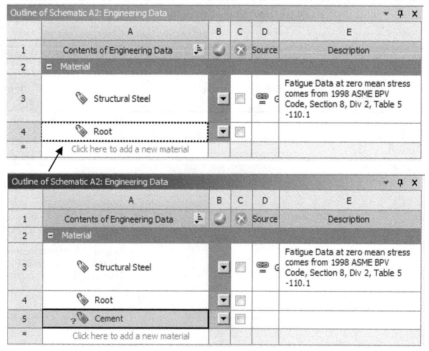

In Tool Box, expand Physical Properties. Double-click Isotropic Secant Coefficient of Thermal Expansion. The Properties window appears. The software system is waiting for the user to enter the value of thermal expansion coefficient. Specify 6e-05. Also enter 20 as the reference temperature.

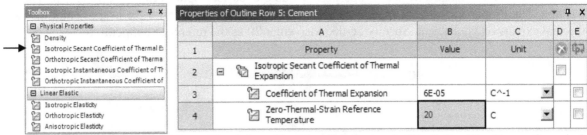

In Tool Box, expand Linear Elasticity. Double-click Isotropic Elasticity. The Properties window appears. The software system is waiting for the user to enter the value of Elasticity (Young's modulus). Specify 6.0e+9 as the Young's modulus value. Also enter 0.33 as the Poisson's Ratio value. Note that other material properties, such as shear modulus equal to 5.882E+9 Pa, and bulk modulus equal to 2.2556E+09 Pa, have default values that are calculated from the entered Young's modulus value and the Poisson's ratio value.

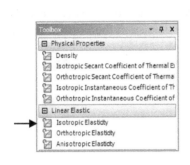

We need to repeat this process one more time to define the material properties for the glass layer. Click the box below Cement, and type Glass as the name of new material type.

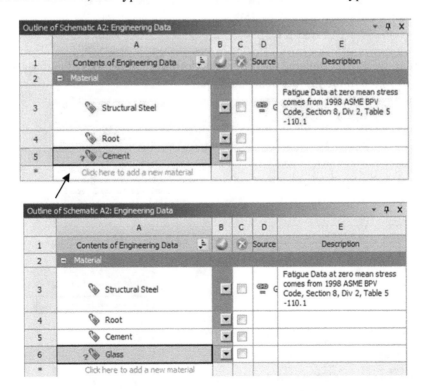

In Tool Box, expand Physical Properties. Double-click Isotropic Secant Coefficient of Thermal Expansion. The Properties window appears. The software system is waiting for the user to enter the value of thermal expansion coefficient. Specify 5e-05. Also enter 20 as the reference temperature.

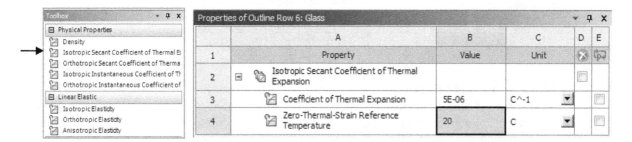

In Tool Box, expand Linear Elasticity. Double-click Isotropic Elasticity. The Properties window appears. The software system is waiting for the user to enter the value of Elasticity (Young's modulus). Specify 8.0e+9 as the Young's modulus value. Also enter 0.27 as the Poisson's Ratio value. Note that other material properties, such as shear modulus equal to 5.7971E+9 Pa, and bulk modulus equal to 3.1496E+09 Pa, have default values that are calculated from the entered Young's modulus value and the Poisson's ratio value.

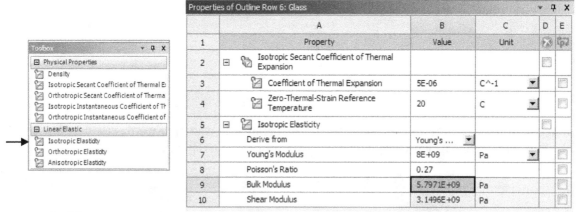

At this moment, we have completed the process of defining 3 new types of materials for the 3 dental layers. Click Project to return to the Project Schematic page.

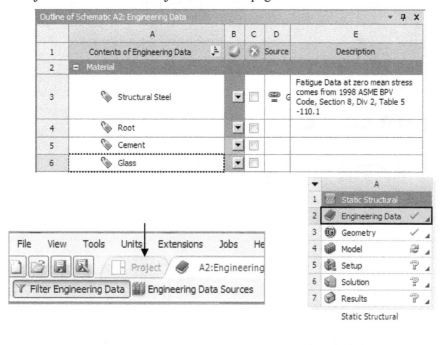

Step 5: Assign Material Types, and Examine the Contact Regions.
  In the Static Structural panel, double click Model. Check the unit system.

From the Project tree, expand Geometry and the three solid models (Root, Cement and Glass) are listed. Highlight **Root**. In the Details of "Root" window, Structural Steel is shown in Assignment. Right-click to pick Root material type.

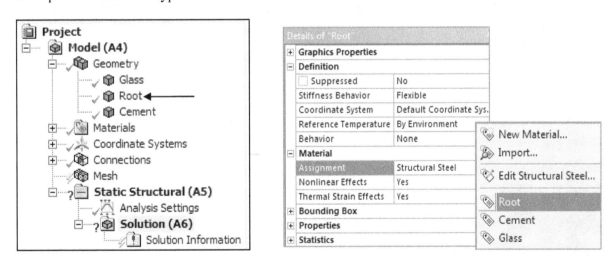

Now let us highlight **Cement** listed in the Project tree. In the Details of "Cement" window, Structural Steel is shown in Assignment. Right-click to pick Cement material type.

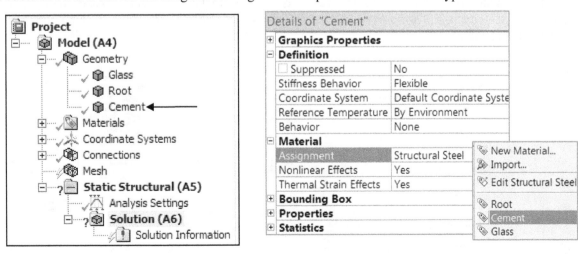

Now let us highlight **Glass** listed in the Project tree. In In the Details of "Glass" window, Structural Steel is shown in Assignment. Right-click to pick Glass material type.

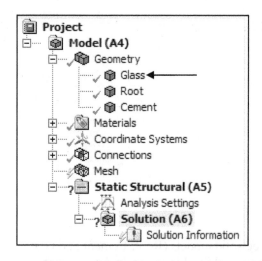 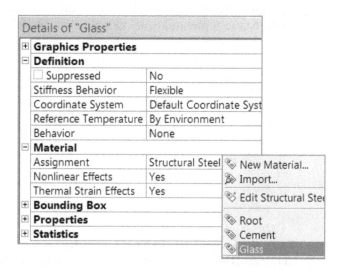

From the Project tree, expand **Connections** > **Contacts**. There are 2 contact regions. Highlight the first contact region. This region is between Root and Cement: Contact Body is Root. Contact Target is Cement, as shown below.

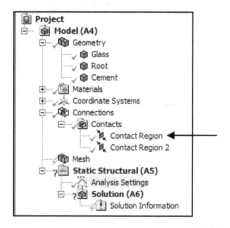

Highlight the first contact region. This region is between Root and Cement: Contact Body is Root. Contact Target is Cement, as shown below. The type of contact region is **Bonded**, meaning no slide or separation between these two faces is allowed.

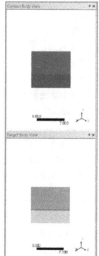

Details of "Contact Region"	
⊟ **Scope**	
Scoping Method	Geometry Selection
Contact	1 Face
Target	1 Face
Contact Bodies	Root
Target Bodies	Cement
⊟ **Definition**	
Type	Bonded
Scope Mode	Automatic
Behavior	Program Controlled
Trim Contact	Program Controlled
Trim Tolerance	5.1539e-002 mm
Suppressed	No

Highlight the second contact region or Contact Region 2. This region is between Cement and Glass. Contact Body is Cement. Contact Target is Glass, as shown below. The type of contact region is **Bonded**, meaning no slide or separation between these two faces is allowed.

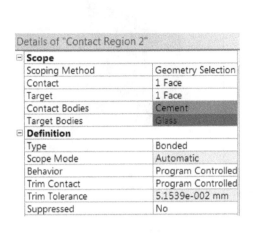

Step 6: Generate Mesh and Define the Boundary Condition.
From the Project tree, Highlight **Mesh**, and right-click to pick **Generate Mesh**.

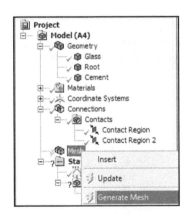

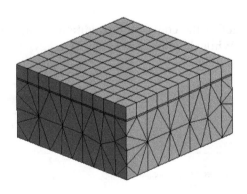

Highlight **Static Structural (A5),** and specify 20 as the environment temperature listed in Details of "Static Structural".

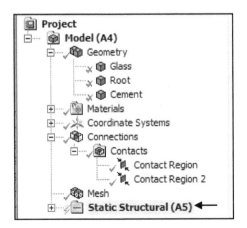

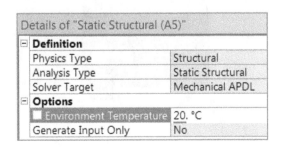

Highlight **Static Structural (A5)**. Pick **Insert** and **Fixed Support**. Pick the imprinted surface on the bottom surface of Root. Pick **Apply**.

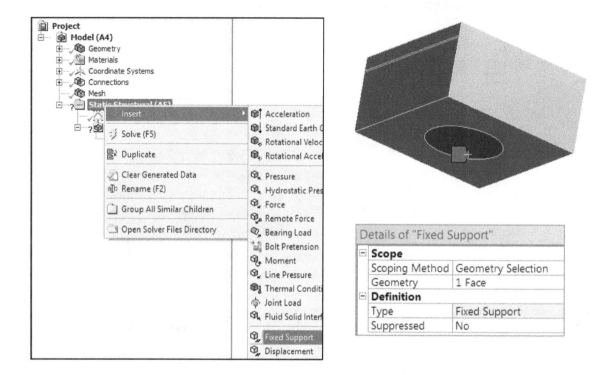

To define the convection condition, highlight **Steady-State Thermal**. Pick **Insert** and **Thermal Condition**. While holding down the Ctrl key, pick the 3 layers (blocks). Specify 40 as the model temperature setting.

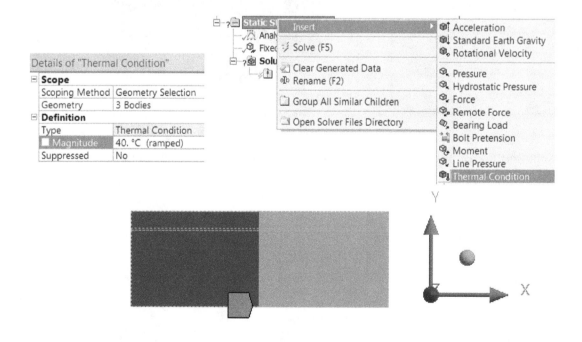

Highlight **Solution (A6)** and pick **Solve**.

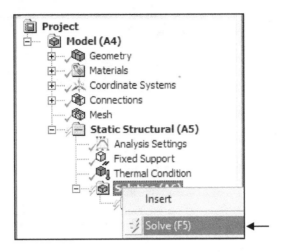

Step 7: Displaying Results

Highlight **Solution (A6)**. Right-click to pick **Insert > Deformation > Total**. Highlight. Right-click to pick **Evaluate all results**. The maximum value of total deformation is 0.004705 mm, located around the cement layer, as shown.

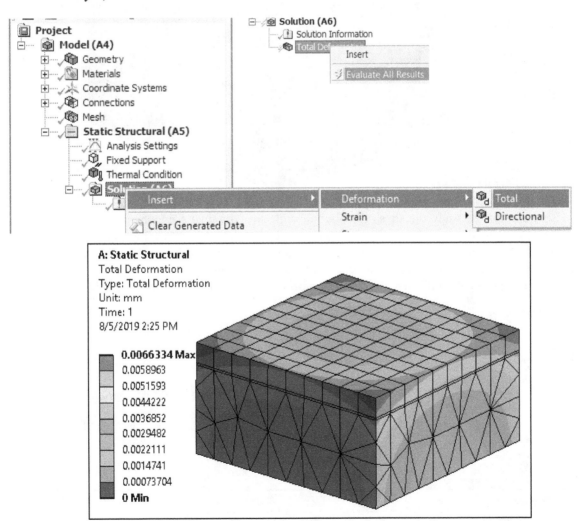

Highlight **Solution (A6)**. Right-click to pick **Insert > Stress > Maximum Principal**. Highlight **Maximum Principal.** Right-click to pick **Evaluate all results**. The maximum value of the maximum principal stress is 4.57 MPa, as shown.

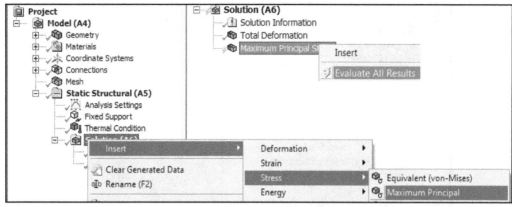

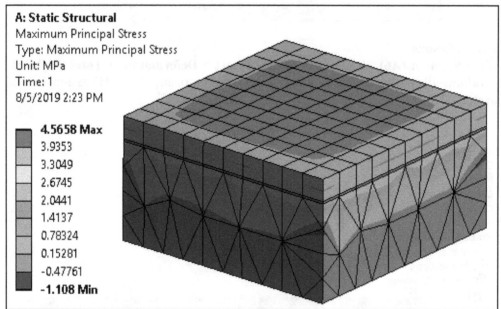

Users are able to study the stress distribution with an individual component, say the maximum principal stress distribution associated with the cement layer. In the Project tree, highlight **Cement** listed under Geometry, right-click to pick Hide All Other Bodies. Afterwards, highlight Maximum Principal Stress listed under **Solution (A6)**. User may click the icon of **Max** and/or **Probe** to view the location and numerical value of stress. As indicated, the maximum principal stress at the corner is 1.01 MPa.

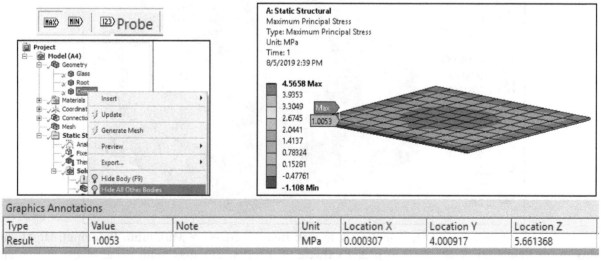

Type	Value	Note	Unit	Location X	Location Y	Location Z
Result	1.0053		MPa	0.000307	4.000917	5.661368

Click **File** and **Close Mechanical** to go back to **Project Schematic** and click **Save**.

3.5   FEA with a Gear Box

The displayed drawing represents a gearbox product. The drawing illustrates that there are two components: gearbox upper and gearbox lower. We first create a 3D solid model for this gearbox assembly. Afterwards we perform FEA. The bottom surface of the gearbox is subjected to a fixed support condition. Four (4) bearing loads act on the 4 bearing locations and two axial forces act on the gearbox as well. The magnitudes of those loads are listed in the table with their units.   Assume that the material type is Gray Cast Iron for both components. We use this example to show the Slice operation for creating an assembly.

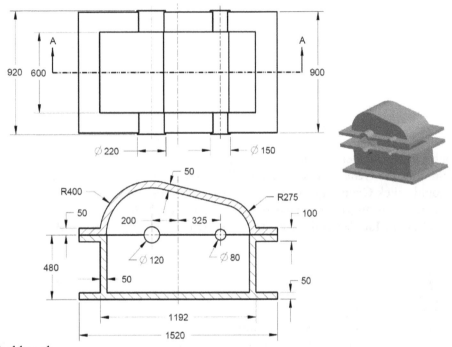

Step 1: Open Workbench.

From the start menu, click **ANSYS 19 > Workbench 19**. Users may directly click Workbench 19 when the symbol of Workbench 19.0 is on display.

Workbench 19.2

From the top menu, click the icon of **Save Project**. Specify **gearbox** as the file name, and click **Save**.

Step 2: From **Toolbox**, highlight **Static Structural** and drag it to the main screen.

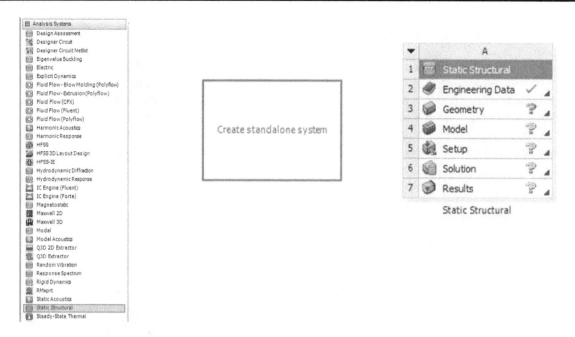

Step 3: Double click Geometry to start Design Modeler.

Right-click the **Geometry** cell and pick **New DesignModeler Geometry** to start **Design Modeler**. Users may double click **Geometry** to directly enter **Design Modeler** if **New SpaceClaim Geometry** is not present. Click the **Units** drop down on the top menu, and select **Millimeter,** double click **Geometry** to enter the screen of **Design Modeler,** and select the unit system. In this case, select Millimeter.

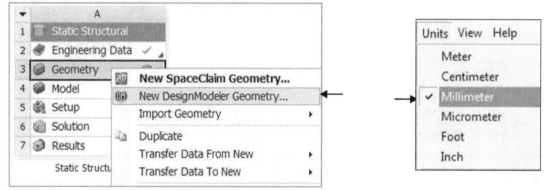

In **Tree Outline**, select YZPlane as the sketching plane. Click Look At to orient the selected sketching plane.

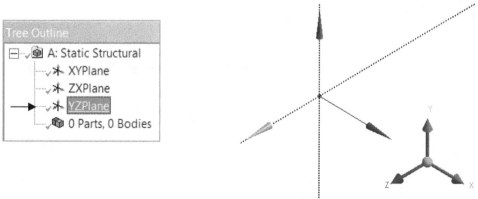

To orient the selected sketching plane, click the X axis. Note do not click Look At for orientation in this case study.

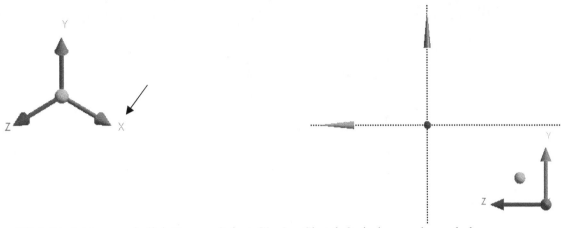

Click **Sketching**, and click **Draw**. Select **Circle**. Sketch 2 circles, as shown below.

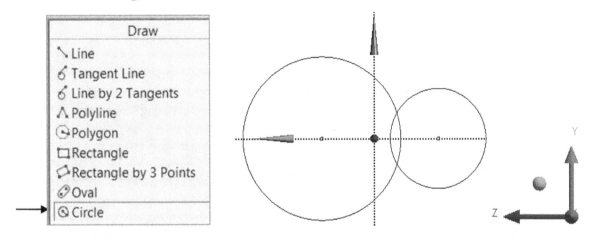

Click **Dimensions** and from the **Dimensions** panel, select Radius and specify 400 and 275. Select **General**. Specify 200 and 325.

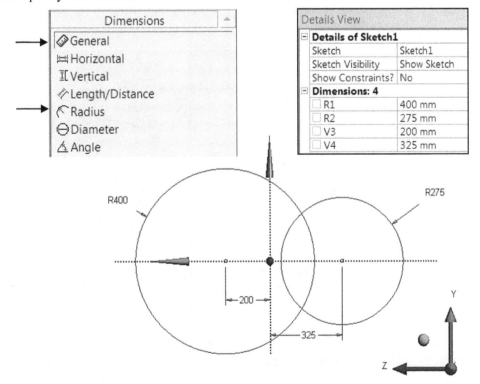

From **Draw** to select **Line**. Sketch 3 lines, as shown. Note one line is an inclined line.

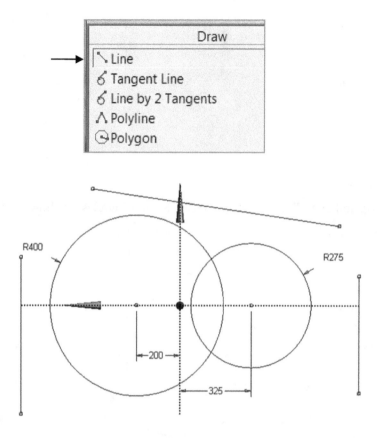

Click **Constraints**, and select **Tangent**. Click the inclined line and click one of the 2 sketched circles. Click the inclined line, again and click the other circle so that the inclined line is tangent to both circles. Follow the same procedure to make the two vertical lines tangent to the two circles, respectively.

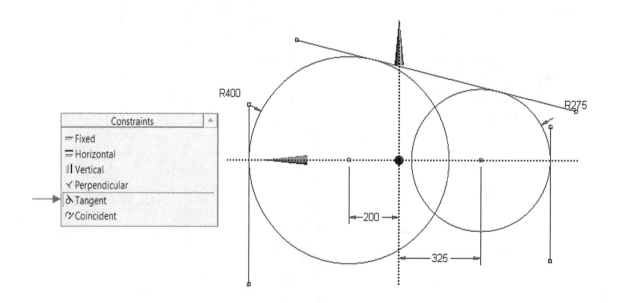

Click **Constraints**, and select **Trim**. Delete those not-needed line segments and arcs, as shown.

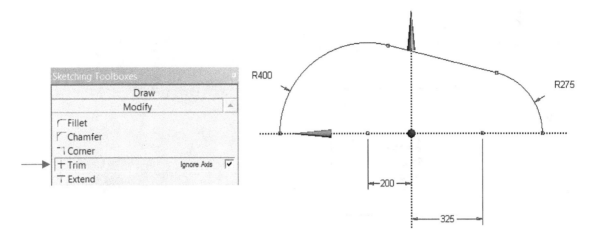

From **Draw** to select **Line**. Sketch a line so that the sketch is a closed sketch, as shown.

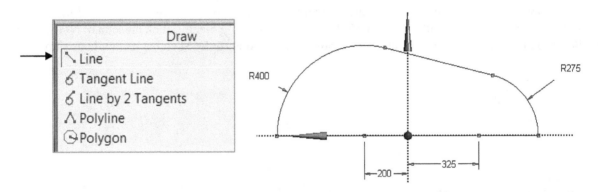

Upon completing the sketch, click the icon of **Extrude**. Click the created sketch or sketch 1 from the Tree Outline, and click **Apply**. For Direction, select **Both-Symmetric** and specify 300 as the distance (both = 600). Click the icon of **Generate** to obtain the 3D solid model, as shown.

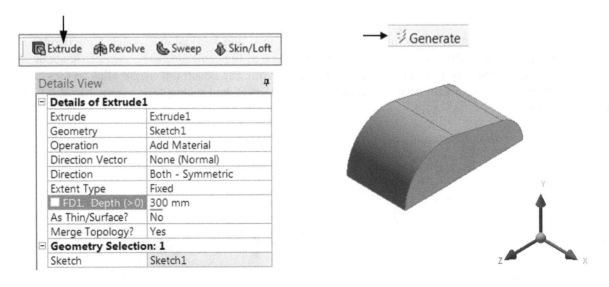

To remove some of the material from the 3D solid model, click **Thin/Surface**. In Details View, select Faces to Remove. Afterwards, pick the bottom surface and click Apply. Keep **Inward**, specify 0 as the thickness value temporarily. Afterwards, change 0 to 50. Click **Generate**.

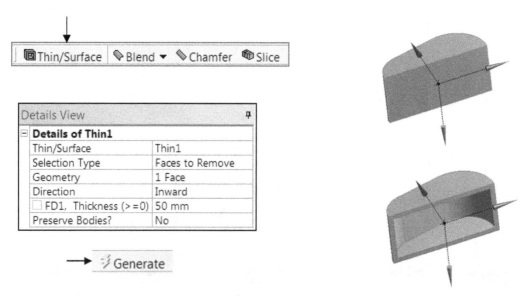

Now let us create the second feature: add a rectangular block feature. The 3 size dimensions are 1520x900x100. In the Tree Outline, select ZXPlane as the sketching plane. , click the icon of **New Sketch** > **Look At** to orient the sketching plane. From **Draw** to select **Rectangle**. Sketch a rectangle as shown.

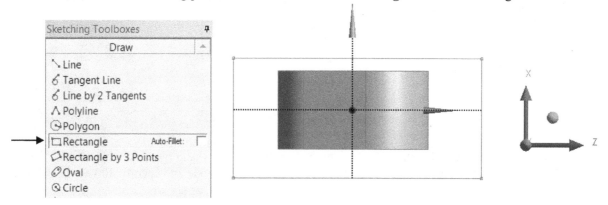

To add 2 symmetrical constraints about the horizontal and vertical axes, click **Constraints** and select **Symmetry**. Using Right-click, pick the horizontal axis as the centerline to be symmetric about, and pick the top and bottom sides of the rectangle to add the symmetric constraint. Right click to pick Select new symmetry axis. Follow the same procedure so that the sketched rectangle is symmetric about both axes, as shown.

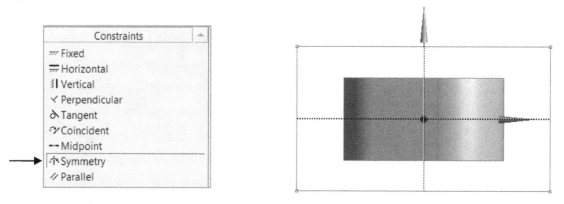

Click **Dimensions,** specify 1520 and 900 as the 2 size dimensions of the rectangle.

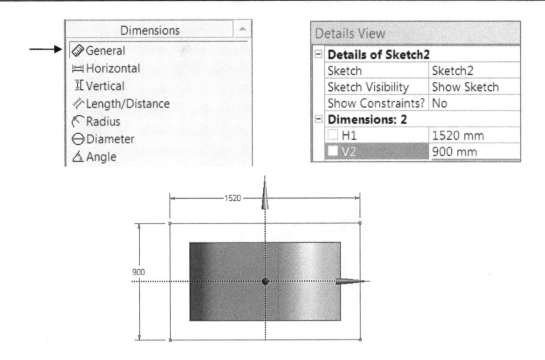

From **Draw** to select **Rectangle**. Sketch a new or second rectangle as shown.

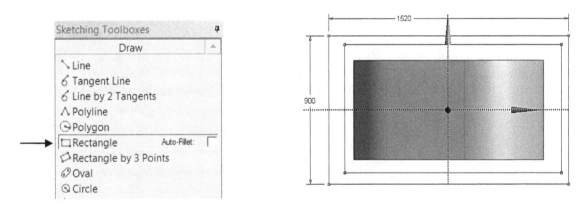

To add 2 symmetrical constraints about the horizontal and vertical axes, click **Constraints** and select **Symmetry**. Using Right-click, pick the horizontal axis as the centerline to be symmetric about, and pick the top and bottom sides of the rectangle to add the symmetric constraint. Follow the same procedure so that the sketched rectangle is symmetric about both axes, as shown.

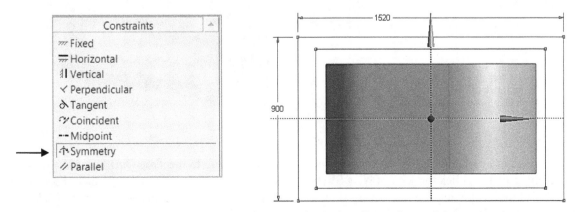

Click **Dimensions**, specify 1100 and 500 as the 2 size dimensions of the rectangle.

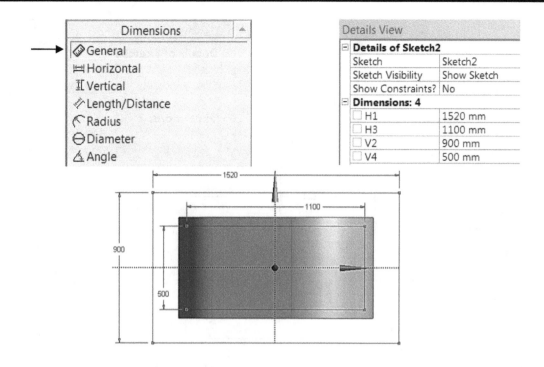

Upon completing the sketch, click the icon of **Extrude**. Click the created sketch or sketch 2 from the Tree Outline, and click **Apply**. For Direction, select **Both-Symmetric** and specify 50 as the extrusion distance (both = 100). Click the icon of **Generate** to obtain the 3D solid model of the cylindrical feature, as shown.

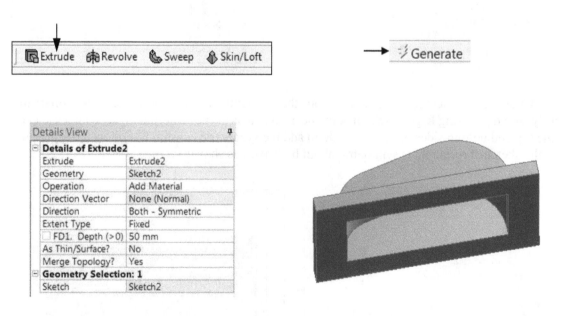

Now let us create the third feature: a box-shaped solid model. In the Tree Outline, select ZXPlane as the sketching plane. , click the icon of **New Sketch** > **Look At** to orient the sketching plane. From **Draw** to select **Polyline**. Following the previously sketched rectangle, sketch a rectangle, as shown. Make sure the Closed End is selected.

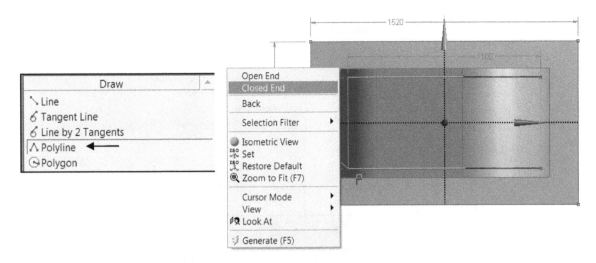

From **Draw** to select **Rectangle**. Sketch a new or second rectangle as shown.

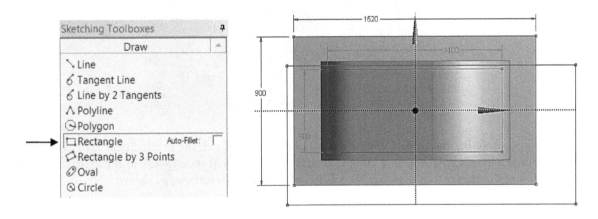

To add 2 symmetrical constraints about the horizontal and vertical axes, click **Constraints** and select **Symmetry**. Using Right-click, pick the horizontal axis as the centerline to be symmetric about, and pick the top and bottom sides of the rectangle to add the symmetric constraint. Follow the same procedure so that the sketched rectangle is symmetric about both axes, as shown.

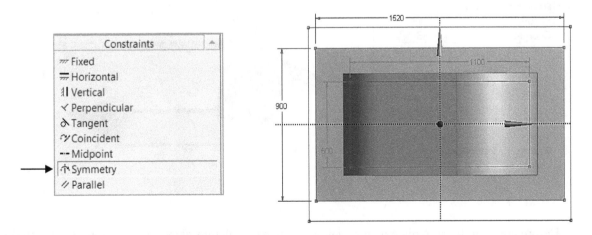

Click **Dimensions** and from the **Dimensions** panel, specify 1200 and 600 as the 2 size dimensions of the rectangle.

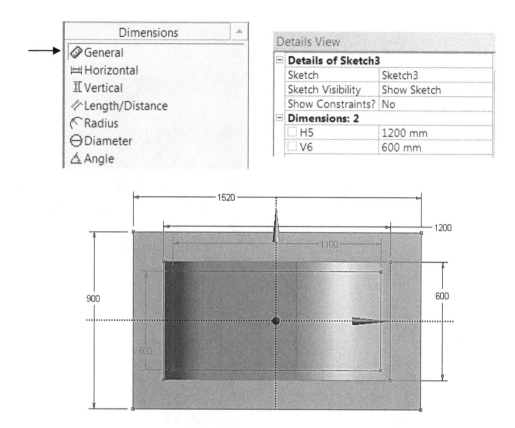

Upon completing the sketch, click the icon of **Extrude**. Click the created sketch or sketch 3 from the Tree Outline, and click **Apply**. For Direction, select Reverse, and specify 430 as the extrusion distance. Click the icon of **Generate** to obtain the 3D solid model of the box feature, as shown.

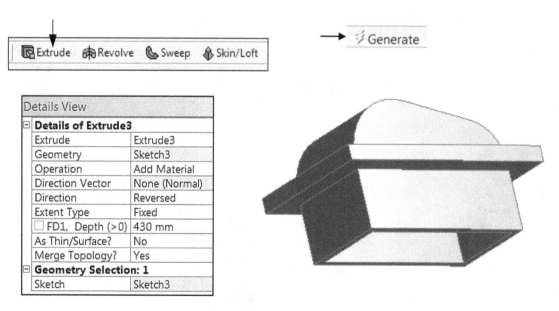

Before creating the base plate feature, let us create a new datum plane. In Tree Outline, highlight ZXPlane. And click New Plane. In Details View, select Offset Z and specify -430 as the offset value. Click Generate. Plane4 is generated and listed in Tree Outline.

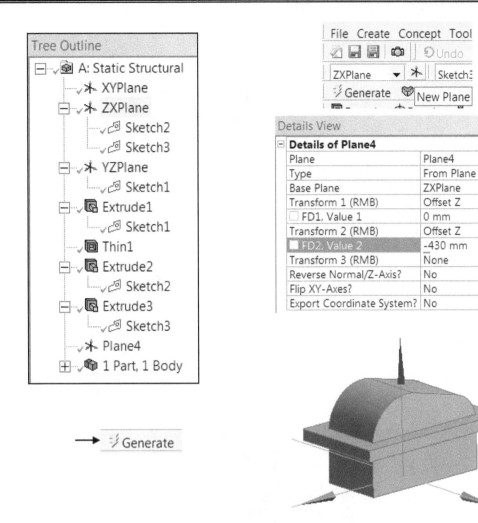

Now highlight Plane4 and click the icon of New Sketch > Look At. Select Rectangle from Draw, and sketch a rectangle, as shown.

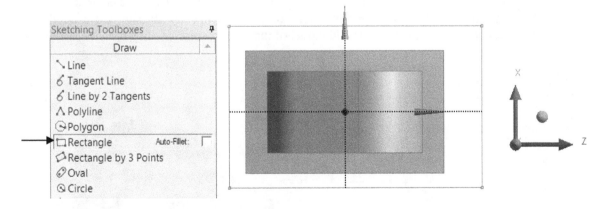

To add 2 symmetrical constraints about the horizontal and vertical axes, click **Constraints** and select **Symmetry**. Using Right-click, pick the horizontal axis as the centerline to be symmetric about, and pick the top and bottom sides of the rectangle to add the symmetric constraint. Follow the same procedure so that the sketched rectangle is symmetric about both axes, as shown.

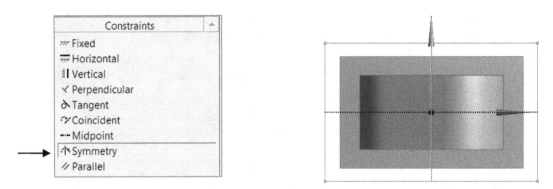

Click **Dimensions** and from the **Dimensions** panel, specify 1520 and 900 as the 2 size dimensions of the rectangle.

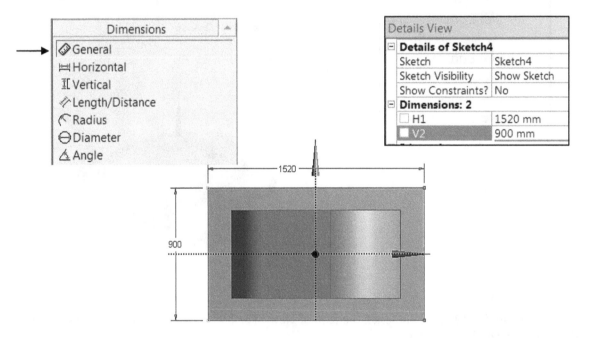

Upon completing the sketch, click the icon of **Extrude**. Click the created sketch or sketch 4 from the Tree Outline, and click **Apply**. For Direction, select **Reverse**, and specify 50 as the extrusion distance. Click the icon of **Generate** to obtain the 3D solid model of the plate feature, as shown.

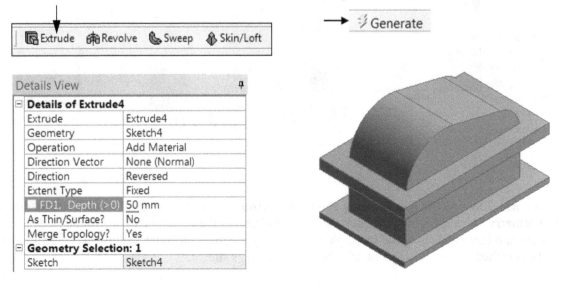

Now let us examine the features listed in Tree Outline. Four (4) features are listed. They are Extrude 1 (include Thin1), Extrude 2, Extrude 3 and Extrude 4. 1 Part, 1 Body with one Solid is also listed, indicating the currently developed solid model is a component model.

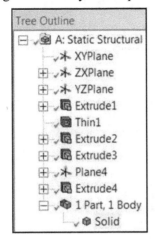

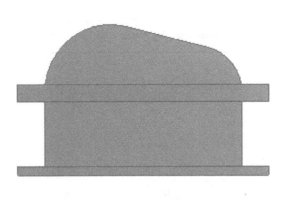

Now let us add the last feature: bearing support feature. Now highlight YZPlane, and click the icon of New Sketch. Click the X-axis for orientation. Select Circle from Draw, and sketch 2 circles, as shown.

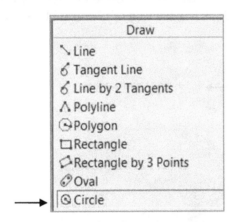

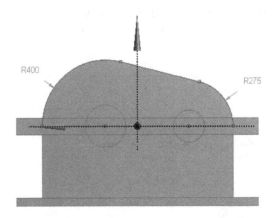

Click **Dimensions,** select **General**, and specify 220 and 150 as the 2 diameter values of the 2 sketched circles.

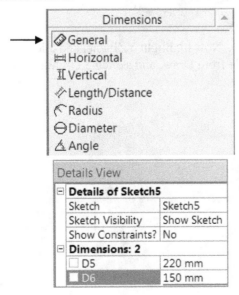

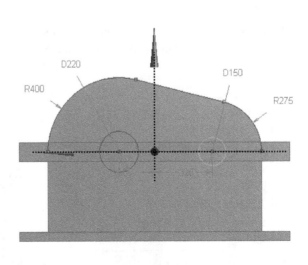

Upon completing the sketch, click the icon of **Extrude**. Click the created sketch or sketch 5 from the Tree Outline, and click **Apply**. For Direction, select **Both-Symmetry**, and specify 470 as the extrusion distance (total = 940). Click the icon of **Generate** to obtain the 3D solid model of the bearing support feature, as shown.

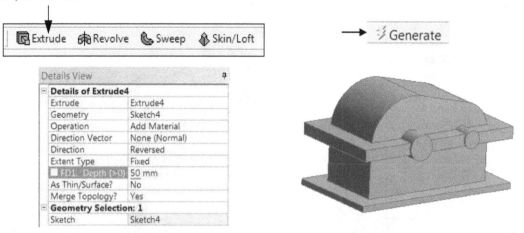

We need to have a cut operation. Click the icon of **Extrude**. Click the created sketch or sketch 5 from the Tree Outline, and click **Apply**. For Operation, select Cut Material. For Direction, select **Both-Symmetry**, and specify 250 as the extrusion distance (total = 500). Click the icon of **Generate** to remove the material within the box.

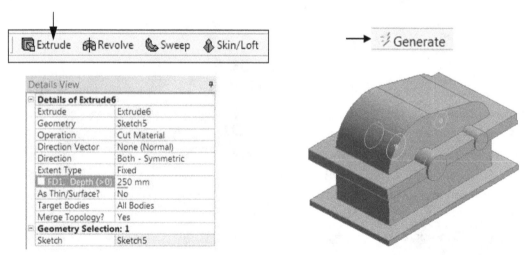

Now let us add the last feature: bearing support feature. Now highlight YZPlane, and click the icon of New Sketch. Click the X-axis for orientation. Select Circle from Draw, and sketch 2 circles, as shown.

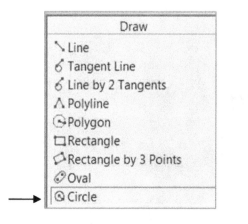

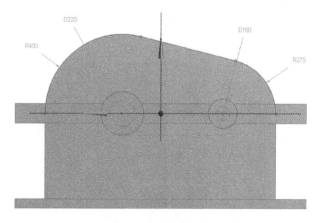

Click **Dimensions** and from the **Dimensions** panel, specify 120 and 80 as the 2 diameter values of the 2 sketched circles.

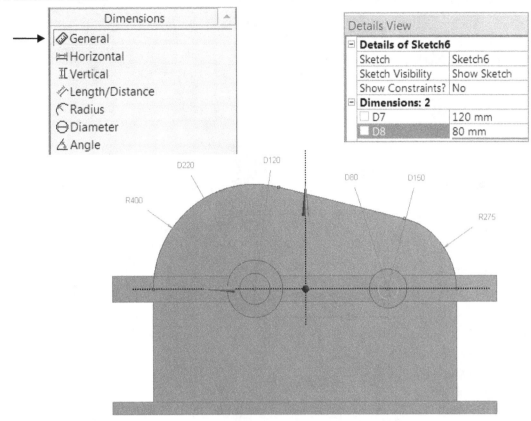

Upon completing the sketch, click the icon of **Extrude**. Click the created sketch or sketch 6 from the Tree Outline, and click **Apply**. Select the Cut Material. For Direction, select **Both-Symmetry**, and specify 470 as the extrusion distance (total = 940). Click the icon of **Generate** to obtain the 3D solid model of the holes with the bearing support feature, as shown.

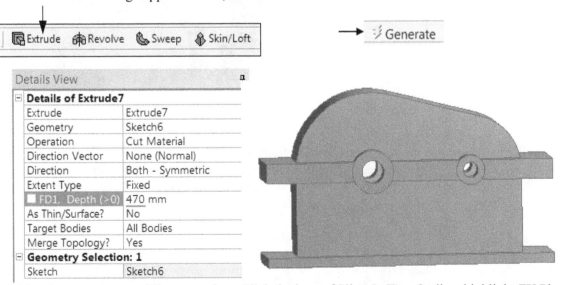

Now let us carry out a Slice operation. Click the icon of Slice. In Tree Outline, highlight ZX Plane. In Details View, click Apply to accept the selected ZX Plane as the slicing plane. Click Generate. The slicing operation cuts the entire model into two parts along the ZXPlane. Now we have an assembly model with two parts as indicated in Tree Outline where the item called 2 Parts, 2 Bodies is listed.

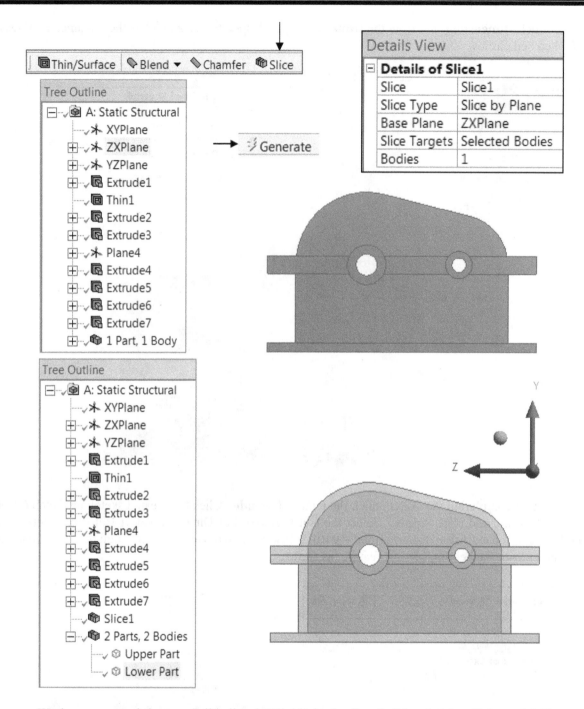

We have renamed the two Solids listed. Highlight the first Solid and right-click to pick Rename. Specify Upper Part. Highlight the second Solid listed and right-click to pick **Rename**. Specify Lower Part. Check the geometry of the completed assembly model, as shown below.

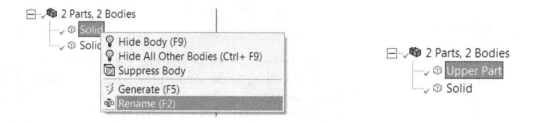

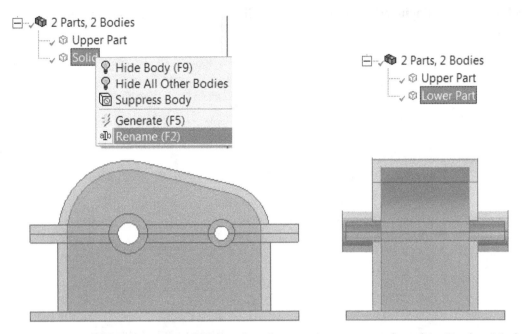

Upon completing the process of creating the geometry, we need to close Design Modeler. Click **File** > Close **Design Modeler**, and go back to **Project Schematic**. Click **Save Project**.

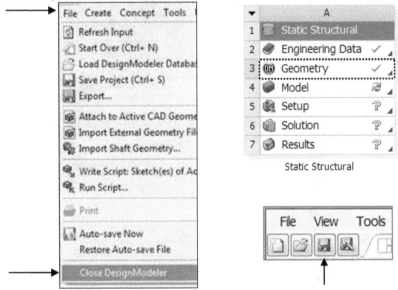

Step 4: Activate the Gray Cast Iron in the Engineering Data Sources.

In Static Structural, double-click the Engineering Data Sources cell. We enter the material library mode, or Engineering Data Sources.

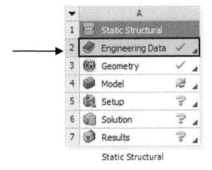

Static Structural

Click the icon of **Engineering Data Sources**. Click **General Materials**. We are able to locate Gray Cast Iron.

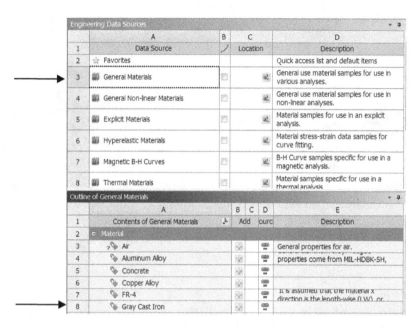

Click the plus symbol nearby Gray Cast Iron to activate it. Afterwards, click the icon of Engineering Data Sources and click the icon of Project to go back to Project Schematic.

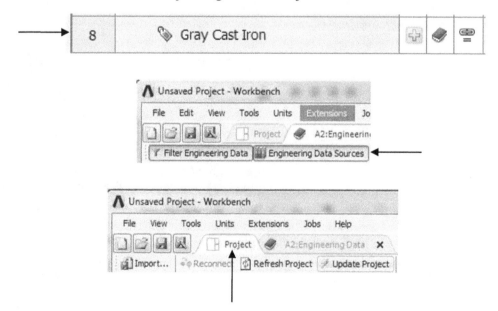

Step 5: Assign Gray Cast Iron to Both Upper and Lower Parts.

Double-click the icon of Model. We enter the Mechanical Mode. Check the unit system, and pick Metric (mm, t, N, s, mV, mA).

Now let us do the material assignment. From the Project tree, expand Geometry. While holding down the **Ctrl** key, pick Upper Part and Lower Parts. In Details of "Multiple Selection", click material assignment, pick Gray Cast Iron to replace Structural Steel, which is the default material assignment.

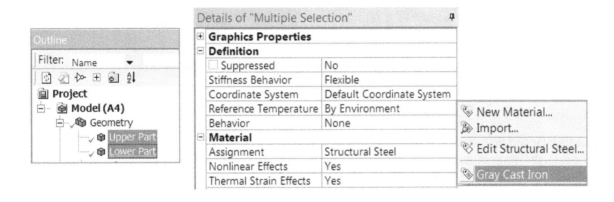

Step 6: Verify the Bonded Configuration in the Contact Setting.

In the Project tree, Expand Connections > Contacts. Click Contact Region. The Upper Part model serves as Contact Body and the Lower Part model serves as Target Body. The type of contact is **Bonded**.

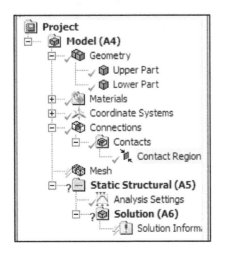

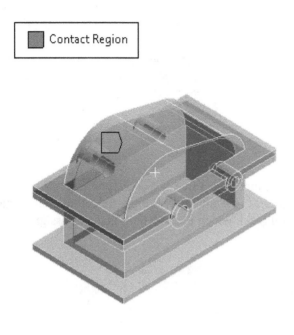

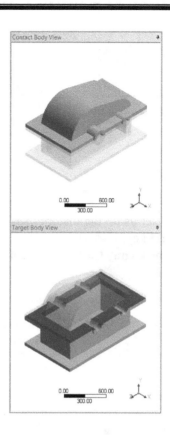

Step 7: Mesh Generation.

From the project model tree, highlight **Mesh.** In Details of "Mesh", expand **Sizing**, and set Resolution to 4. Right-click **Mesh** again to pick **Generate Mesh**.

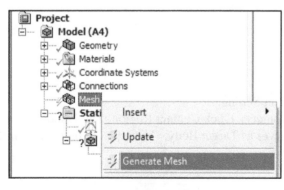

Details of "Mesh"	
Display	
Display Style	Body Color
Defaults	
Physics Preference	Mechanical
Element Order	Program Controlled
Element Size	Default
Sizing	
Use Adaptive Sizing	Yes
Resolution	6
Mesh Defeaturing	Yes
Defeature Size	Default
Transition	Fast
Span Angle Center	Coarse
Initial Size Seed	Assembly
Bounding Box Diagonal	483.2 mm
Average Surface Area	3875.4 mm²
Minimum Edge Length	10.213 mm
Quality	

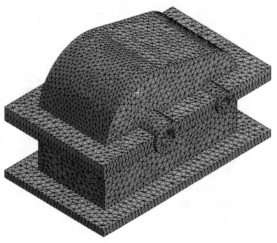

Step 8: Define the boundary condition and load condition.

Highlight **Static Structural (A5)** and right-click to pick **Insert**. Afterwards, select **Fixed Support**. From the screen, pick the surface at the bottom, as shown. Click **Apply**.

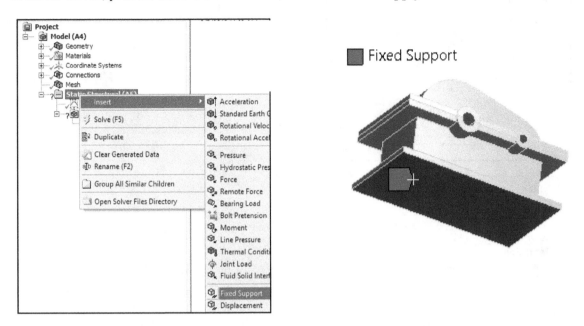

As a gearbox case study, defining the condition of bearing loads is important. Pay attention to the orientation, as shown. Let us first name each of the 4 bearing locations, as illustrated in the following figure. The 4 bearing locations are named B, C, D, and E. The bearing load at each of the 4 locations has Y and Z components, which is listed in the table. The X component is zero for all the 4 locations, assuming that only spur gears are used in the operation.

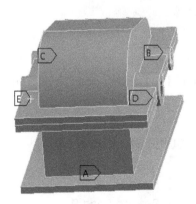

Bearing Location	X Component	Y Component	Z component
Location B	0	15000 N	10000 N
Location C	0	15000 N	-10000 N
Location D	0	-15000 N	10000 N
Location E	0	-15000 N	-10000 N

To define the bearing loads at Location B and Location C, highlight **Static Structural (A5)**. Pick **Insert** and **Bearing Load.** Pick the upper half cylindrical surface at Location B and click **Apply**. Specify 15000 N as Y component and 10000 N as Z component. Keep 0 as X component. Repeat this process at Location C. Make sure the Z component is -10000 N.

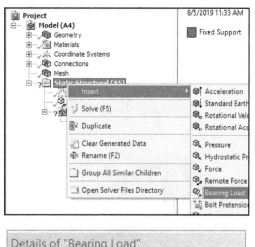

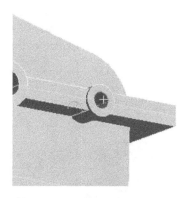

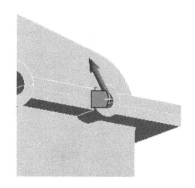

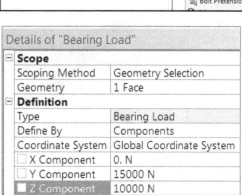

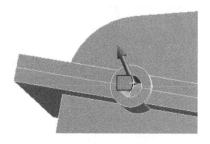

Let us rename Bearing Load to Bearing Load B and Bearing Load 1 to Bearing Load C, listed under Static Structural (A5).

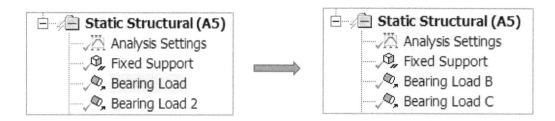

To define the bearing loads at Location D and Location E, highlight **Static Structural (A5)**. Pick **Insert** and **Bearing Load.** Pick the lower half cylindrical surface at Location D and click **Apply**. Specify -15000 N as Y component and 10000 N as Z component. Keep 0 as X component. Repeat this process at Location E. Make sure the Z component is -10000 N.

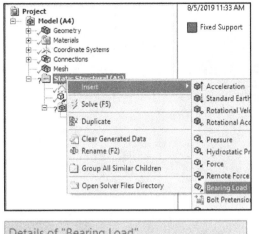

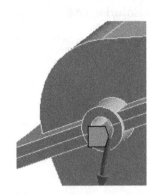

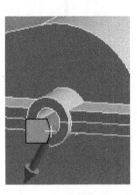

Details of "Bearing Load"

Scope	
Scoping Method	Geometry Selection
Geometry	1 Face
Definition	
Type	Bearing Load
Define By	Components
Coordinate System	Global Coordinate System
X Component	0. N
Y Component	-15000 N
Z Component	10000 N
Suppressed	No

Details of "Bearing Load 2"

Scope	
Scoping Method	Geometry Selection
Geometry	1 Face
Definition	
Type	Bearing Load
Define By	Components
Coordinate System	Global Coordinate System
X Component	0. N
Y Component	-15000 N
Z Component	-10000 N
Suppressed	No

Let us rename Bearing Load to Bearing Load D and Bearing Load 2 to Bearing Load E, listed under Static Structural (A5).

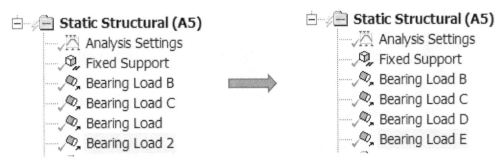

Let highlight Static Structural (A5), the fixed support constraint and the 4 bearing loads are all listed, as shown below.

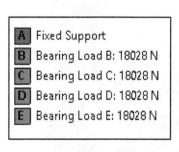

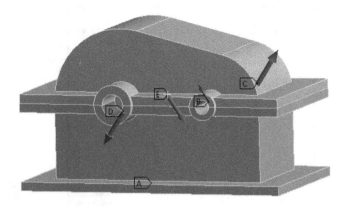

Step 6: Run FEA program

Highlight **Solution (A6)** and right-click to pick **Solve**.

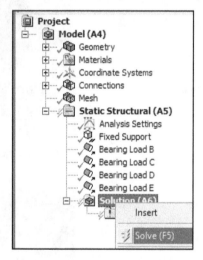

Step 7: Displaying Results

Highlight **Solution (A6)**. Right-click to pick **Insert > Deformation > Total**.

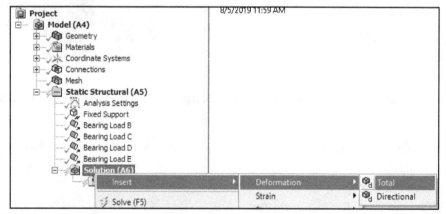

Highlight **Total Deformation** under **Solution (A6)**. Right-click to pick **Evaluation All Results**. The maximum value of deformation is 0.011 mm.

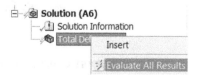

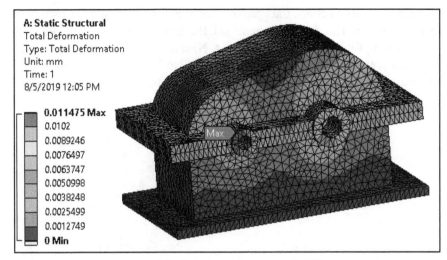

Highlight **Solution (A6)**. Right-click to pick **Insert > Stress > Equivalent (von-Mises)**.

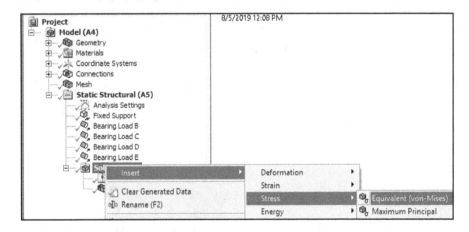

Highlight **Equivalent Stress** under **Solution (A6)**. Right-click to pick **Evaluation All Results**. The maximum value of von Mises stress is 2.71 MPa.

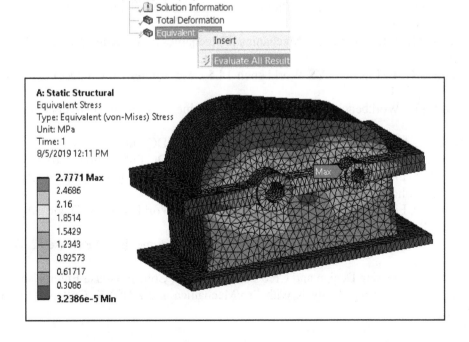

It is important for users to check the graphic screen before closing a case study. At the bottom of the graphic screen, there is a Tabular Data area. At the bottom of this area, the unit system used in performing the current analysis is listed, Metric (mm, t, N, s, mV, mA) Degrees, rad/s Celsius. Also at the bottom of the graphic screen, the Message Board indicates "No Message", indicating the FEA running did not account any unexpected process, namely, the FEA run was successful.

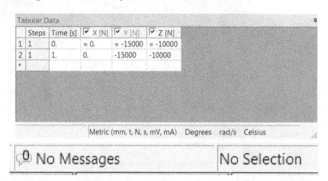

Click **File** and **Close Mechanical** to go back to Project Schematic and click **Save**.

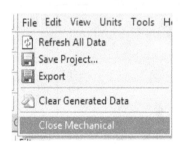

3.6   References

1. ANSYS Parametric Design Language Guide, Release 15.0, Nov. 2013, ANSTS, Inc.
2. ANSYS Workbench User's Guide, Release 15.0, Nov. 2013, ANSYS, Inc.
3. X. L. Chen and Y. J. Liu, Finite Element Modeling and Simulation with ANSYS Workbench, 1st edition, Barnes & Noble, January 2004.
4. J. W. Dally and R. J. Bonnenberger, Problems: Statics and Mechanics of Materials, College House Enterprises, LLC, 2010.
5. J. W. Dally and R. J. Bonnenberger, Mechanics II Mechanics of Materials, College House Enterprises, LLC, 2010.
6. X. S. Ding and G. L. Lin, ANSYS Workbench 14.5 Case Studies (in Chinese), Tsinghua University Publisher, Feb. 2014.
7. G. L. Lin, ANSYS Workbench 15.0 Case Studies (in Chinese), Tsinghua University Publisher, October 2014.
8. K. L. Lawrence. ANSYS Workbench Tutorial Release 13, SDC Publications, 2011.
9. K. L. Lawrence. ANSYS Workbench Tutorial Release 14, SDC Publications, 2012
10. Huei-Huang Lee, Finite Element Simulations with ANSYS Workbench 14, Theory, Applications, Case Studies, SDC Publications, 2012.
11. Huei-Huang Lee, Finite Element Simulations with ANSYS Workbench 16, Theory, Applications, Case Studies, SDC Publications, 2015.
12. Jack Zecher, ANSYS Workbench Software Tutorial with Multimedia CD Release 12, Barnes & Noble, 2009.
13. G. M. Zhang, Engineering Design and Creo Parametric 3.0, College House Enterprises, LLC, 2014.
14. G. M. Zhang, Engineering Analysis with Pro/Mechanica and ANSYS, College House Enterprises, LLC, 2011.

3.7 Exercises

1. A simple deck beam bridge is shown below. The unit used in the drawing is meter. The bridge consists of a deck beam and 3 piles. Assume that the deck beam is constructed of Structural Steel and the piles is made of Concrete. The three (3) size dimensions of the deck beam are 24x3x0.2 meter. The three (3) size dimensions of the piles are $\phi 1.5 \times \phi 1.0 \times 10$ meter. Assume that the bottom surfaces of the 3 piles are fixed to the ground. Also assume that the two (2) ends of the deck beam are fixed to the ground through the two surface regions of 3X0.2 meter. Assume the pressure load acting on the deck beam is 0.01 MPa. Evaluate the deflection of the bridge due to the load and span.

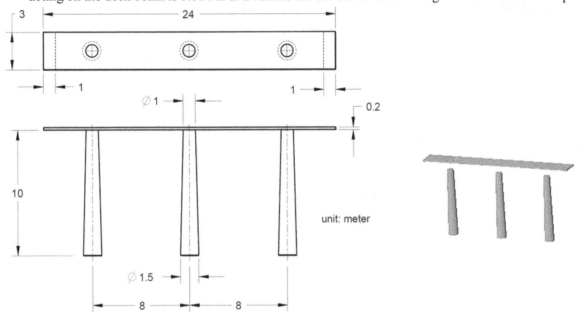

The following steps are suggested for creating a 3D solid model of this bridge in DesignModeler. Step 1: select Meter unit. Select ZXPlane as the sketching plane to create a plate feature. Afterwards create the Imprint Faces, as shown. It is important that you click Freeze from Tools listed on the top menu.

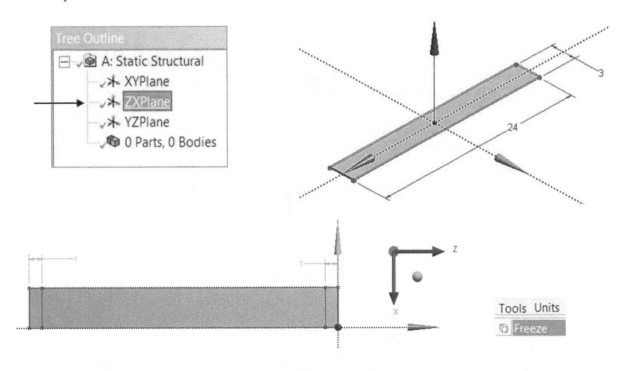

Step 2: select YZPlane as the sketching plane to create the pile feature. Use Create > Body Transformation > Translate to create the other 2 piles, as shown.

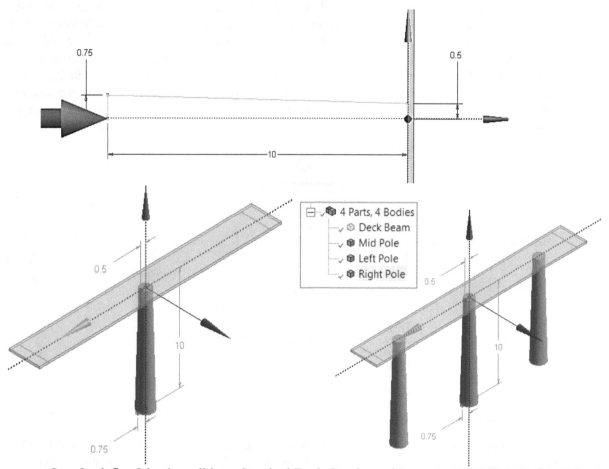

Step 3: define 2 load conditions: Standard Earth Gravity and Pressure (0.01MPa). Define 2 fixed support constraints, and run FEA.

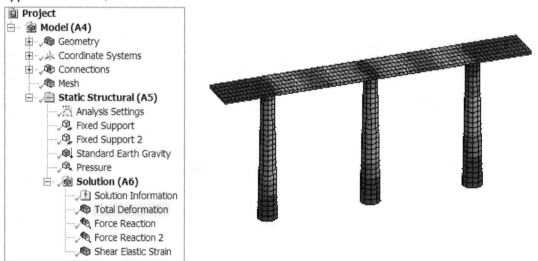

2.   The engineering drawing of a cam-follower assembly is shown below. The cam component is assembled with the shaft component, as shown below. The dimensions of the cam and shaft components are shown in the drawing. The unit is millimeter. The material type of the cam component is stainless steel. The material type of the shaft component is structural steel. The two components are bonded together.

Cam-Follower Mechanism

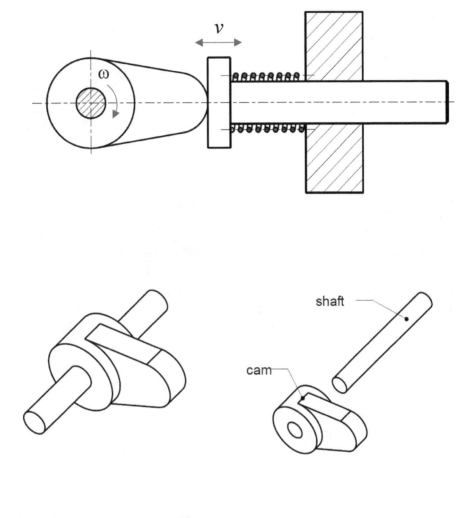

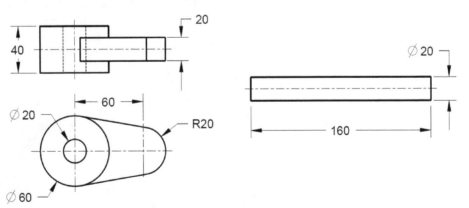

The following steps are suggested for working with this cam-shaft assembly in DesignModeler.

Step 1: select the Millimeter unit. Select XYPlane as the sketching plane to create the first feature. Afterwards create 2 more features, as shown. It is important that you click Freeze from Tools listed on the top menu before getting on the creation of shaft.

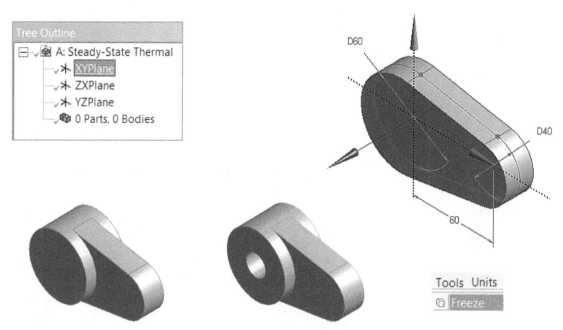

Step 2: select XYPlane as the sketching plane, click New Sketch to create the shaft feature, as shown. Afterwards, you need to create imprint surfaces at the two ends for defining the fixed support constraints. Use ZXPlane to create the following sketch. Afterwards use these 2 sketches to create 2 surface regions.

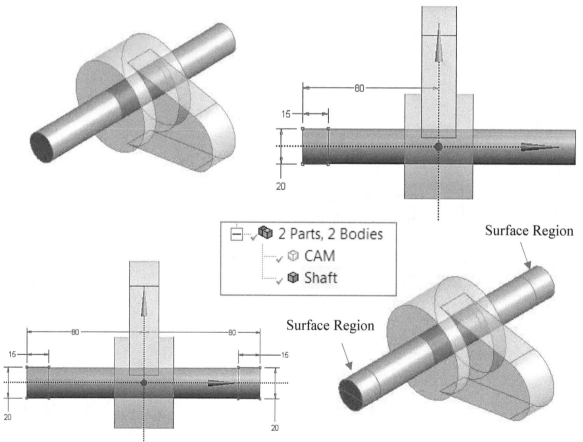

Step 3: define 2 load conditions: Standard Earth Gravity and Pressure (0.01MPa). Define 2 fixed support constraints, and run FEA.

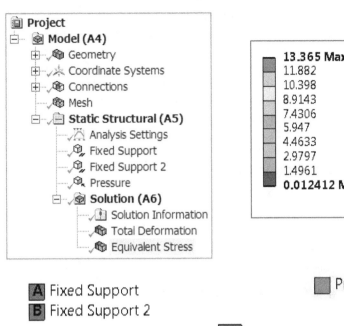

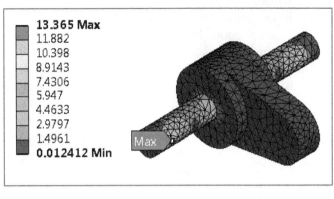

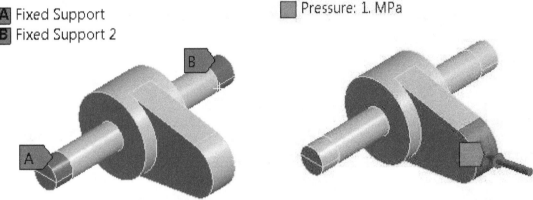

3. The engineering drawing of a bearing-shaft assembly is shown below. The bearing material is stainless steel, and the shaft material is structural steel. Assume that the outer cylindrical surface of bear is fixed to the ground. The outer cylindrical surface of shaft is subjected to a pressure load. Its magnitude is 10 MPa. Construct the 3D solid model for the assembly and perform FEA.

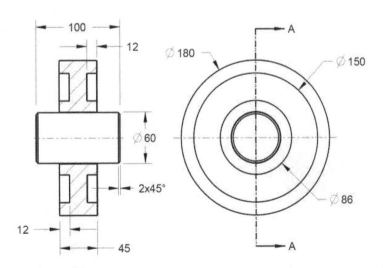

4.   The following drawing shows an assembly consists of two layers. The upper layer material is structural steel and the lower layer material is copper alloy. These two types of material are listed in the library in Workbench. Both layers are bonded. The right end has a constraint condition: Fixed Support. The left end is set free. A pressure load acts on a rectangular surface region located on the top surface, as shown. The pressure magnitude is 0.20 MPa. The rectangular area is 6x2 mm^2 with a position dimension equal to 36 mm, as shown. The assembly is also subjected to temperature load. The environment temperature is 5°C, and the temperature of the two layers of the assembly is 40°C.

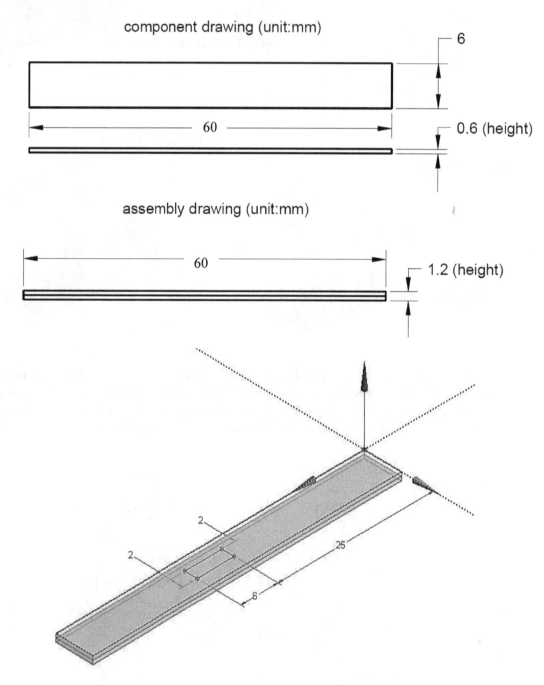

component drawing (unit:mm)

6

60

0.6 (height)

assembly drawing (unit:mm)

60

1.2 (height)

2

2

25

6

Use Mesh > Insert > Method and select Hex Dominant. Specify 0.30 mm in Element Size. Click Generate Mesh. Determine the max deformation of this assembly and observe its location. Determine the max value of von Mises stress.

CHAPTER 4

Steady-State Thermal Analysis

4.1   Introduction

In Chapter 2 and Chapter 3, the focuses of discussion are Static Structural Analysis. Important concepts of the creation of geometric and assembly models are detailed. The focus of discussion in Chapter 4 is Thermal Analysis. In this chapter, we present seven case studies using thermal analysis. All these 7 case studies are related to steady-state thermal analysis. Section 4.2 presents the first case study of a firewall structure. We review those basic concepts of geometry creation discussed in Chapter 2 and chapter 3 with the focus on applying thermal loads. Section 4.3 is a case study illustrating the procedure of defining the boundary conditions related to pre-scribed temperatures and thermal convection conditions. Section 4.4 presents a heat sinker design with performance evaluation. Section 4.5 and Section 4.6 are two case studies where a thermal steady-state analysis is combined with a static structural analysis. The case study presented in Section 4.5 starts with thermal analysis and export the temperature distribution as an input to the static structural analysis. The case study presented in Section 4.6 starts with the static structural analysis first, and thermal analysis afterwards. The obtained temperature distribution is fed back to the static structural analysis. Section 4.7 introduces the concept of creating an assembly system with two components, namely, a door frame and an inserted glass window. A steady-state thermal analysis illustrates the temperature distribution developed within the door-window assembly when subjected to 2 convection conditions. Section 4.8 presents the design of an assembly structure using a pattern operation. We demonstrate the importance of using the operation called Form a New Part where independent solid models are combined in the process of forming a new part. This operation creates a bonded configuration on the contact surfaces so that no need to construct the contact regions when entering the Mechanical mode.

4.2   Steady-State Thermal Analysis of a Firewall Structure

A firewall model is shown below. The model is a part of a long firewall. The 3 dimensions of the firewall model are 1000 x 500 x 2000 mm. To control the temperature distribution, there are 3 water channels designed to dissipate excessive heat confined within the firewall. The diameter of each of those 3 holes is 180 mm. The material type is brick. The thermal heat conductivity of this brick material is 0.47 W/m/°C.

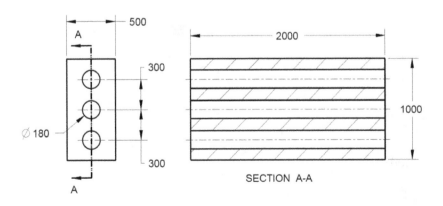

SECTION A-A

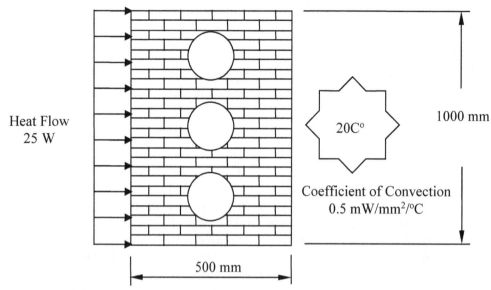

Heat Flow
25 W

20C°

1000 mm

Coefficient of Convection
0.5 mW/mm²/°C

500 mm

Step 1: Open Workbench.

From the start menu, click **ANSYS 19** > **Workbench 19**. Users may directly click Workbench 19 when the symbol of Workbench 19 is on display.

Workbench 19.2

Step 2: From **Toolbox**, highlight **Steady-State Thermal** and drag it to the main screen.

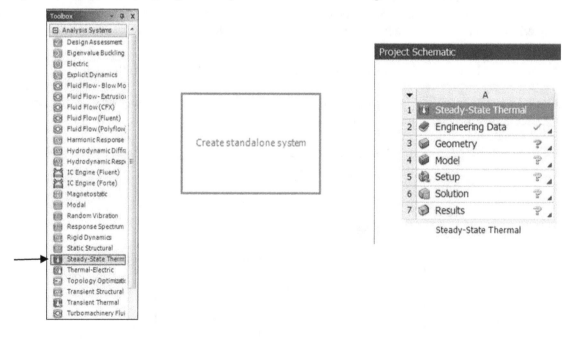

Step 3: Start Design Modeler.

Right-click the **Geometry** cell and pick **New DesignModeler Geometry** to start **Design Modeler**. Users may double click **Geometry** to directly enter **Design Modeler** if New **SpaceClaim Geometry** is not present. Click the icon of **Units** listed on the top menu, and select **Millimeter**.

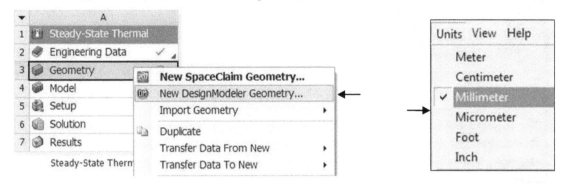

In Tree Outline, highlight **XYPlane** and select it as the sketching plane. Click **Look At** to orient the sketching plane for preparing a sketch.

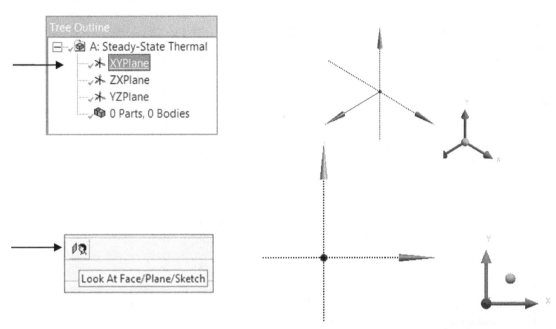

Click **Sketching**, and click **Draw**. In the Draw window, click **Rectangle**. Sketch a rectangle, as shown below.

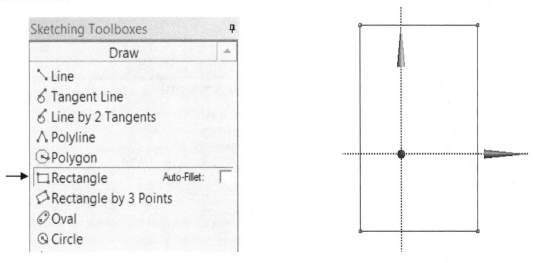

Add 2 constraint conditions: **Symmetry**. Click the vertical axis first, and pick the 2 vertical lines of the rectangle. Afterwards, pick the horizontal axis, and pick the 2 horizontal lines of the rectangle.

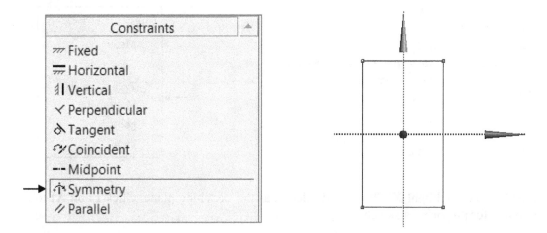

Click **Dimensions** and from the **Dimensions** panel, select **General**. Add 2 size dimensions: 1000 and 500, as shown.

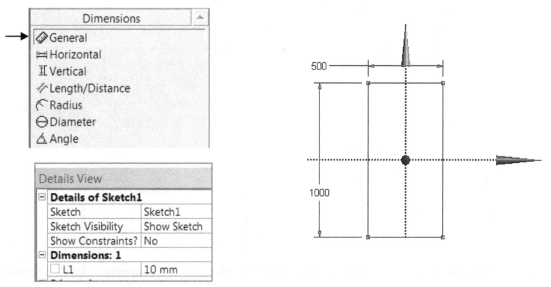

Click **Draw**. In the Draw window, click **Circle**. Sketch 3 circles, and add the diameter dimension of 180 and two (2) position dimensions of 300, as shown below.

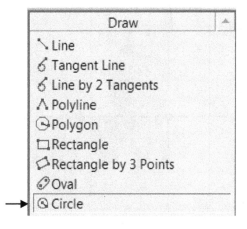

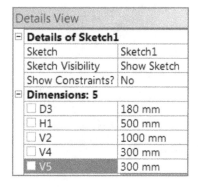

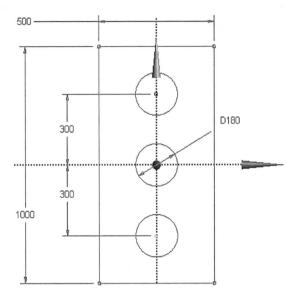

Upon completing the sketch, click the icon of **Extrude**. Click the created sketch or sketch 1 from the Tree Outline, and click **Apply**. For Operation, select **Add Material** and specify 2000 as the extrusion value. Click the icon of **Generate** to obtain the 3D solid model of the firewall, as shown.

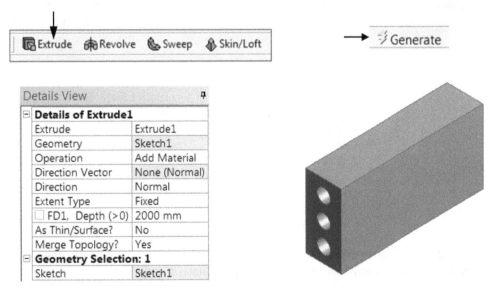

Expand the model tree, highlight Solid listed under 1 Part 1 Body, right-click to pick Rename. Change Solid to Firewall.

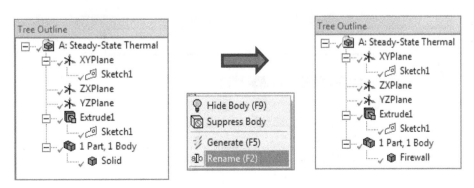

Upon completing the process of creating the geometry, click **File** > Close **Design Modeler**, and go back to **Project Schematic**. Specify **Firewall** as the file name. Click **Save**.

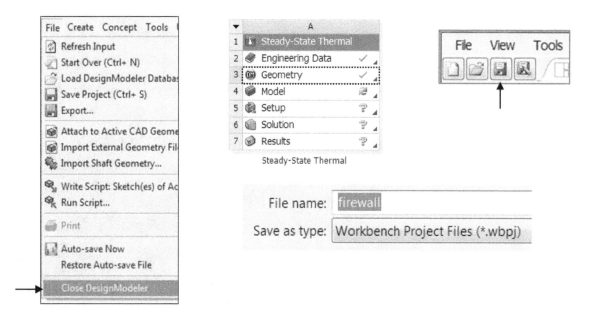

Step 4: Define a New Type of Material for Brick in the Engineering Data Sources

In the Steady-State Thermal panel, double click Engineering Data. We enter the Engineering Data Mode.

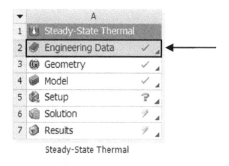

Steady-State Thermal

Click the icon of **Engineering Data Sources**. There is box called **Click here to add a new material**. Click this box and type Brick as the name of new material type.

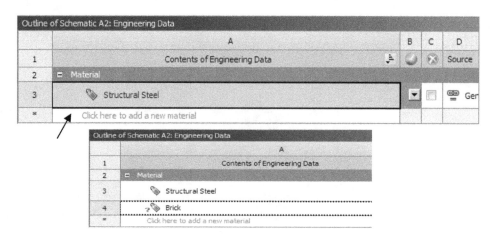

In Tool Box, expand Thermal. Double-click Isotropic Thermal Conductivity. The Properties window appears. The software system is waiting for the user to enter the value of thermal conductivity. Specify 0.47. Set temperature to 20° C. Click Project to return to the Project Schematic page.

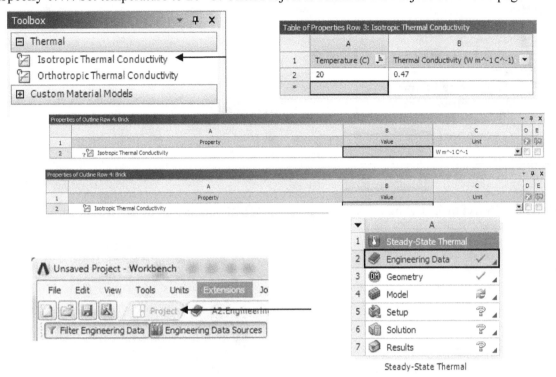

Step 5: Assign Brick, Generate Mesh and Define the Boundary Conditions to Perform FEA
    In the Steady-State Thermal panel, double click Model. Check the unit system.

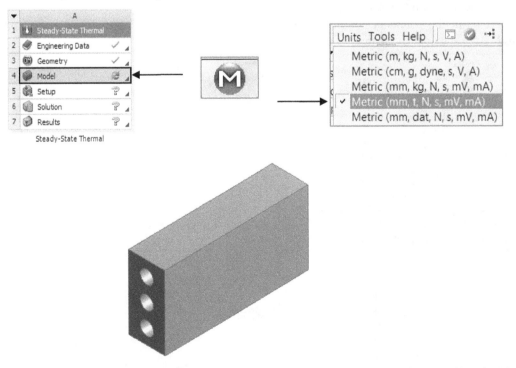

From the Project tree, expand Geometry and highlight **Firewall**. In the Details of "Firewall" window, Structural Steel is shown in Assignment. Right-click to pick Brick.

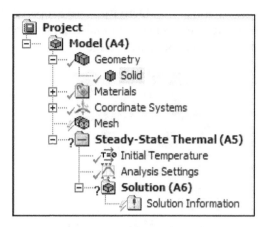

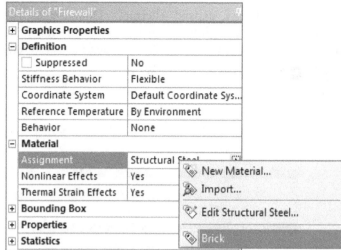

From the Project tree, Highlight **Mesh**. Right-click to pick **Generate Mesh**.

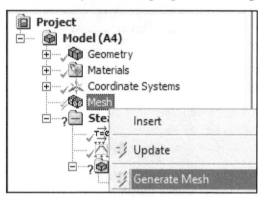

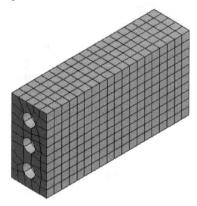

Step 6: Define the boundary condition.

To define the boundary conditions, first highlight **Initial Temperature** listed under **Steady-State Thermal (A5)**, and specify 20 degree as the initial temperature setting.

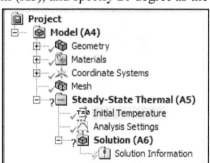

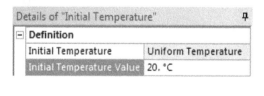

Highlight **Steady-State Thermal (A5)**. Pick **Insert** and **Heat Flow**. Pick the surface at the right side of firewall. Click **Apply**. Specify 25 Watt the magnitude of **Heat Flow**.

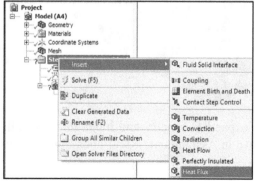

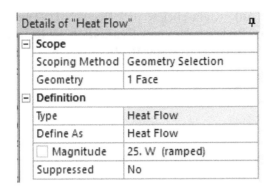

Details of "Heat Flow"	中
⊟ **Scope**	
Scoping Method	Geometry Selection
Geometry	1 Face
⊟ **Definition**	
Type	Heat Flow
Define As	Heat Flow
☐ Magnitude	25. W (ramped)
Suppressed	No

Again, highlight **Steady-State Thermal (A5)**. Pick **Insert** and **Convection**. Pick the surface at the right side of firewall, as shown. Click **Apply**. Specify 20 as the environment temperature setting, and specify 0.5 (mW/°C/mm²) as the coefficient of convection.

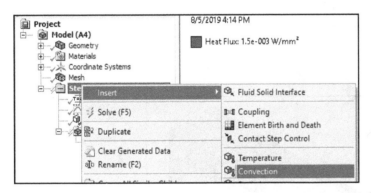

Details of "Convection"	中
⊟ **Scope**	
Scoping Method	Geometry Selection
Geometry	1 Face
⊟ **Definition**	
Type	Convection
☐ Film Coefficient	5.e-004 W/mm².°C ... ▸
☐ Ambient Temperature	20. °C (ramped)
Convection Matrix	Program Controlled
Suppressed	No

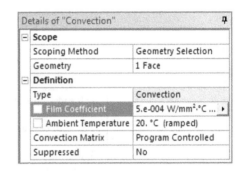

Step 7: Run FEA program
   Highlight **Solution (A6)** and pick **Solve**.

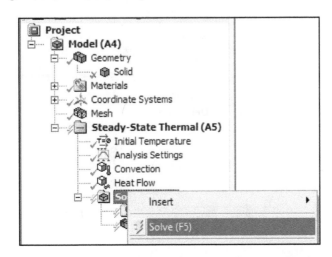

To examine the obtained results, highlight **Solution (A6)**. Right-click to pick **Insert > Thermal > Temperature**. Highlight **Temperature** under **Solution (A6)**. Right-click to pick **Evaluation All Results**. The maximum and minimum values of temperature are 38.9 °C and 20 °C, respectively.

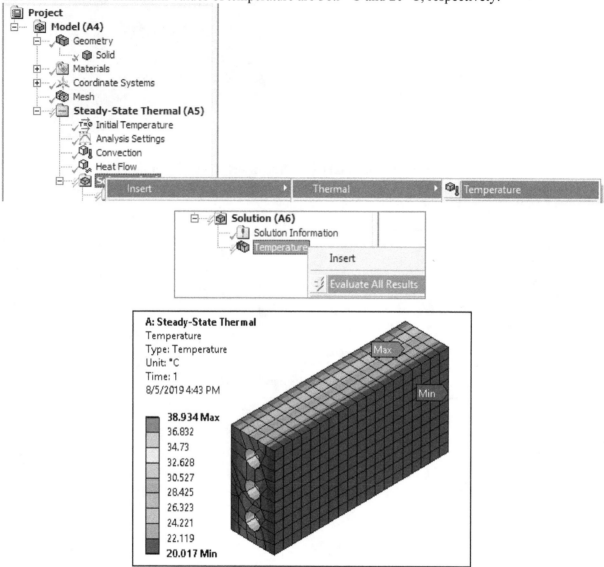

In the Project tree, highlight **Heat Flow**. Modify 25 Watt to 40 Watt to increase the heat flow strength. Click **Solve** to run FEA. Examining the FEA result of temperature distribution, the maximum and minimum values of temperature are 50 °C and 20 °C, respectively.

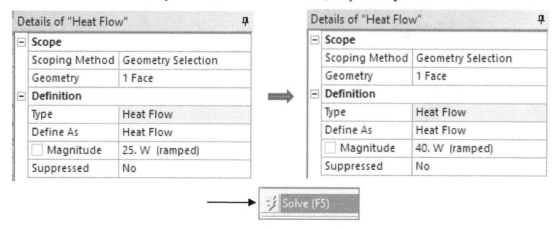

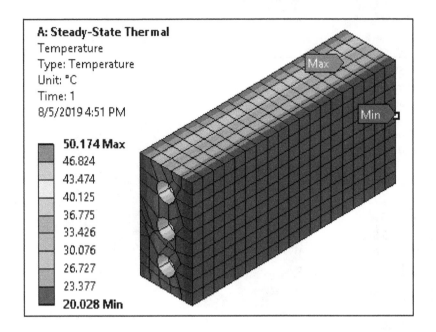

Let us add a new convection condition so that the temperature distribution will go lower side. To do so, highlight **Steady-State Thermal (A5)**. Pick **Insert** and **Convection**. While holding down the **Ctrl** key, pick the 3 cylindrical surfaces from the 3 circular-shaped channels, as shown. Click **Apply**. Specify 20 as the environment temperature setting, and specify 2.0 (mW/°C/mm²) as the coefficient of convection. Click **Solve** and run FEA. Plot the temperature distribution. The maximum and minimum values of temperature are 28,7 °C and 20.0 °C, respectively.

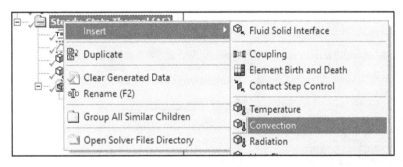

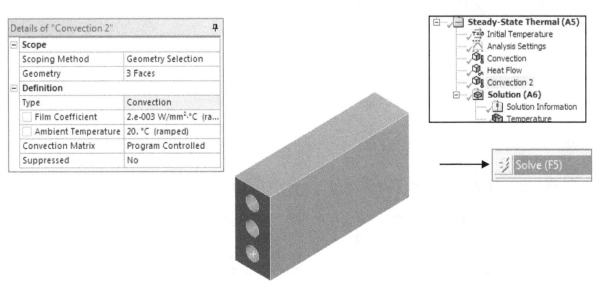

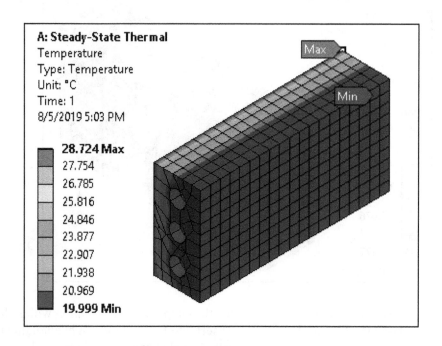

Click **File** and **Close Mechanical** to go back to the **Project Schematic** screen. The displayed Steady-State Thermal panel shows all check marks, indicating that the FEA run was successful. From the top menu, click the icon of **Save Project** to complete this study.

4.3   Thermal Analysis for a Plate with Three Holes

The displayed drawing represents a plate component with 3 holes. In this study, we perform a steady-state thermal analysis. Assume that the material type is steel, and the environmental temperature setting is 10 °C. The left side of the plate component is subjected to a pre-defined temperature condition. The temperature setting is 200 °C. The surface on the right side is subjected to convection condition. The temperature setting is 10 °C, and coefficient is 5 mW/mm²°C.

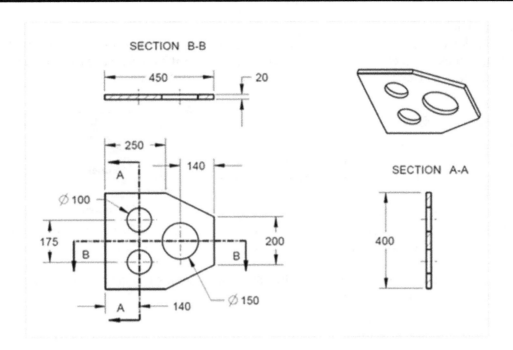

Step 1: Open Workbench.

From the start menu, click **ANSYS 19** > **Workbench 19**. Users may directly click Workbench 19 when the symbol of Workbench 19 is on display.

 Workbench 19.2

Step 2: From **Toolbox**, highlight **Steady-State Thermal** and drag it to the main screen.

Step 3: Start Design Modeler.

Right-click the **Geometry** cell and pick **New DesignModeler Geometry** to start **Design Modeler**. Users may double click **Geometry** to directly enter **Design Modeler** if New **SpaceClaim Geometry** is not present. Click the icon of **Units** listed on the top menu, and select **Millimeter**.

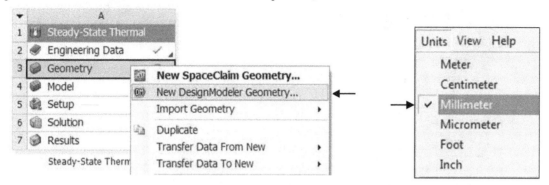

Select **XYPlane** as the sketching plane. Click the icon of **Look At** so that the sketching plane is parallel to the screen and ready for preparing a sketch

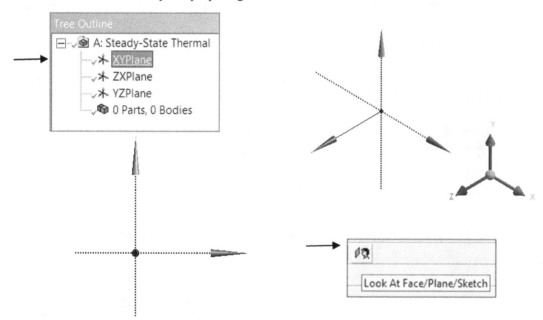

Click **Sketching**, and click **Draw**. In the Draw window, click **Polyline**. Make the following sketch. At the end of making this sketch, right-click to pick **Closed Sketch**.

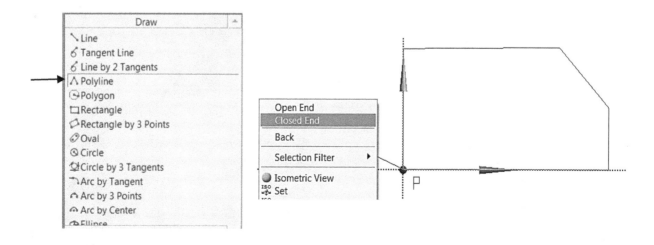

Click **Dimensions**, and select **General**. Specify the 4 size dimensions: 200, 450, 250 and 100.

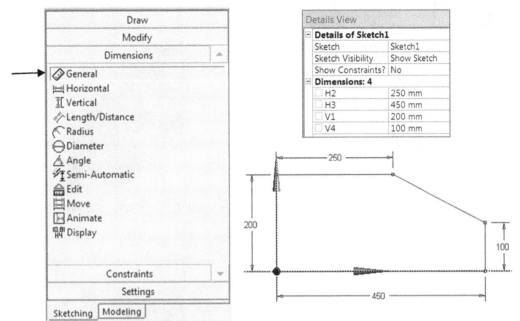

Upon completing the sketch, click the icon of **Extrude**. Click the created sketch or sketch 1 from the Tree Outline, and click **Apply**. Specify 20 as the thickness value. Click the icon of **Generate** to obtain the 3D solid model of the plate feature, as shown.

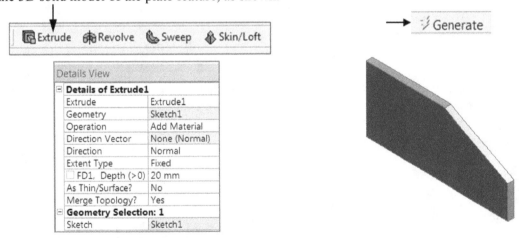

Now let us create the second feature: add a mirrored part of the created plate feature. From the top menu, click **Create > Body Transformation > Mirror**. From the screen pick the created body and click Apply. Afterwards, Click Mirror Plane, pick ZX Plane as the Mirror Plane, and Click **Generate**.

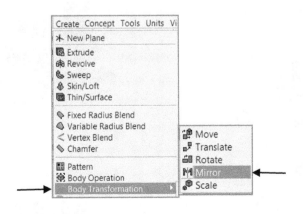

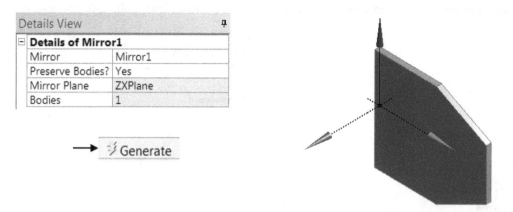

To create the third feature of adding 3 holes, first highlight XY Plane and click **New Sketch**. Afterwards, click **Look At**.

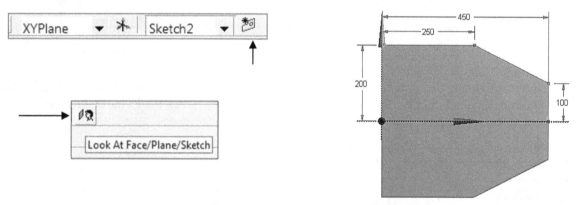

Click **Sketching**, and click **Draw**. In the Draw window, click **Circle**. Sketch a circle, as shown below. Add a diameter dimension of 150 and a position dimension of 140.

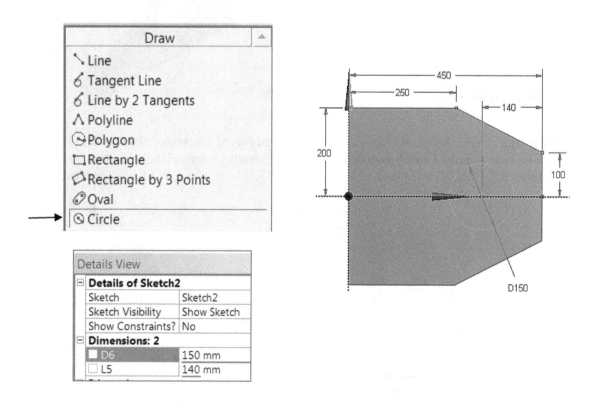

Sketch 2 more circles, as shown below. Add a diameter dimension of 100 and 2 position dimensions of 140 and 175. Note that the 2 circles are symmetric about the horizontal axis.

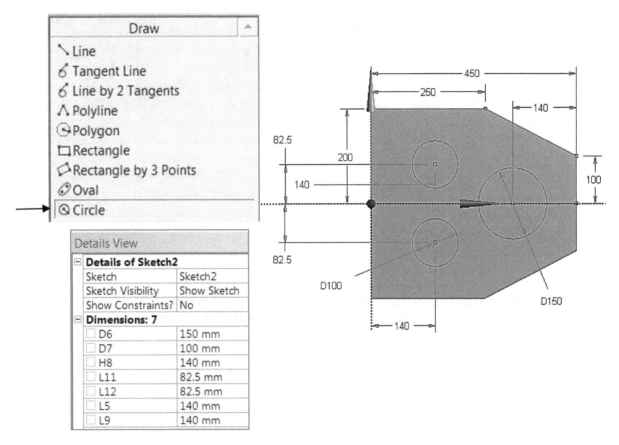

Upon completing the sketch, click the icon of **Extrude**. Click the created sketch or sketch 2 from the Tree Outline, and click **Apply**. For Operation, click Cut Material. For Extent Type, click Through All. Click the icon of **Generate** to obtain the 3D solid model of the plate with 3 holes, as shown.

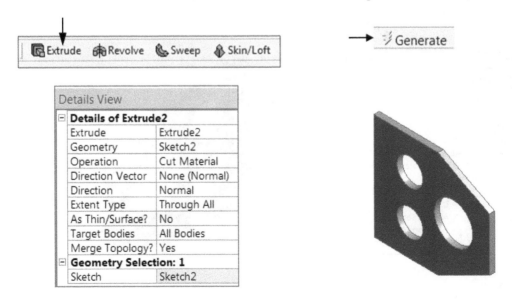

Upon completing the process of creating the geometry, we need to close Design Modeler. Click **File** > Close **Design Modeler**, and go back to **Project Schematic**. Specify **Thermal Plate Three Holes** as the file name. Click **Save**.

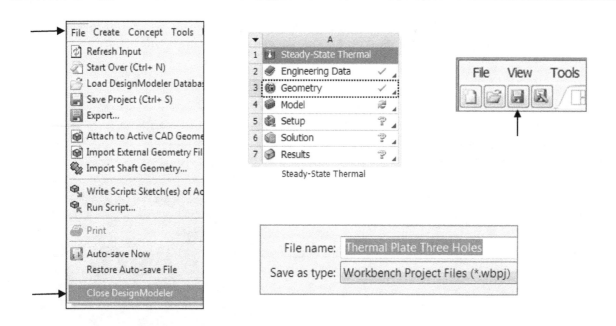

Step 4: Accept the Default Material Assignment of Structural Steel and Perform FEA.
Double-click the icon of **Model**. We enter the **Mechanical Mode**. Check the unit system.

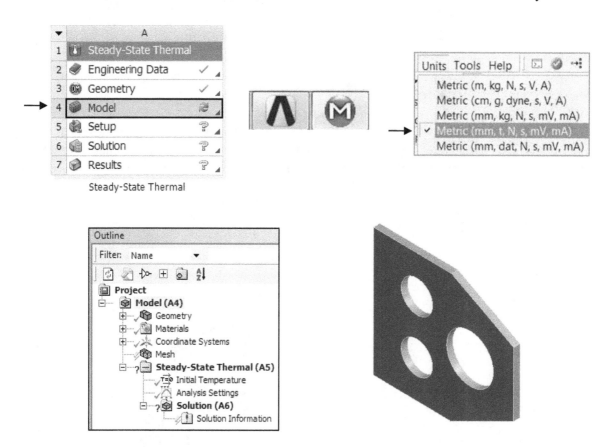

From the Project tree, Highlight **Mesh**. In the Details of "Mesh" window, click **Sizing**. Set Resolution to 4. Right-click **Mesh** again to pick **Generate Mesh**.

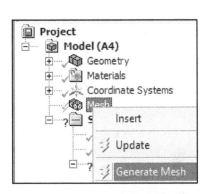

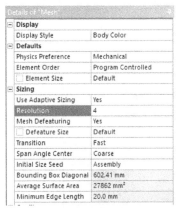

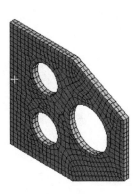

Step 5: Define the boundary condition.

Highlight **Initial Temperature** listed under **Steady-State Thermal (A5)**, and specify 10 degree as the initial temperature setting.

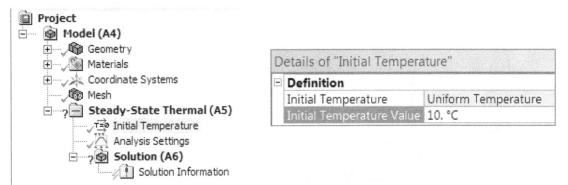

Highlight **Steady-State Thermal (A5)**. Pick **Insert** and **Temperature**. Pick the surface on the left side, as shown. Pick **Apply**. Specify 200 as the magnitude of pre-defined temperature.

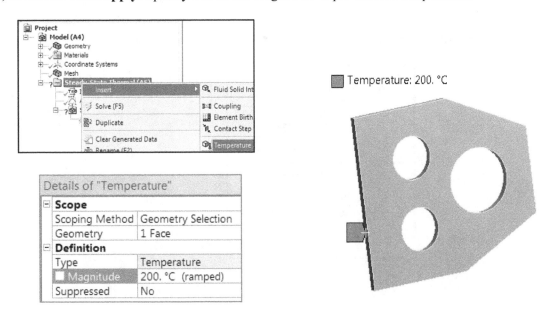

To define the convection condition, highlight **Steady-State Thermal**. Pick **Insert** and **Convection**. Pick the surface on the right side, as shown. Click **Apply**. Specify 10 °C as the **Ambient** or environmental temperature, and 0.005 (mW/°C/mm^2) as the **Film Coefficient**.

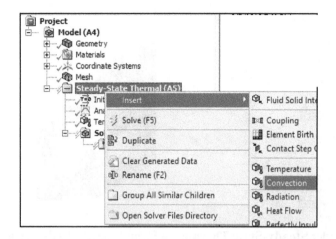

Convection: 10. °C, 5.e-003 W/mm²·°C

Details of "Convection"	
Scope	
Scoping Method	Geometry Selection
Geometry	1 Face
Definition	
Type	Convection
Film Coefficient	5.e-003 W/mm²·°C (ra.
Ambient Temperature	10. °C (ramped)
Convection Matrix	Program Controlled
Suppressed	No

Highlight **Solution (A6)** and pick **Solve**.

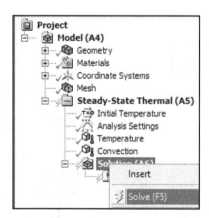

Step 6: Displaying Results

Highlight **Solution (A6)**. Right-click to pick **Insert > Thermal > Temperature**.

Highlight **Temperature** under **Solution (A6)**. Right-click to pick **Evaluation All Results**. The maximum and minimum values of temperature are 200°C and 13.891 °C, respectively.

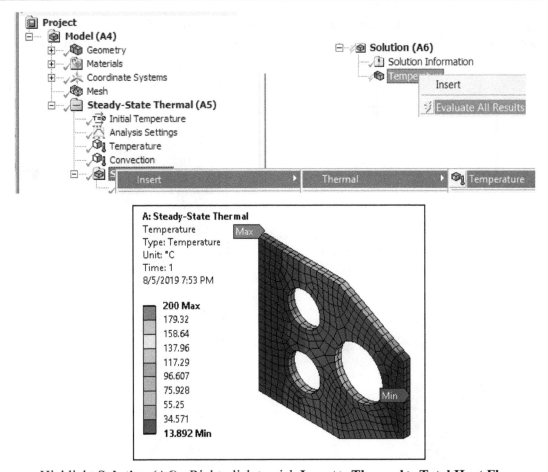

Highlight **Solution (A6)**. Right-click to pick **Insert > Thermal > Total Heat Flex**.

Highlight **Total Heat Flux** under **Solution (A6)**. Right-click to pick **Evaluation All Results**. The maximum and minimum values of total hear flux are 0.045 W/mm² and 0.00010 W/mm², respectively.

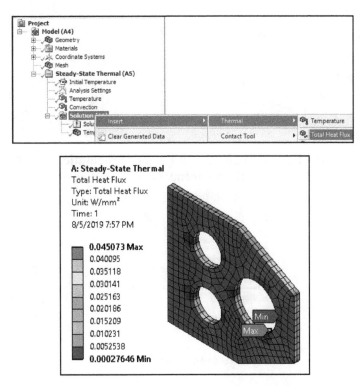

To simulate the heat flux flow, click the Vector Symbol, as shown. Afterwards, click the triangle symbol. Users may adjust the number of frames and speed to have a better view of the heat flux flow.

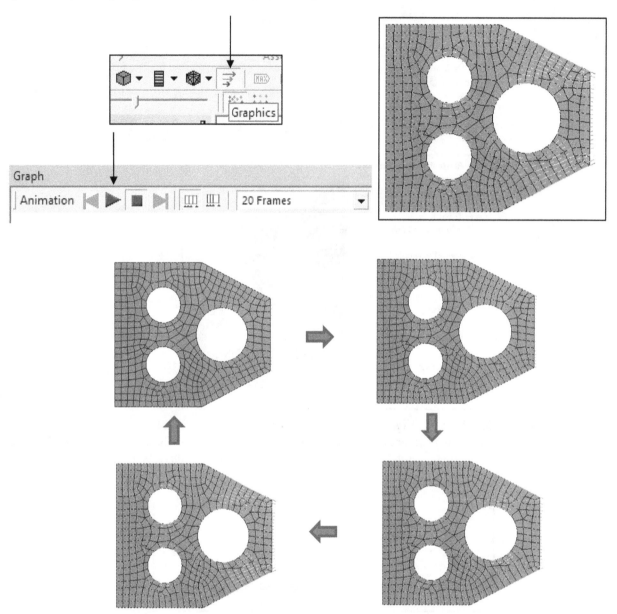

Click **File** and **Close Mechanical** to go back to **Project Schematic**. The displayed Steady-State Thermal panel shows all check marks. From the top menu, click **Save Project** to complete this study.

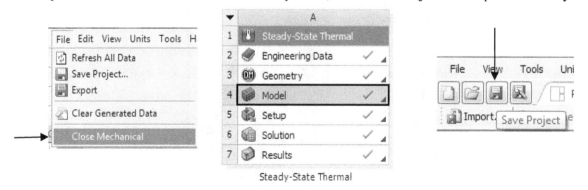

Steady-State Thermal

4.4 Thermal Analysis for a Heat Sinker

The engineering drawing of a heat sink is shown below. Heat sinks are widely used and assembled on the CPU system, taking heat away from the areas closed to the CPU site, which functions as a heat flow load. The material type is aluminum alloy. Perform three (3) case studies. Identify the temperature distribution with the maximum value and its location.

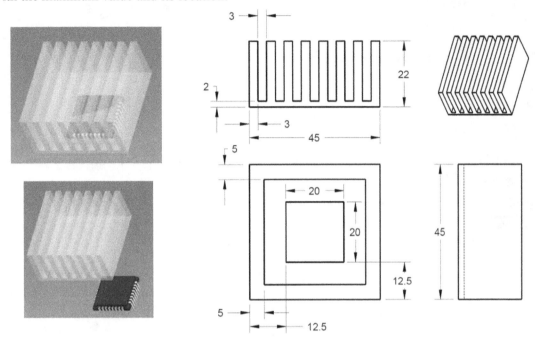

Step 1: Open Workbench.

From the start menu, click **ANSYS 19** > **Workbench 19**. Users may directly click Workbench 19 when the symbol of Workbench 19 is on display.

Workbench 19.2

Step 2: From **Toolbox**, highlight **Steady-State Thermal** and drag it to the main screen.

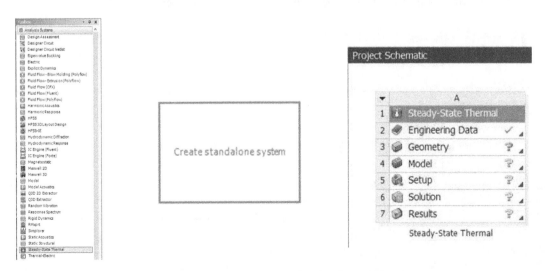

Step 3: Start Design Modeler.

Right-click the **Geometry** cell and pick **New DesignModeler Geometry** to start **Design Modeler**. Users may double click **Geometry** to directly enter **Design Modeler** if New **SpaceClaim Geometry** is not present. Click the icon of **Units** listed on the top menu, and select **Millimeter**.

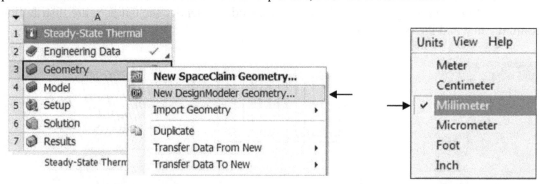

Select **XYPlane** as the sketching plane. Click the icon of **Look At** so that the sketching plane is parallel to the screen and ready for preparing a sketch

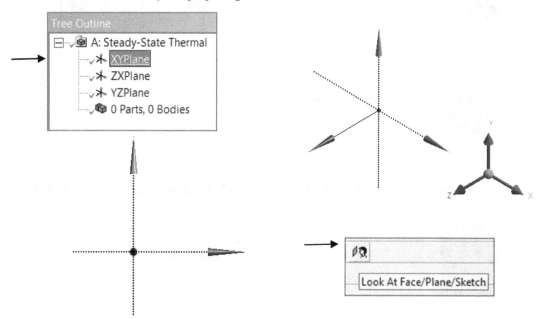

Click **Sketching**, and click **Draw**. In the Draw window, click **Rectangle.** Make the following sketch. At the end of making this sketch, right-click to pick **Closed Sketch**.

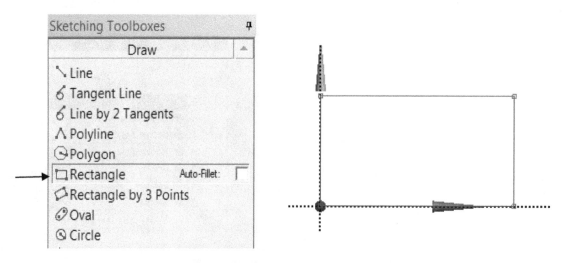

Click **Dimensions**, and select **General**. Specify the 4 size dimensions 22 and 45.

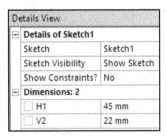

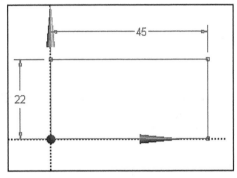

Upon completing the sketch, click the icon of **Extrude**. Click the created sketch or sketch 1 from the Tree Outline, and click **Apply**. Specify 45 as the distance value. Click the icon of **Generate** to obtain the 3D solid model of the plate feature, as shown.

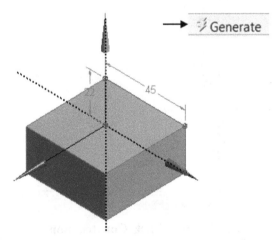

Highlight **XYPlane**. Click **New Sketch**. Click **Look At**.

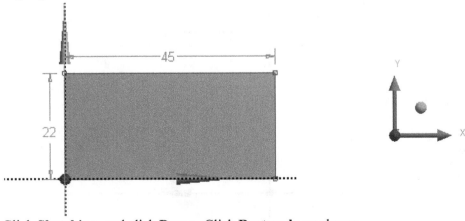

Click **Sketching**, and click **Draw**. Click **Rectangle,** as shown.

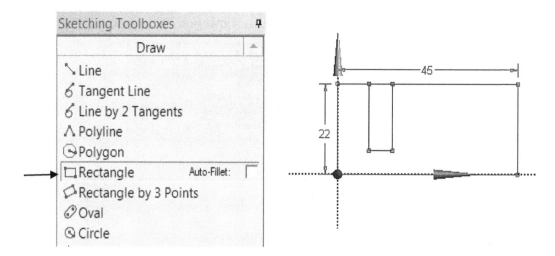

Click **Dimensions**, and select **General**. Specify 3 as the width of the rectangle. Select **Horizontal** to specify 3 as the position dimension. Select **Vertical** and specify 2 as the position dimension.

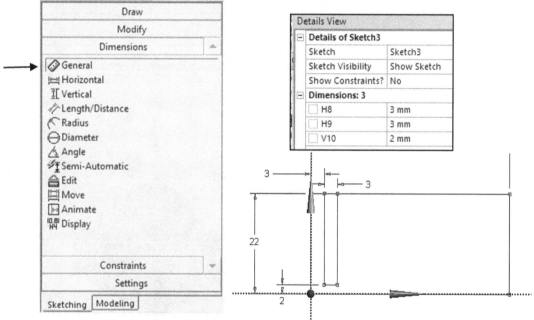

Click **Draw**. Click **Construction point**. On the X-axis, create a point and specify 9 as its coordinate, as shown.

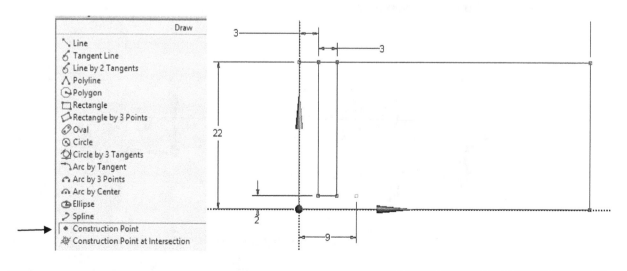

Click **Modify**. Select **Copy**. While holding down the **Ctrl** key and pick the rectangle (4 sides) and the construction point. Click **Paste**. Move forward and make a left when P is on display. Repeat this process for 6 times.

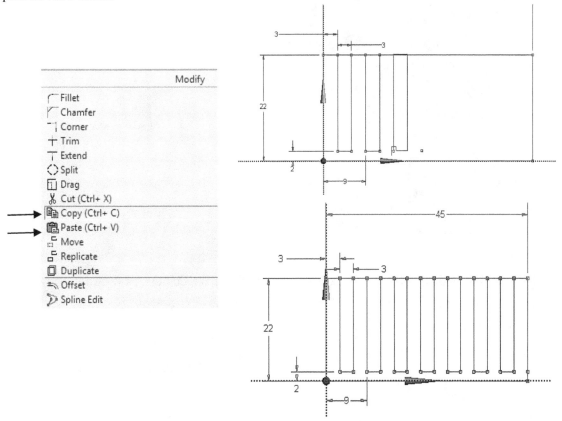

Upon completing the sketch, click the icon of **Extrude**. Click the created sketch or sketch 2 from the Tree Outline, and click **Apply**. For Operation, click Cut Material. For Extent Type, click Through All. Click the icon of **Generate** to obtain 7 slots, as shown.

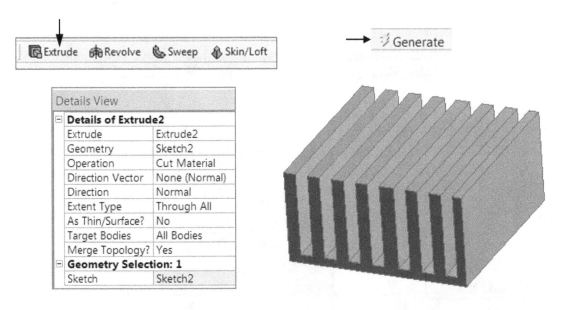

Highlight ZXPlane. Click the icon of **New Plane**. Click **Look At**.

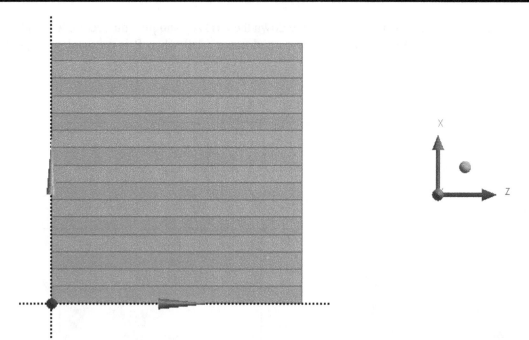

Click **Sketching**, and click **Draw**. Click **Rectangle** to sketch a rectangle, as shown.

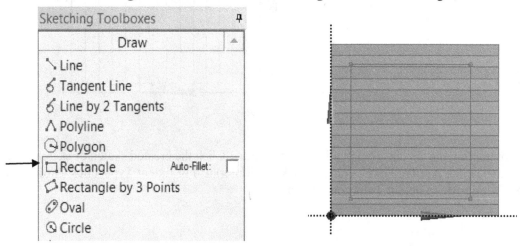

Click **Dimensions**, and select **General**. Specify 35 and 35, as shown. Select **Horizontal** to specify 5, as the position dimension. Select **Vertical** and specify 5 as the position dimension.

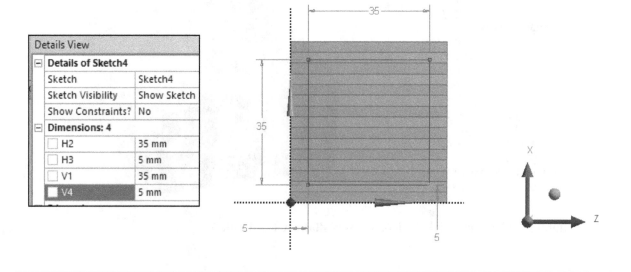

Click **Modeling** > **Extrude**. Click the created sketch. Click **Apply**. Select **Imprint Faces** from the **Operation** field. Click **Generate**.

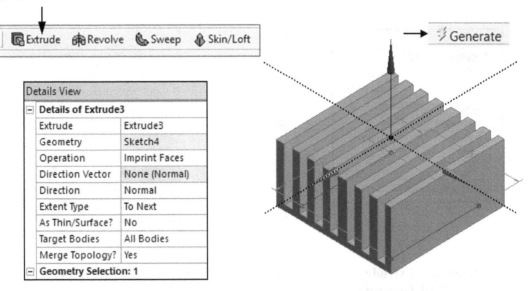

Details View	
Details of Extrude3	
Extrude	Extrude3
Geometry	Sketch4
Operation	Imprint Faces
Direction Vector	None (Normal)
Direction	Normal
Extent Type	To Next
As Thin/Surface?	No
Target Bodies	All Bodies
Merge Topology?	Yes
Geometry Selection: 1	

Repeat this process to create the second surface region. Highlight **ZXPlane**. Click the icon of **New Plane**. Click **Look At**.

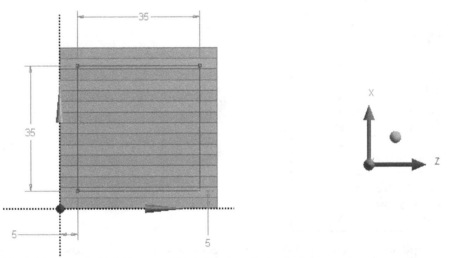

Click **Sketching**, and click **Draw**. Click **Rectangle** to sketch a rectangle, as shown.

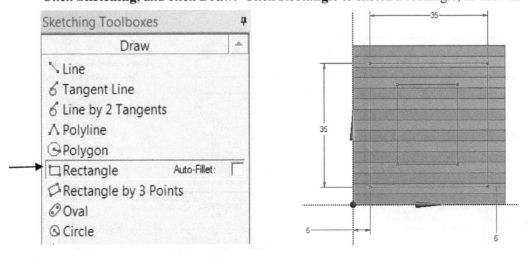

Click **Dimensions**, and select **General**. Specify 20 and 20, as shown. Select **Horizontal** to specify 12.5, as the position dimension. Select **Vertical** and specify 12.5 as the position dimension.

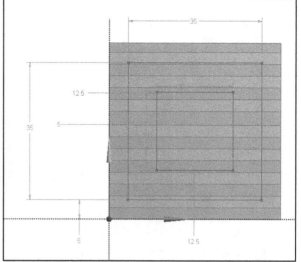

Click **Modeling** > **Extrude**. Click the created sketch. Click **Apply**. Select **Imprint Faces** from the **Operation** field. Click **Generate**.

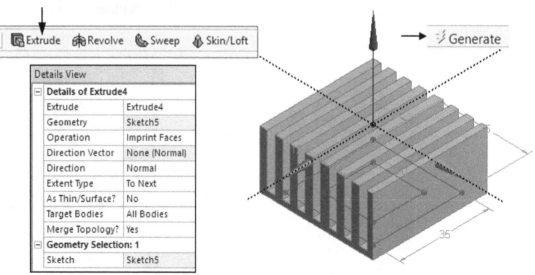

Upon completing the process of creating the geometry, we need to close Design Modeler. Click **File** > Close **Design Modeler**, and go back to **Project Schematic**. Specify **Heat Sinker** as the file name. Click **Save**.

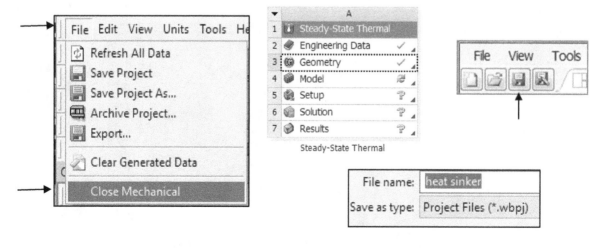

Step 4: Material Assignment of Aluminum Alloy and Perform FEA.

In the Steady-state thermal panel, double click Engineering Data. We enter the Engineering Data Mode.

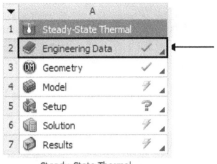

Steady-State Thermal

Click the icon of **Engineering Data Sources**. Click **General Materials**. We are able to locate Aluminum Alloy.

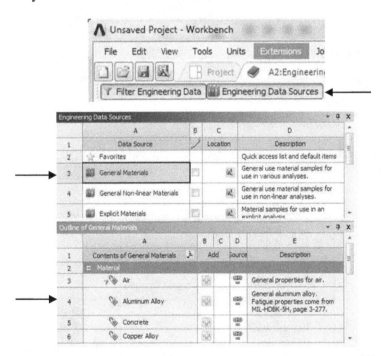

Click the plus symbol nearby Aluminum Alloy to activate it. Afterwards, click the icon of **Engineering Data Sources** and click the icon of Project to go back to **Project Schematic**.

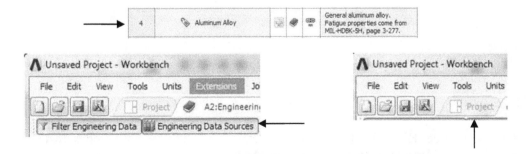

Double-click the icon of **Model**. We enter the **Mechanical Mode**. Check the unit system.

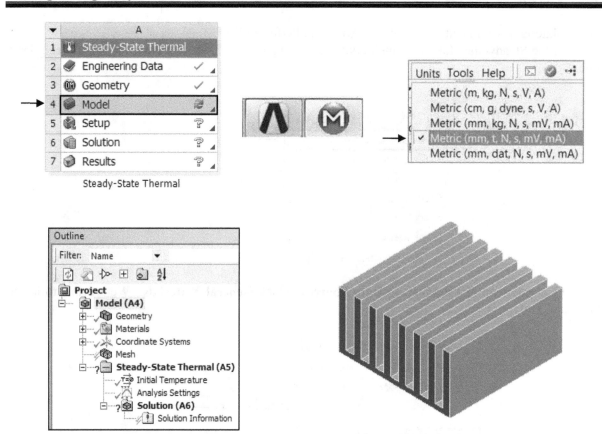

Steady-State Thermal

Let us do the material assignment. From the Project tree, expand Geometry. Highlight Solid. In Details of "Solid", pick Aluminum.

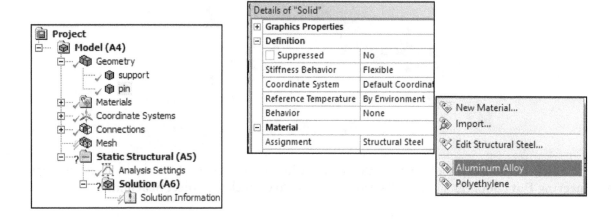

From the Project tree, Highlight **Mesh**. In the Details of "Mesh" window, click **Sizing**. Set Resolution to 4. Right-click **Mesh** again to pick **Generate Mesh**.

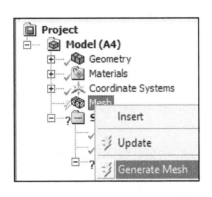

Step 5: Define the boundary condition.

Highlight **Initial Temperature** listed under **Steady-State Thermal (A5)**, and specify 25 degree as the initial temperature setting.

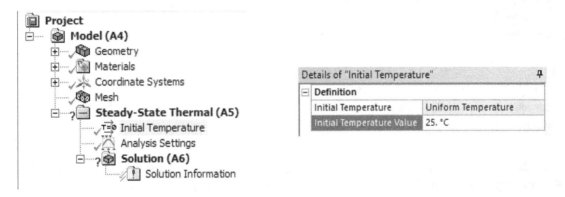

Highlight **Steady-State Thermal (A5)**. Pick **Insert** and **Temperature**. Pick the surface region on the bottom, as shown. Pick **Apply**. Specify 20 as the magnitude of pre-defined temperature.

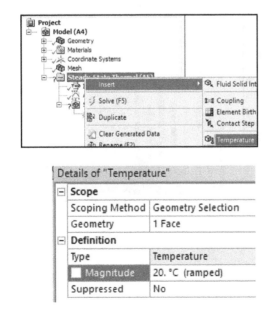

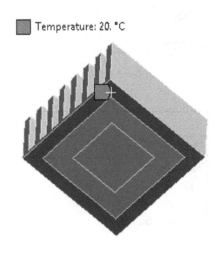

To define the convection condition, highlight **Steady-State Thermal**. Pick **Insert** and **Convection**. Pick the five surfaces (outside surfaces), as shown. Click **Apply**. Specify 25 °C as the **Ambient** or environmental temperature, and 0.005 (mW/°C/mm²) as the **Film Coefficient**.

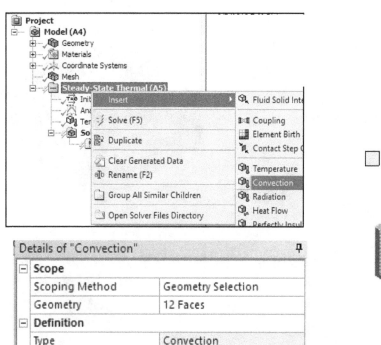

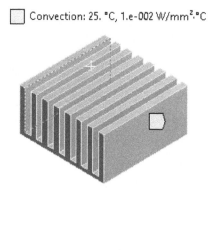

Convection: 25. °C, 1.e-002 W/mm².°C

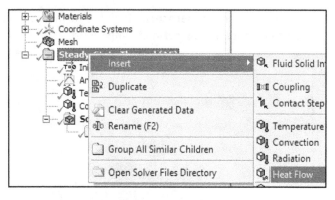

To define the heat flow, highlight **Steady-State Thermal**. Pick **Insert** and **Heat Flow**. Pick the small surfaces region at the bottom, as shown. Specify 5 W. Click **Apply**.

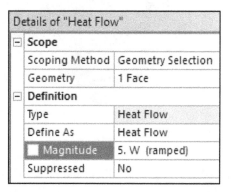

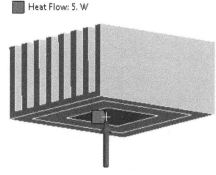

Heat Flow: 5. W

Highlight **Solution (A6)** and pick **Solve**.

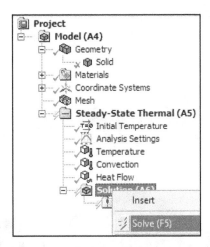

Step 6: Displaying Results

  Highlight **Solution (A6)**. Right-click to pick **Insert** > **Thermal** > **Temperature**.

  Highlight **Temperature** under **Solution (A6)**. Right-click to pick **Evaluation All Results**. The maximum and minimum values of temperature are 25.8°C and 20 °C, respectively.

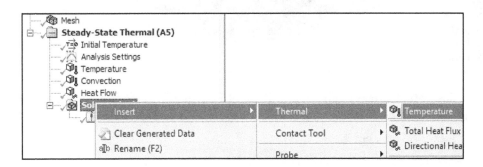

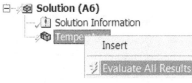

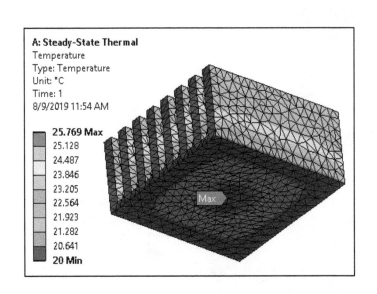

  Highlight **Solution (A6)**. Right-click to pick **Insert** > **Thermal** > **Total Heat Flex**.

  Highlight **Total Heat Flux** under **Solution (A6)**. Right-click to pick **Evaluation All Results**. The maximum and minimum values of total hear flux are 0.15 W/mm² and 0.00 W/mm², respectively.

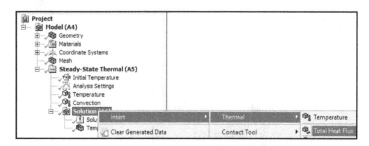

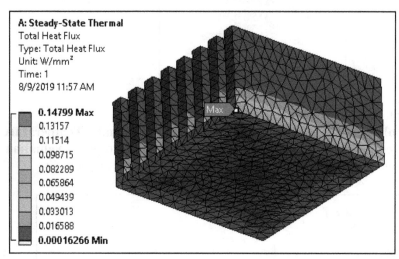

Click **File** and **Close Mechanical** to go back to **Project Schematic**. The displayed Steady-State Thermal panel shows all check marks. From the top menu, click **Save Project** to complete this study.

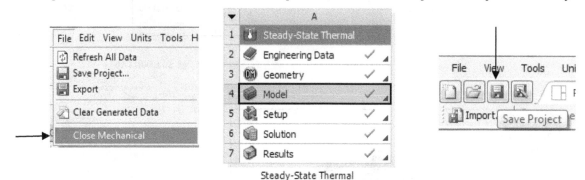

The above analysis is termed as operating at a normal condition. The maximum temperature is located at the CPU location. Its magnitude is just above the ambient temperature 25°. Let us create an overheating operating condition. Right click Steady-State Thermal, and click Duplicate. Change the name to overheating. Users may change the name of the previous one to Normal.

Double click Setup. Highlight Temperature and change the value from 20 to 40.

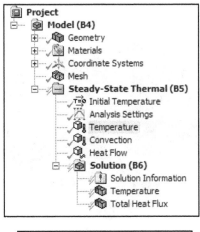

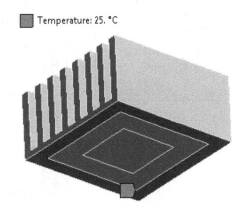

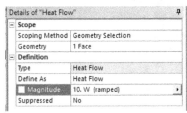

Highlight Heat Flow and change the value from 5 to 10.

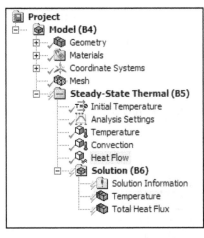

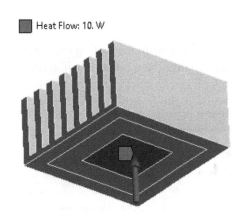

Highlight Solution and click Solve. Plot the temperature distribution. The max temperature is increased from 25.8 to 31.5 degrees Celsius. Its location is at CPU. Under such a circumstance, a cooling fan is automatically to operate to reducing the temperature in the CPU location.

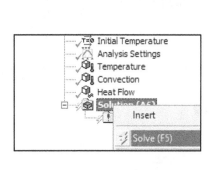

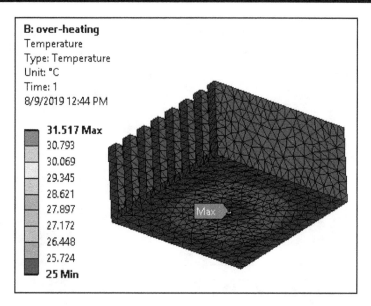

Click **File** and **Close Mechanical** to go back to **Project Schematic**. From the top menu, click **Save Project** to complete this study of over-heating case.

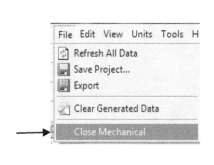

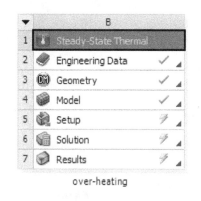

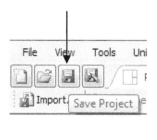

over-heating

Now let us working with the cooling off operating condition. Right click Steady-State Thermal, and click Duplicate. Change the name to cooling off.

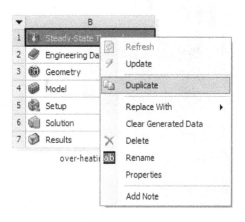

cooling-off

Double click **Setup**. Add a new convection condition or Convection 2. While holding down the **Ctrl** key. Pick the 3 surfaces of the channel for all 7 channels, as shown. Set the ambient temperature to 15 and film coefficient to 100 W/mm^2/^{o}C.

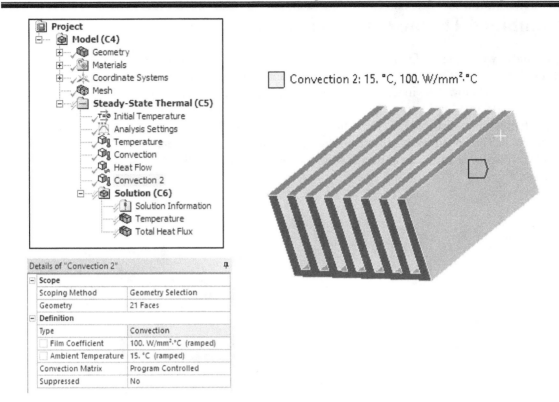

Highlight **Solution** and click **Solve**. Plot the temperature distribution. The max temperature is 25 degrees **Celsius**. Its location is NOT at CPU. The temperature at the CPU location is about 14.8 degrees Celsius.

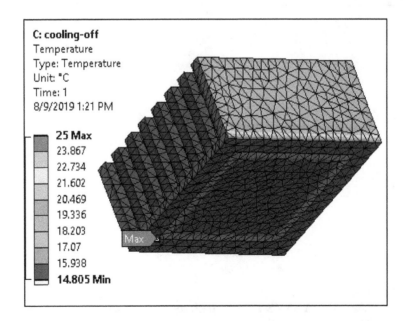

The results listed in the following table summarize the three case studies.

	Case 1: Normal	Case 2: Over-Heating	Case 3: CoolingOff
Max Temperature at CPU (Celsius)	25	25	14.8

4.5 Combined Thermal and Structural Analysis I

In this case study, we create a 3D model for the ring component, as shown below. The material type is structural steel (system default). For a steady-state thermal analysis, the 2 pre-scribed temperature conditions are shown below. The inner part of ring is subjected to a convection condition. The coefficient of convection is 0.25 (mW/°C/ mm²). Determine the temperature distribution and heat flux flow. After obtaining the temperature distribution. Input this temperature distribution to a static structural analysis as a temperature load. The boundary conditions of the static-structural analysis are the top end is "fixed to the ground" and the displacement of the low end is set to zero in the Z direction, and is set to free for both X and Y direction.

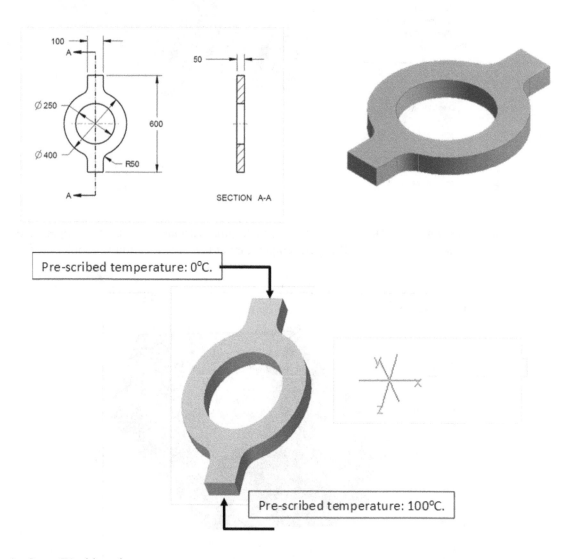

Step 1: Open Workbench.

From the start menu, click **ANSYS 19 > Workbench 19**. Users may directly click Workbench 19 when the symbol of Workbench 19 is on display.

 Workbench 19.2

Step 2: From **Component Systems,** highlight **Geometry** and drag it to the main screen.

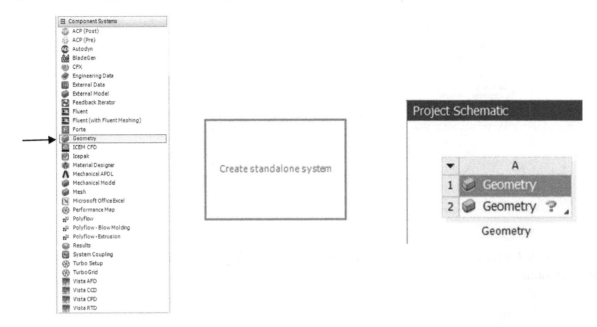

Step 3: Start Design Modeler.

Right-click the **Geometry** cell and pick **New DesignModeler Geometry** to start **Design Modeler**. Users may double click **Geometry** to directly enter **Design Modeler** if New **SpaceClaim Geometry** is not present. Click the icon of **Units** listed on the top menu, and select **Millimeter**.

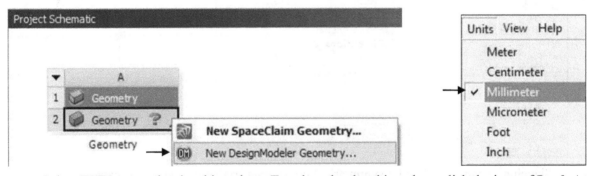

Select **ZXPlane** as the sketching plane. To orient the sketching plane, click the icon of **Look At.**

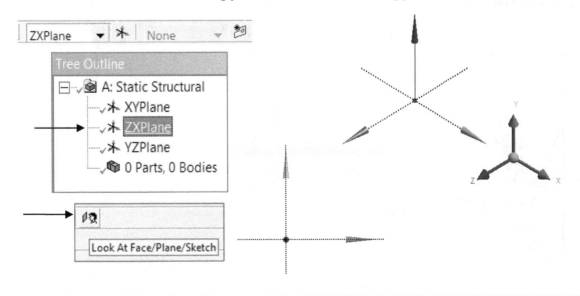

Click **Sketching**, and click **Draw**. Click Circle. Sketch 2 circles, as shown below.

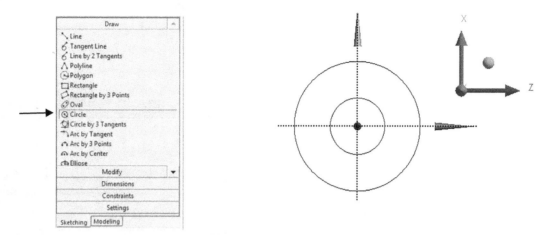

Click **Dimensions** and from the **Dimensions** panel, select **General**. Specify 250 and 400 as the diameter values.

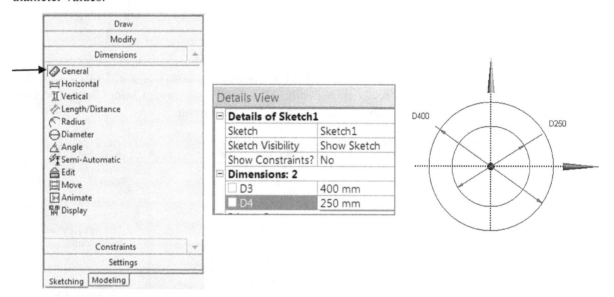

Sketch a rectangle and add 2 symmetrical constraints so that the sketched rectangle is symmetric about both X- and Z- axes.

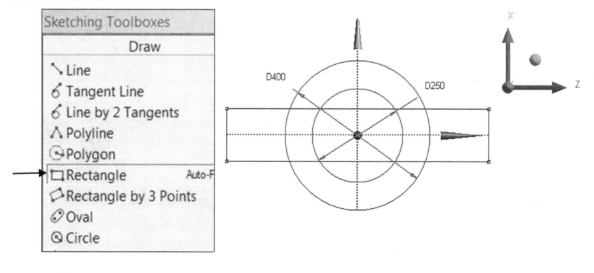

Specify the 2 size dimensions. They are 100 and 600, respectively.

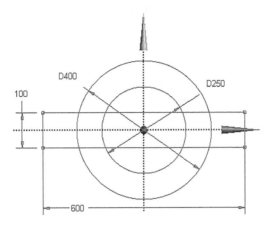

Click **Modify** and select **Trim**. Delete several line segments, as shown.

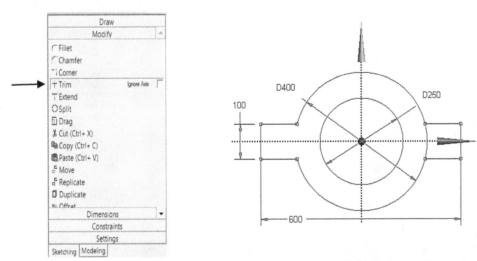

Upon completing the sketch, click the icon of **Extrude**. Click the created sketch or sketch 1 from the Tree Outline, and click **Apply**. For Direction, select **Both-Symmetric** and specify 25 as the thickness value (both = 50). Click **Generate** to obtain the 3D solid model of the cylindrical feature, as shown.

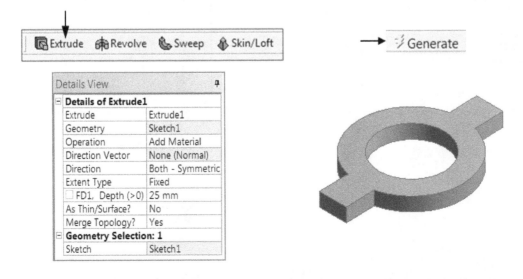

From the top menu, click **Blend > Fixed Radius**. Specify 50 as the radius value and pick the 4 vertical edges. Click **Generate**.

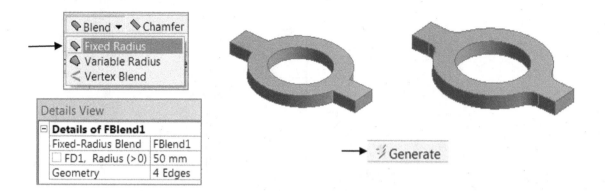

Click **File** and **Close Design Modeler**. We go back to the **Workbench** screen, or Project Schematic.

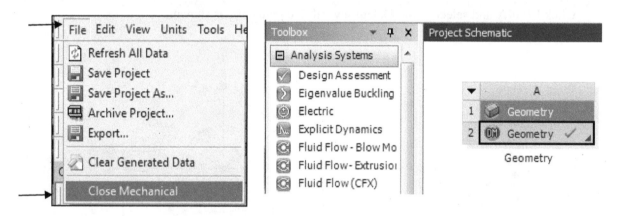

Step 4: From Analysis Systems, highlight Steady-state thermal and drag it to Window A. When Share A2 appears, release the drag button. Window B appears.

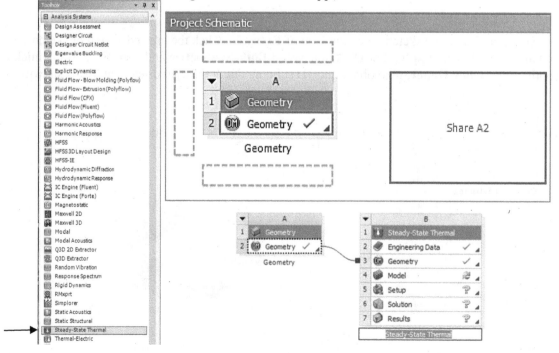

For the material properties assignment, we accept the system default assignment, assuming that the material type is Structural Steel. Now double-click **Model** to Start **Mechanical** Analysis.

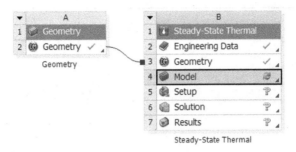

From the Project tree, Highlight **Mesh**. In the Detailed of "Mesh" window, click **Sizing**. Set Resolution to 6. Right-click **Mesh** again to pick **Generate Mesh**.

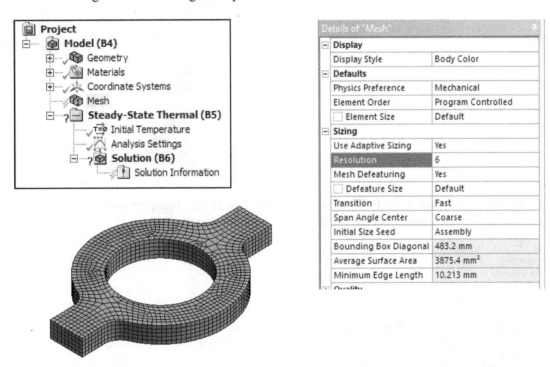

Step 5: Define the boundary condition.

Highlight **Initial Temperature** listed under **Steady-State Thermal (B5)**, and specify 0 degree as the initial temperature setting.

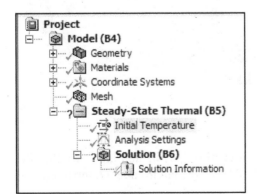
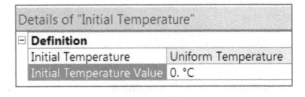

Highlight **Steady-State Thermal (B5)**. Pick **Insert** and **Temperature**. Pick the surface at the end, as shown. Pick **Apply**. Specify 0 as the magnitude of pre-defined temperature.

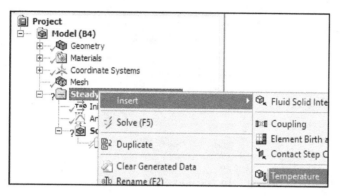

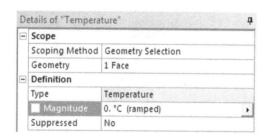

Again, highlight **Steady-State Thermal (B5)**. Pick **Insert** and **Temperature**. Pick the surface at the front, as shown. Pick **Apply**. Specify 100 as the magnitude of temperature.

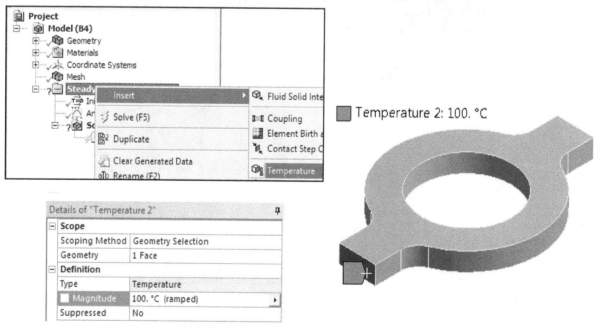

Step 7: Define the convection condition.

Highlight **Steady-State Thermal**. Pick **Insert** and **Convection**. While holding down the Ctrl key, pick the 2 half surfaces, as shown. Click **Apply**. Specify 0 °C as the environment temperature, and 0.25 (mW/°C/mm^2) as the Film Coefficient.

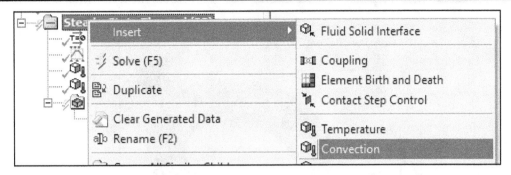

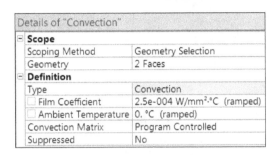

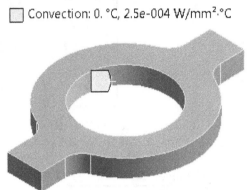

Step 8: Run FEA program

Highlight **Solution (B6)** and pick **Solve**.

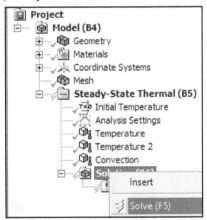

Step 9: Displaying Results

Highlight **Solution (B6)**. Right-click to pick **Insert > Thermal > Temperature**.

Highlight **Temperature** under **Solution (B6)**. Right-click to pick **Evaluation All Results**. The maximum and minimum values of temperature are 100°C and 0°C, respectively.

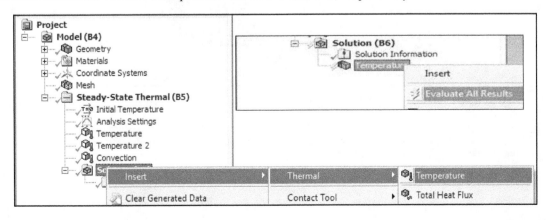

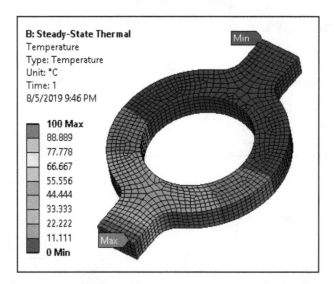

Highlight **Solution (B6)**. Right-click to pick **Insert** > **Thermal** > **Total Heat Flux**.

Highlight **Total Heat Flux** under **Solution (B6)**. Right-click to pick **Evaluation All Results**. The pattern of total hear flux is shown. The maximum and minimum values of heat flux are 34694 and 165 W/m², respectively.

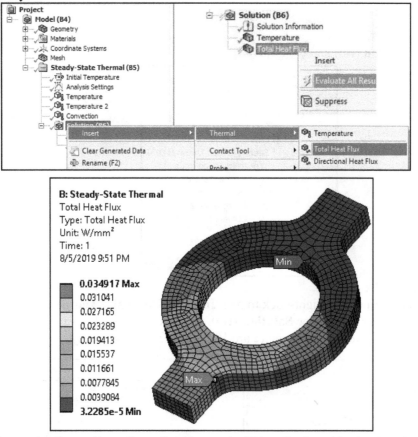

To animate the flow of heat flux, click the symbol Play displayed in the Graph window.

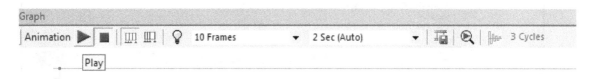

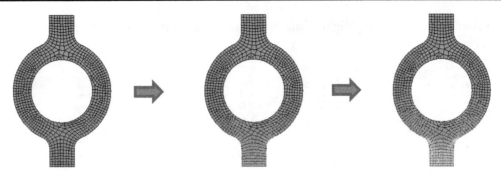

Click **File** and **Close Mechanical** to go back to the **Analysis Systems** screen. The displayed Windows A and B show all check marks, indicating that the FEA run was successful. From the top menu, click Save Project. Specify ring thermal as the file name and click **Save**.

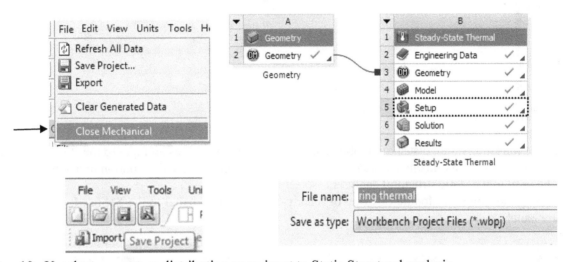

Step 10: Use the temperature distribution as an input to Static Structural analysis.

Highlight Steady-State Thermal and right-click to pick Duplicate and get copy of Steady-State Thermal Panel. Click **Setup** to go to Mechanical. Click **Solve** to update the database. Click **File** and **Close Mechanical** to go back to the **Analysis Systems** screen.

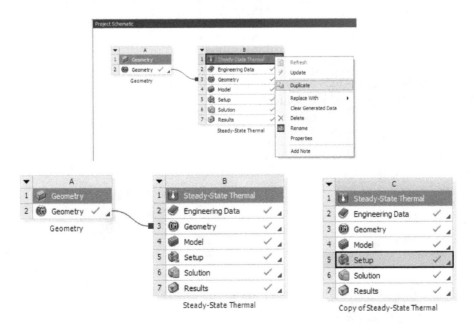

From Analysis Systems, highlight Static Structure and drag it to Geometry panel so that a Static Structure panel is created.

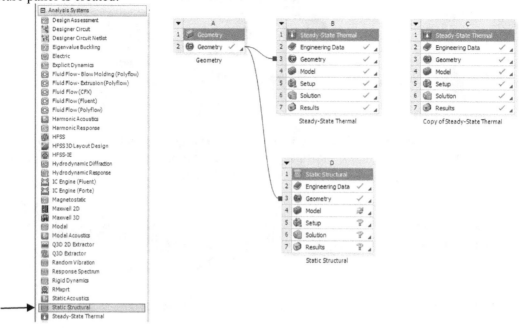

Use the temperature distribution obtained from thermal analysis as an input to the static structural analysis. The ANSYS Workbench system will convert such a temperature distribution to a temperature load to be used in the static structural analysis.

Drag the Solution Module of Copy of Steady-State Thermal panel to the Setup Module of Static Structure panel.

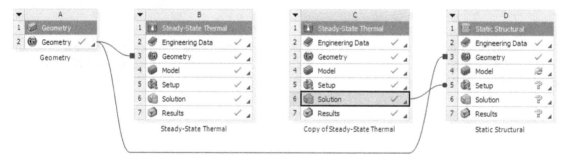

Step 11: Work on Static Structural Analysis with the Static Structural Panel.
Double-click **Model** to get into Mechanical.

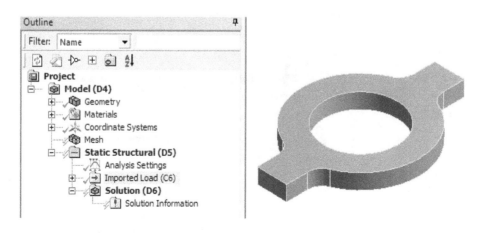

From the Project tree, Highlight **Mesh**. In the Detailed of "Mesh" window, click **Sizing**. Set Resolution to 6. Right-click **Mesh** again to pick **Generate Mesh**.

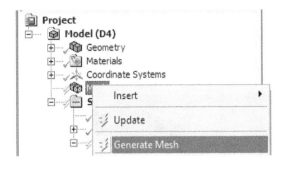
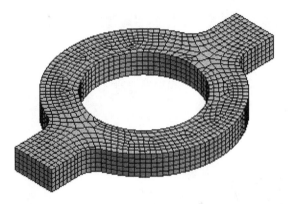

Highlight **Static Structural (D5).** Change the environment Temperature to 0 °C to match the thermal condition used in the previous steady-state thermal analysis.

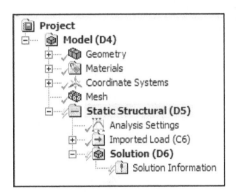
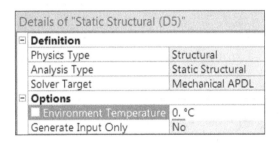

Highlight **Static Structural (D5)**. Pick Insert and **Fixed Support**. Pick the surface at the end, as shown. Pick **Apply**.

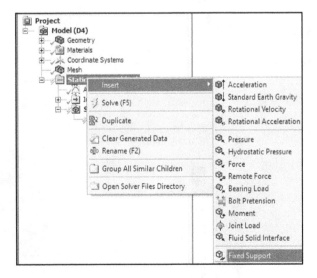

Highlight **Static Structural (D5)**. Pick Insert and **Displacement**. Pick the surface at the front end, as shown. Pick **Apply**. Specify 0 as the Z component and keep Free for both X and Y components.

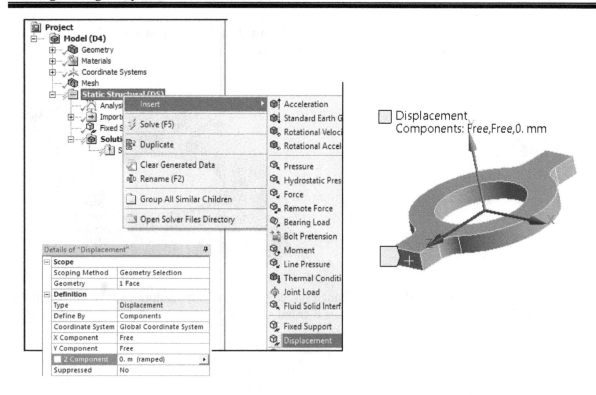

Expand **Imported Load (C6)**. Highlight Imported Body Temperature and right-click to pick Import Load.

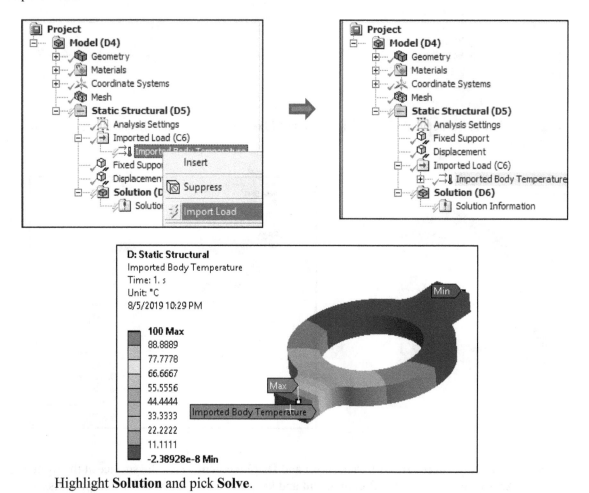

Highlight **Solution** and pick **Solve**.

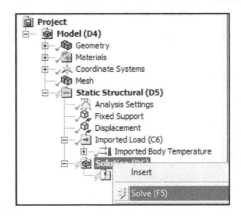

To display the results, highlight **Solution**. Right-click to pick **Insert** > **Deformation** > **Total**. Highlight **Total Deformation** under **Solution (D6)**. Right-click to pick **Evaluation All Results**. The maximum displacement value is 0.13785 mm.

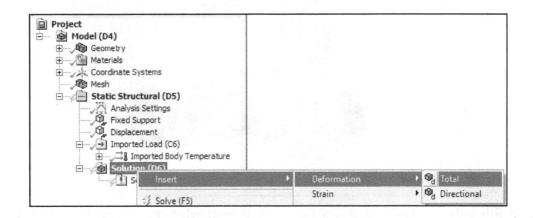

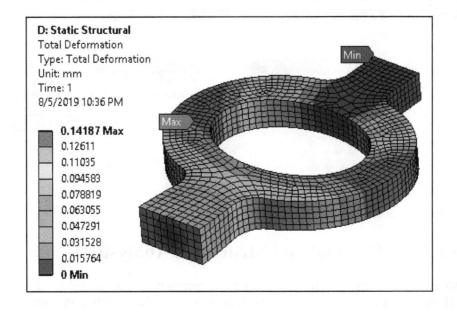

Highlight **Solution (D6)**. Right-click to pick **Insert** > **Stress** > **Equivalent (von-Mises)**. Right-click to pick **Evaluation All Results**. The maximum value of von Mises stress is 95.1 MPa.

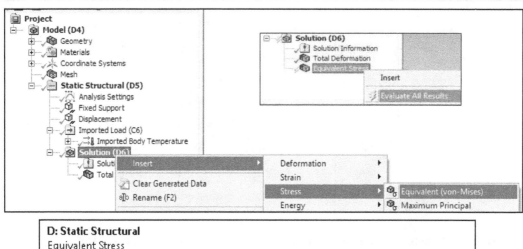

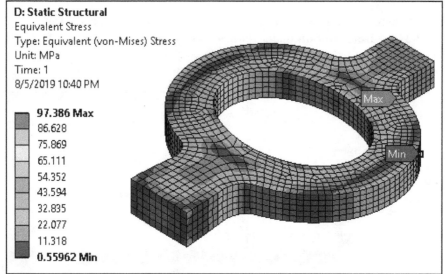

D: Static Structural
Equivalent Stress
Type: Equivalent (von-Mises) Stress
Unit: MPa
Time: 1
8/5/2019 10:40 PM

97.386 Max
86.628
75.869
65.111
54.352
43.594
32.835
22.077
11.318
0.55962 Min

Click **File** and **Close Mechanical** to go back to the **Project Schematic** screen. The displayed Static Structural panel D shows all check marks, indicating that the FEA run was successful.

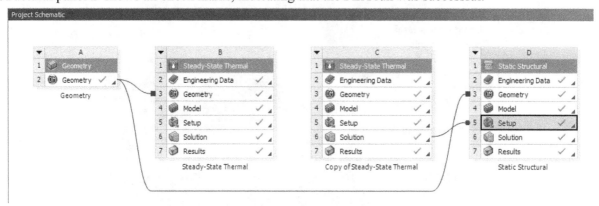

From the top menu. Click **Save**.

4.6   Combined Thermal and Structural Analysis II

In this case study, we first create a 3D model for the component, as shown below. The material type is structural steel (system default). Afterwards, perform a static analysis with a pressure loads of 10 MPa. The displacement constraints are the right side is fixed to the ground. The left side is constrained along the X axis direction (horizontal direction). Then we perform a steady-thermal analysis. A heat flow acts on the inner cylindrical surface, and a convection condition is added on the both ends. Finally, perform a combined thermal-structural analysis with 2 loads: pressure load and thermal-mechanical load.

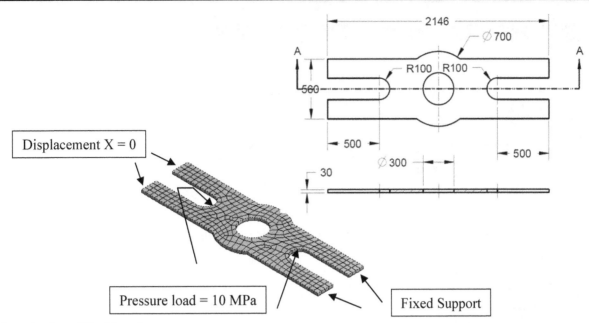

Displacement X = 0

Pressure load = 10 MPa

Fixed Support

Step 1: Open Workbench.

From the start menu, click **ANSYS 19** > **Workbench 19**. Users may directly click Workbench 18 when the symbol of Workbench 19 is on display.

Workbench 19.2

Step 2: From **Component Systems,** highlight **Geometry** and drag it to the main screen.

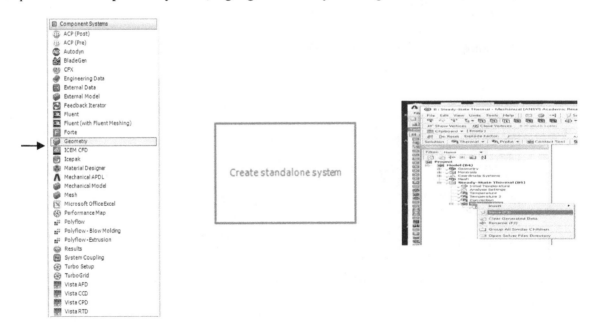

Step 3: Start Design Modeler.

Right-click the **Geometry** cell and pick **New DesignModeler Geometry** to start **Design Modeler**. Users may double click **Geometry** to directly enter **Design Modeler** if New **SpaceClaim Geometry** is not present. Click the icon of **Units** listed on the top menu, and select **Millimeter**.

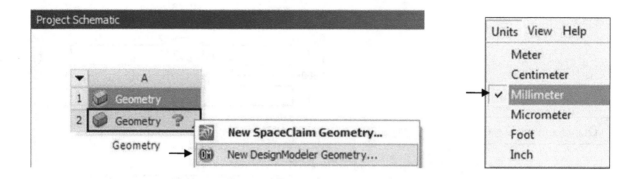

Select **ZXPlane** as the sketching plane. To orient the sketching plane, click the icon of **Look At.**

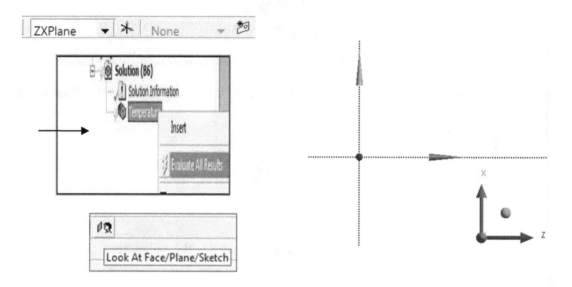

Click **Sketching**, and click **Draw**. In the Draw window, click Circle. Sketch 2 circles, as shown below.

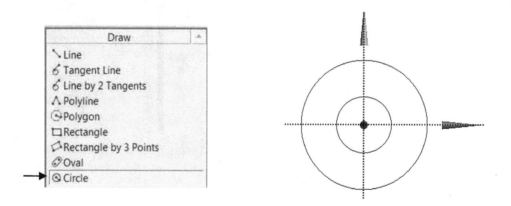

Click **Dimensions** and from the **Dimensions** panel, select **General**. Specify 300 and 700 as the diameter values.

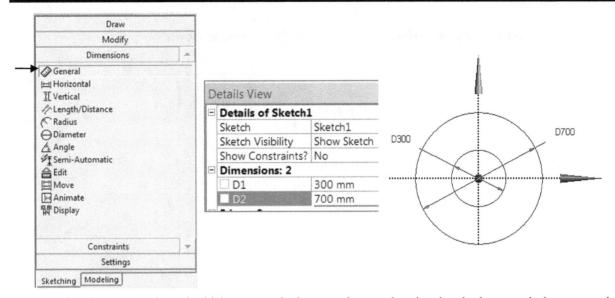

Sketch a rectangle and add 2 symmetrical constraints so that the sketched rectangle is symmetric about both X- and Z- axes.

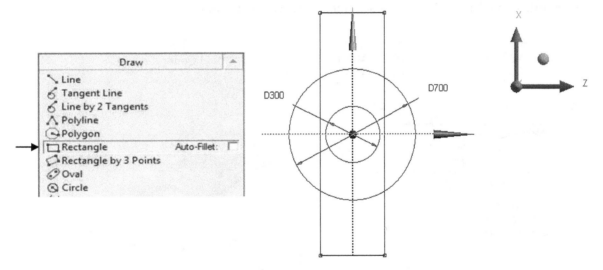

Specify the 2 size dimensions. They are 2146 and 560, respectively.

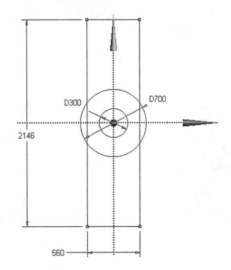

Click **Modify** and select **Trim**. Delete several line segments, as shown.

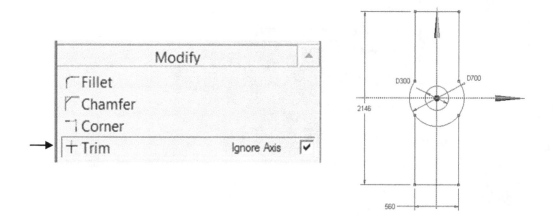

Upon completing the sketch, click the icon of **Extrude**. Click the created sketch or sketch 1 from the Tree Outline, and click **Apply**. For Direction, select **Normal** and specify 30 as the thickness value. Click **Generate** to obtain the 3D solid model of the plate feature, as shown.

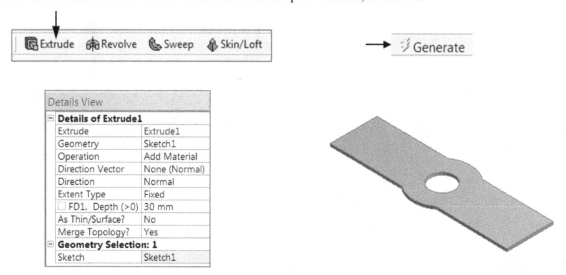

To create the second feature, select ZXPlane and click **New Sketch > Look At**. In **Draw**, click **Arc by Center**, and sketch a half circle. Afterwards, click **Line** and sketch 3 lines to form a closed sketch, as shown. Specify two dimensions: R100 and 500.

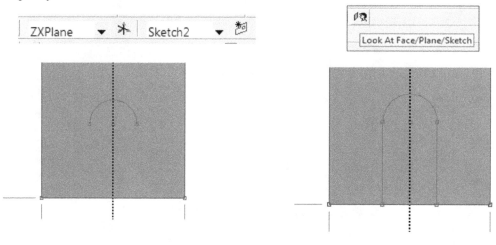

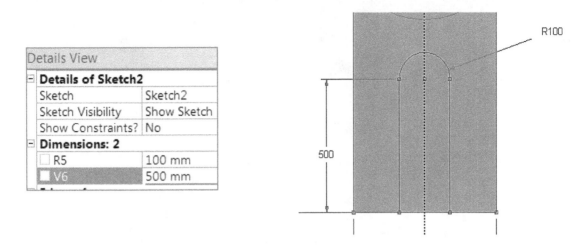

Click **Modify**, and select **Copy**. Pick the four items (an arc and three lines) while holding down the **Ctrl** key. Afterwards, right-click on the screen to pick **End/Use Plane Origin as Handle**. Move your cursor to the plane origin and right-click to pick **Flip vertical**. Make a left click at the plane origin. The picked sketch is shown on the other end, instead of repeating the sketching process.

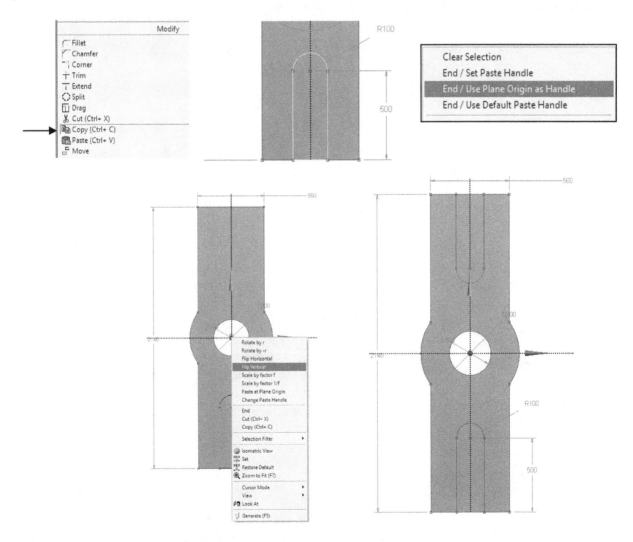

Users may just make the same sketch on the opposite side, as shown, if using the copy method is not working.

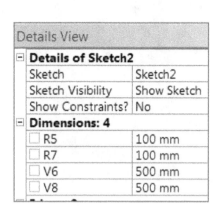

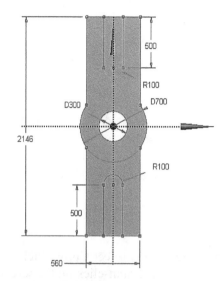

Upon completing the sketch, click the icon of **Extrude**. Click the created sketch or sketch 2 from the Tree Outline, and click **Apply**. In Operation, select Cut Material. In Extend Type, select Through All. Click **Generate** to obtain the 3D solid model of the cut feature, as shown.

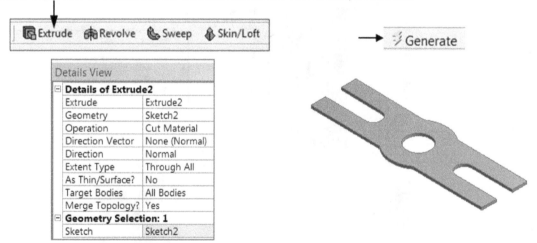

Click **File** and **Close Design Modeler**. We go back to the **Workbench** screen, or Project Schematic. Click **Save Project**. Specify Thermal-Mechanical Combined as the file name and click **Save**.

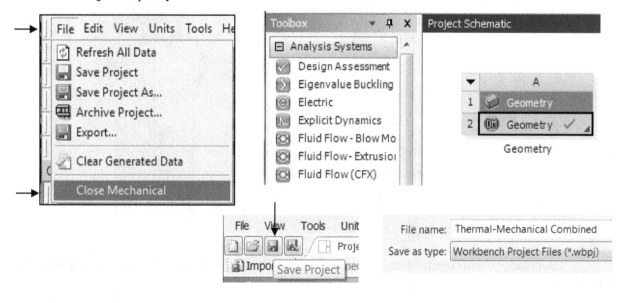

Step 4: From Analysis Systems, highlight Static Structural and drag it to Window A. When Share A2 appears, release the drag button. Window B appears.

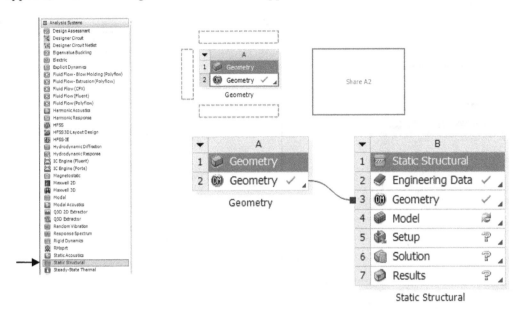

We accept the default material assignment, namely, the material type is Structural Steel. Now double-click **Model** to Start **Mechanical** Analysis. Check the unit system.

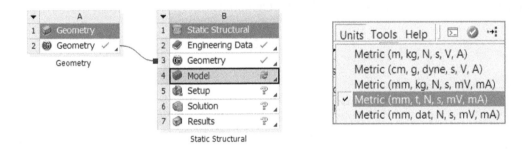

From the Project tree, Highlight **Mesh**. In the Detailed of "Mesh" window, click **Sizing**. Set Resolution to 6. Right-click **Mesh** again to pick **Generate Mesh**.

Step 5: Define the boundary condition.

Highlight **Static Structural (B5)**. Pick Insert and **Fixed Support**. While holding down the **Ctrl** key, pick the 2 surfaces at the end on the right side, as shown. Pick **Apply**.

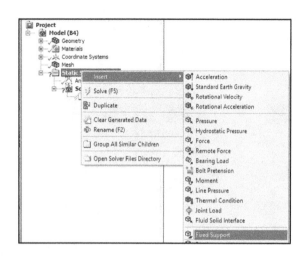

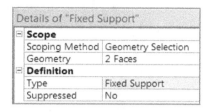

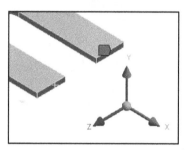

Highlight **Static Structural (B5)**. Pick Insert and **Displacement**. While holding down the **Ctrl** key, pick the 2 surfaces at the end on the left side, as shown. Pick **Apply**. Specify 0 as the value for X Component, and set 0 for the values for both Y and Z Components.

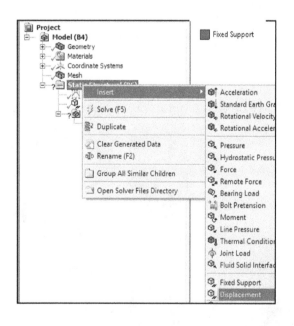

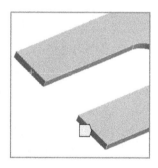

Highlight **Static Structural (B5),** right-click to pick **Insert > Pressure**. While holding down the **Ctrl** key, pick the 2 half-cylindrical surfaces, and click **Apply**. Specify 10 as the value of pressure.

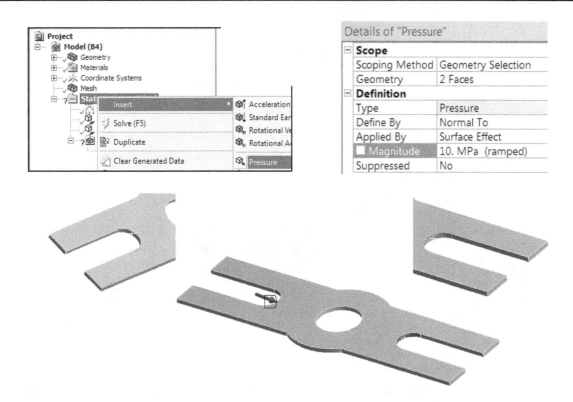

Details of "Pressure"

Scope	
Scoping Method	Geometry Selection
Geometry	2 Faces
Definition	
Type	Pressure
Define By	Normal To
Applied By	Surface Effect
☐ Magnitude	10. MPa (ramped)
Suppressed	No

Highlight **Solution** and pick **Solve**.

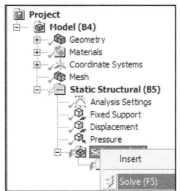

To display the results, highlight **Solution**. Right-click to pick **Insert** > **Deformation** > **Total**.

Highlight **Total Deformation** under **Solution (B6)**. Right-click to pick **Evaluation All Results**. The maximum displacement value is 0.011 mm.

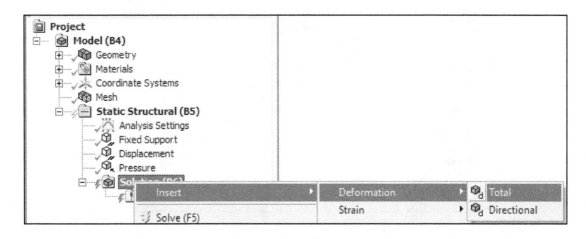

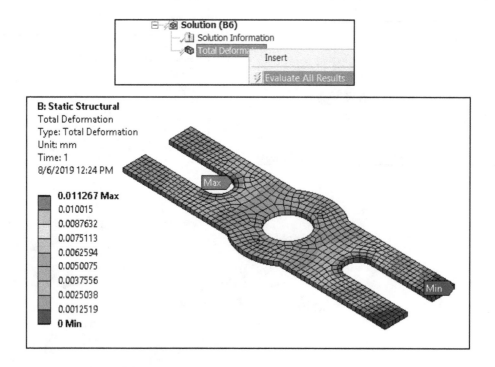

Highlight **Solution (B6)**. Right-click to pick **Insert > Stress > Equivalent (von-Mises)**. Right-click to pick **Evaluation All Results**. The maximum value of von Mises stress is 21.0 MPa.

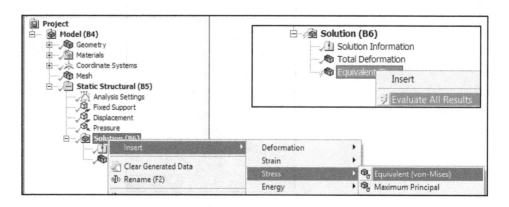

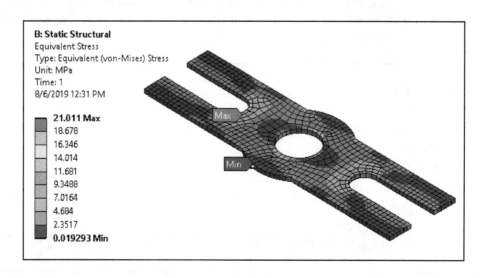

Click **File** and **Close Mechanical** to go back to the **Project Schematic** screen. The displayed Static Structural panel D shows all check marks, indicating that the FEA run was successful. Click the icon of **Save Project**.

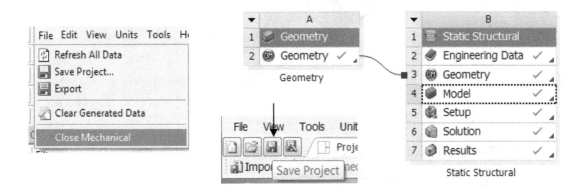

Step 6: From Analysis Systems, highlight Steady-State Thermal and drag it to Window A. When Share A2 appears, release the drag button. Window C appears.

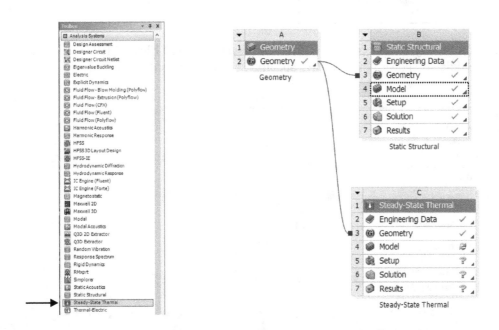

Now double-click **Model** to Start **Mechanical** Analysis. Check the unit system.

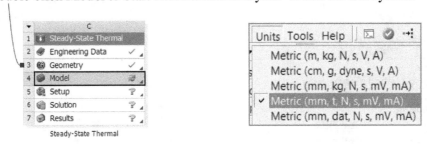

From the Project tree, Highlight **Mesh**. In the Detailed of "Mesh" window, click **Sizing**. Set Resolution to 6 for consistence. Right-click **Mesh** again to pick **Generate Mesh**.

Highlight Initial Temperature. Specify 10 as its value.

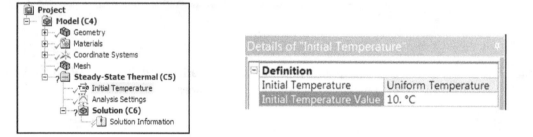

Highlight **Steady-State Thermal (C5)**. Pick **Insert** and **Heat Flow**. Pick the cylindrical surface located at the middle, as shown. Pick **Apply**. Specify 4 as the magnitude of heat flow.

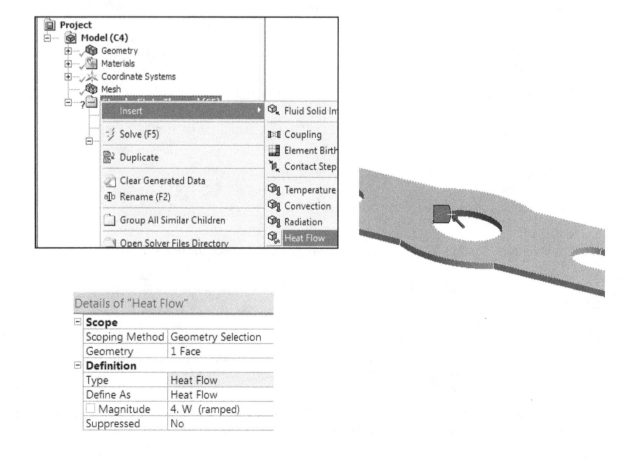

To define the convection condition, highlight **Steady-State Thermal (C5)**. Pick **Insert** and **Convection**. While holding down the **Ctrl** key, pick the 2 surfaces at the right end and the 2 surfaces at the left end, as shown. Click **Apply**. Specify 10 °C as the environment temperature, and 0.1 (mW/°C/mm²) as the Film Coefficient.

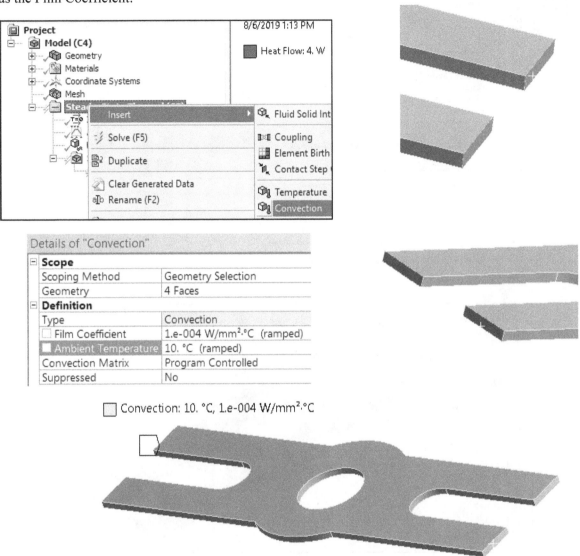

Step 7: Run FEA program
Highlight **Solution (C6)** and pick **Solve**.

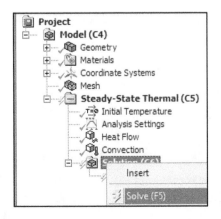

Highlight **Solution (C6)**. Right-click to pick **Insert** > **Thermal** > **Temperature**.

Highlight **Temperature** under **Solution (C6)**. Right-click to pick **Evaluation All Results**. The maximum and minimum values of temperature are 14.5°C and 11.85°C, respectively.

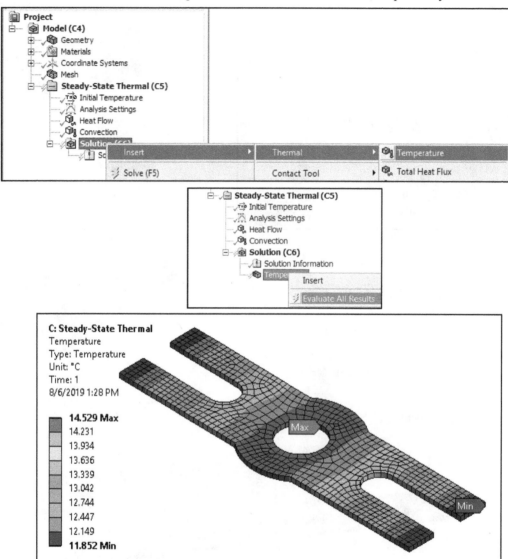

Highlight **Solution (C6)**. Right-click to pick **Insert** > **Thermal** > **Total Heat Flux**.

Highlight **Total Heat Flux** under **Solution (C6)**. Right-click to pick **Evaluation All Results**. The pattern of total hear flux is shown. The maximum and minimum values of heat flux are 24327 and 0.498 W/m², respectively.

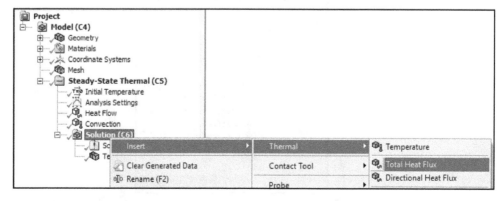

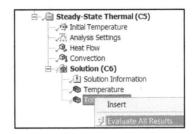

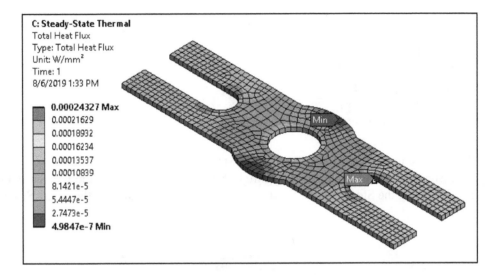

To animate the flow of heat flux, click the symbol Play displayed in the Graph window.

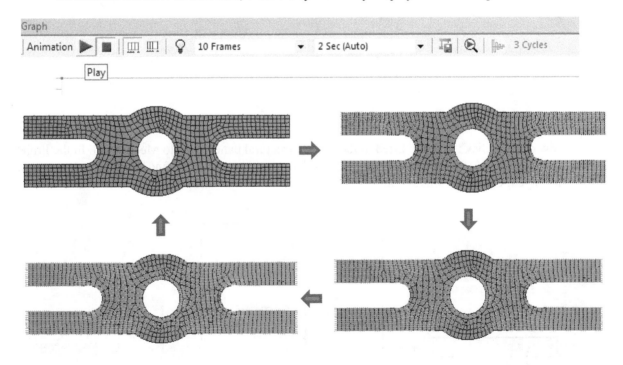

Click **File** and **Close Mechanical** to go back to the **Analysis Systems** screen. The displayed Windows B and C show all check marks, indicating that the FEA run was successful. Click **Save Project**.

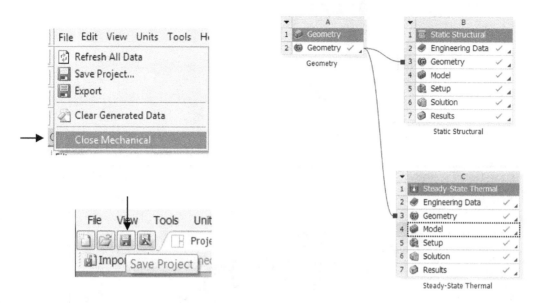

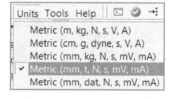

Step 8: Use the temperature distribution as an input to Static Structural analysis.

From the Steady-State Thermal panel, highlight Solution and drag it to Setup listed in the Static Structural panel. Note that the Steady-State Thermal panel is renamed as **B**, and the Static Structural panel is renamed as **C**

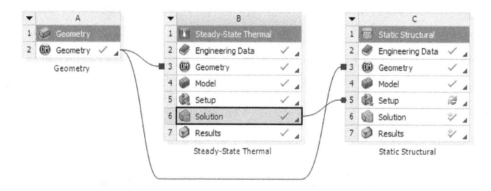

Now double-click the Setup listed in the Static Structural panel to go to Mechanical. In the Project tree, Imported Load (B6) is listed. The pressure load, fixed support constraint and displacement constraint are also displayed. Check the unit system.

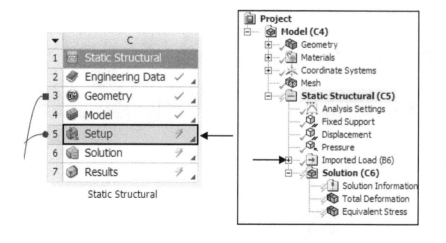

Static Structural
Time: 1. s
3/20/2017 2:58 PM

A Fixed Support
B Displacement
C Pressure: 10. MPa

Highlight Static Structure (C5), specify 10 as the value of environment Temperature.

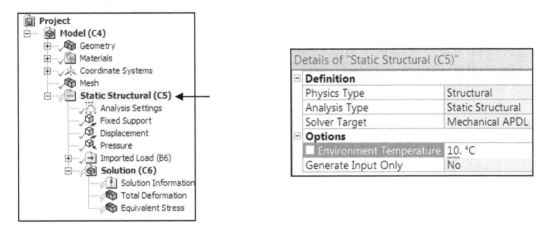

Expand the Imported Load (B6). Highlight Imported Body Temperature and right-click to pick Imported Load.

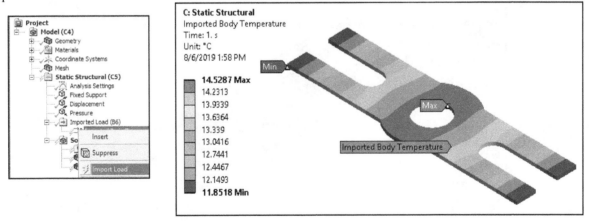

Highlight **Solution (C6)** and right-click to pick **Solve**. Highlight **Total Deformation**. The maximum displacement value is 0.03298 mm.

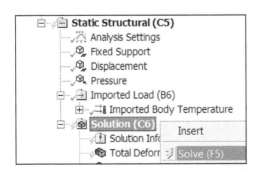

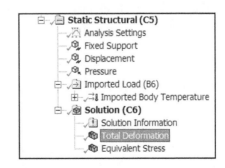

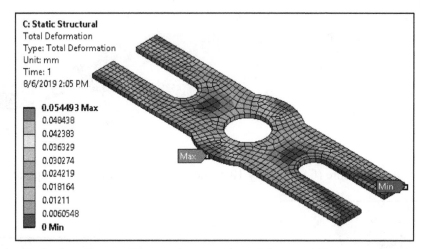

Highlight Equivalent Stress. The maximum value of the von Mises stress is 31.771 MPa.

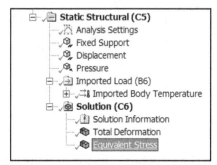

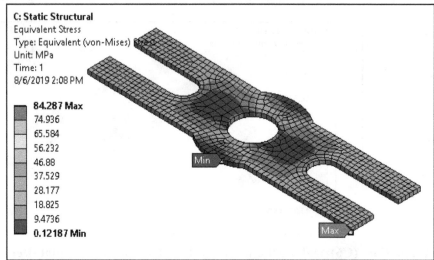

Click **File** and **Close Mechanical** to go back to the **Project Schematic** screen. The displayed Static Structural panel C shows all check marks, indicating that the FEA run was successful.

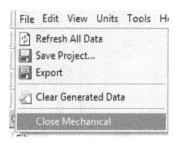

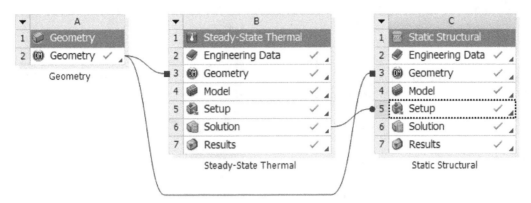

From the top menu, click the icon of **Save Project**.

4.7 Thermal Analysis for a Window and Door Structure

In this case study, we work with an assembly structure. The structure consists of a window and a door. The dimensions and shapes of the window component and door components are shown below. Note the unit used to show those dimensions is foot. The window material is polyethylene and the door material is aluminum alloy. Assume that the environment temperature is 50 °F. There are two thermal convection conditions. On the right-side of the window-door structure, the thermal convection condition is the coefficient of convection equal to 1.46 BTU/s/ft²/°F, which is equivalent to 1.66 W/cm², and the ambient temperature is set to 70 °F, which is equivalent to 21 degrees Celsius representing the room temperature. On the left-side of the window-door structure, the thermal convection condition is the coefficient of convection equal to 10.5 BTU/s/ft²/°F, which is equivalent to 12 W/cm2, and the ambient temperature is set to 50 °F, which is equivalent to 10 degrees of Celsius, representing the outdoor temperature. Perform FEA to obtain the temperature distribution under the thermal steady-state condition.

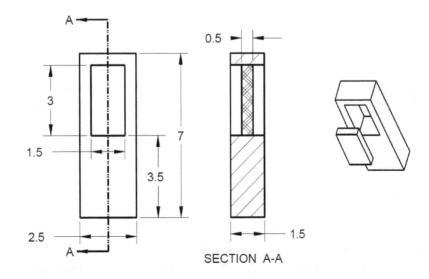

SECTION A-A

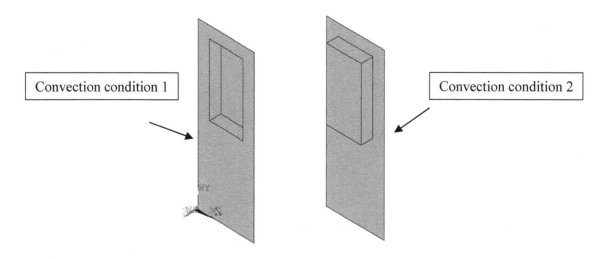

Convection condition 1

Convection condition 2

Step 1: Open Workbench.

From the start menu, click **ANSYS 19** > **Workbench 19**. Users may directly click Workbench 19 when the symbol is on display.

Workbench 19.2

Step 2: From **Toolbox**, highlight Steady-State Thermal and drag it to the main screen.

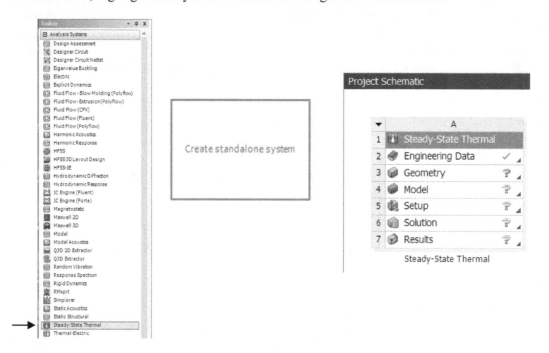

Step 3: Start Design Modeler and create a 2D sketch first.

Right-click the **Geometry** cell and pick **New DesignModeler Geometry** to start **Design Modeler**. Users may double click **Geometry** to directly enter **Design Modeler** if New **SpaceClaim Geometry** is not present. Click the icon of **Units** listed on the top menu, and select **Foot**.

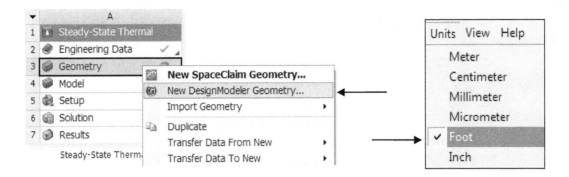

Select XYPlane as the sketching plane. Click Look At to orient the sketching plane.

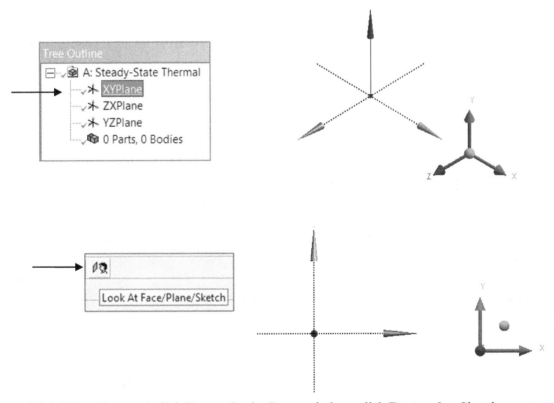

Click **Sketching**, and click **Draw**. In the Draw window, click **Rectangle**. Sketch a rectangle, as shown below.

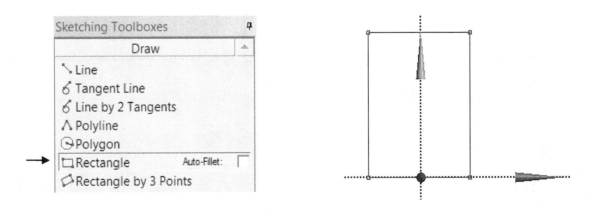

Click Constraints, select Symmetry to add a symmetrical constraint. Click the vertical axis and pick the 2 vertical sides.

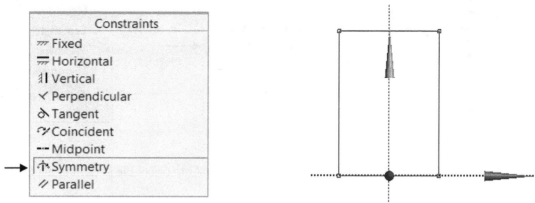

Click **Dimensions** and from the **Dimensions** panel, select **General**. Specify 3 and 1.5, as shown.

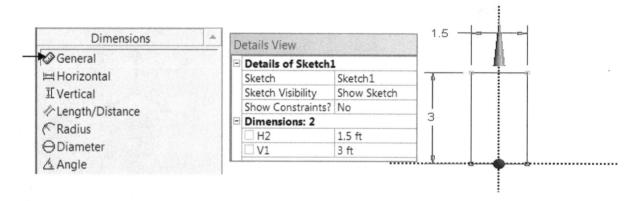

Sketch a rectangle and add a symmetrical constraint so that the sketched rectangle is symmetric about the Y-axis.

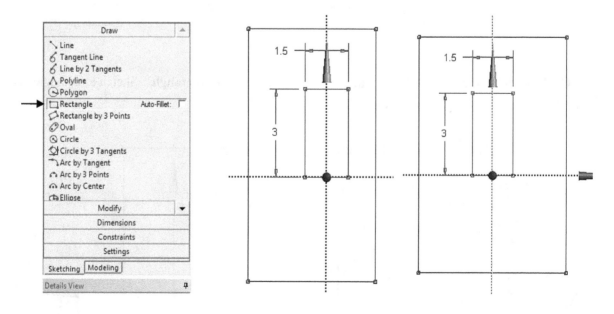

Specify 2 size dimensions, and one position dimension. They are 25, 7 and 3.5, respectively.

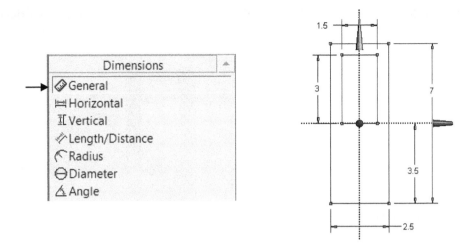

Upon completing the sketch, click the icon of **Extrude**. Click the created sketch or sketch 1 from the Tree Outline, and click **Apply**. Specify 1.5 as the thickness value. Click **Generate** to obtain the 3D solid model of the door component, as shown.

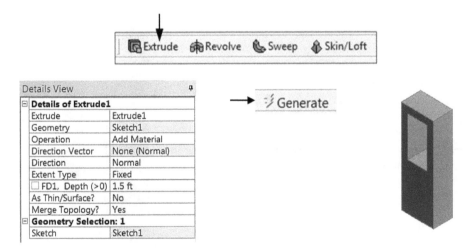

In order to create the window component, we first create a new datum plane. First highlight XY Plane. Afterwards, click the icon of **New Plane**. In Details View, keep the default selection of From Plane. Select Offset Z in Transform. Specify 0.5 as the offset distance. Click Generate. Plane4 is created.

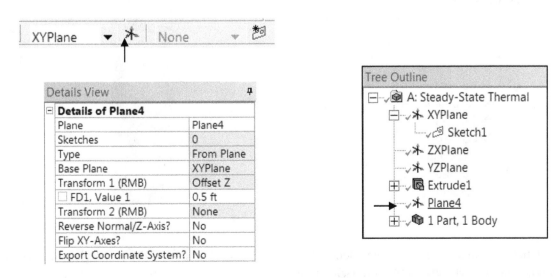

Click the icon of New Sketch, using Plane4 as the sketching plane. Click Look At to orient the sketching plane.

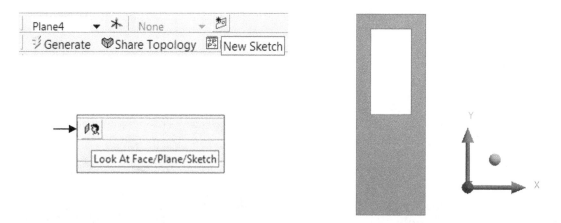

In Draw, click Rectangle and sketch a rectangle along the profile on display.

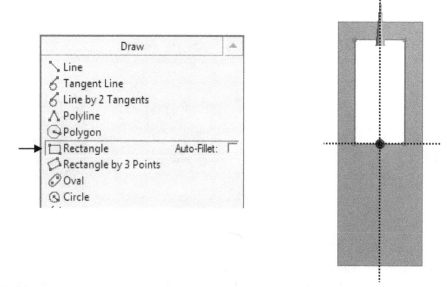

Click **Constraints**, select **Symmetry** to add a symmetrical constraint. Click the vertical axis and pick the 2 vertical sides.

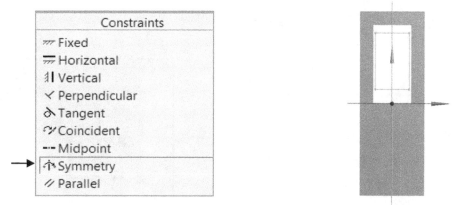

Click **Constraints**, select Coincident. Click the horizontal side from the rectangle and click the X axis to position the rectangle on the X axis, as shown.

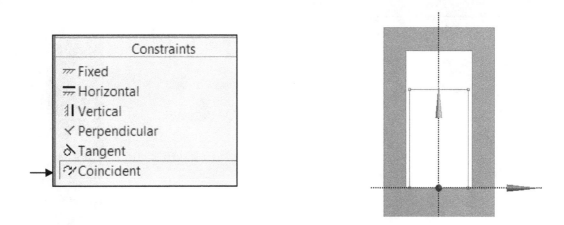

Click **Dimensions** and from the **Dimensions** panel, select **General**. Specify 2 dimensions: 3 and 1.5, as shown.

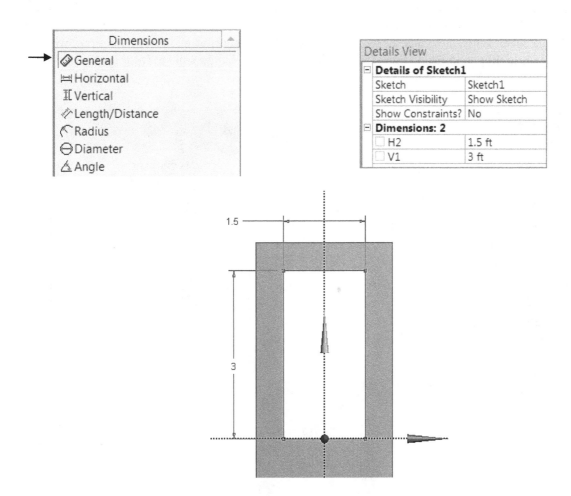

Upon completing the sketch, click the icon of **Extrude**. Click the created sketch or sketch 2 from the Tree Outline, and click **Apply**. In Operation, select **Add Frozen**. Specify 0.5 as the depth value. Click **Generate** to obtain the 3D solid model of the window component, as shown.

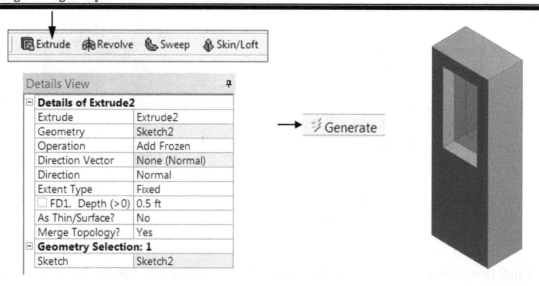

Let us examine the information listed in Tree Outline. We have created a new datum plane: Plane4. We have created 2 sketches: Sketch1 and Sketch2. Using those 2 sketches, we crated 2 extruded features. We use **Add Frozen** function so that those 2 bodies are independent of each other. Therefore, we have 2 parts. Now let us highlight Solid and right-click to pick Rename to change Solid to Door for the first feature, and change the second Solid to Window.

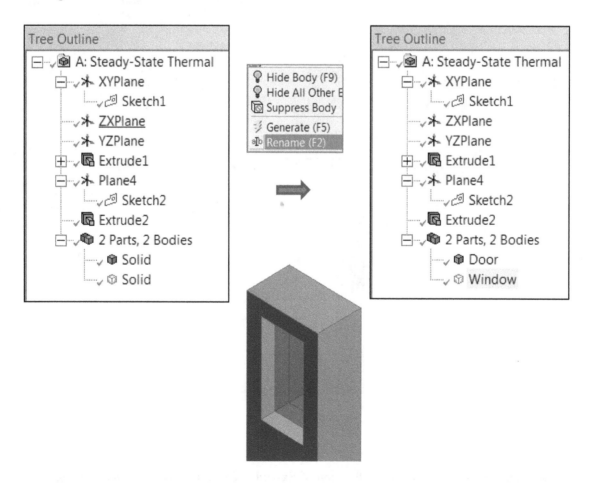

Click **File** and **Close Design Modeler**. We go back to the **Workbench** screen, or Project Schematic. Click **Save Project**. Specify Thermal-Window-Door as the file name and click **Save**.

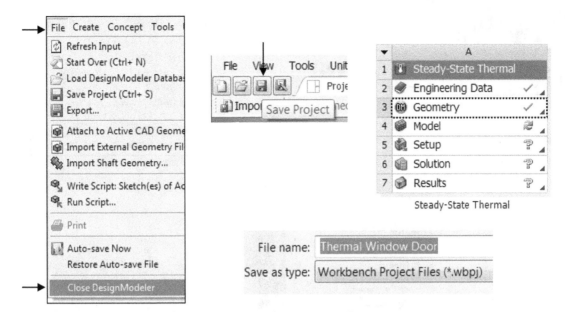

Step 4: Define Two (2) New Types of Material for Window and Door in the Engineering Data Sources
        In the Steady-State Thermal panel, double click **Engineering Data**. We enter the Engineering Data Mode.

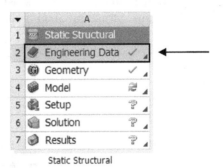

Static Structural

        Click the icon of **Engineering Data Sources**. Click **General Materials**. We are able to locate Aluminum Alloy. Click the plus symbol nearby Aluminum Alloy to activate it. Also locate Polyethylene and activate it, too. Afterwards, click the icon of Engineering Data Sources and click the icon of Project to go back to Project Schematic.

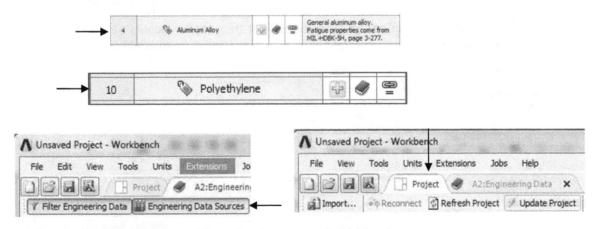

Step 5: Assign Material Types, and Examine the Contact Regions.
        In the Static Structural panel, double click Model. Check the unit system: U. S. Customary (ft, lbm, lbf, °F, s, V, A)

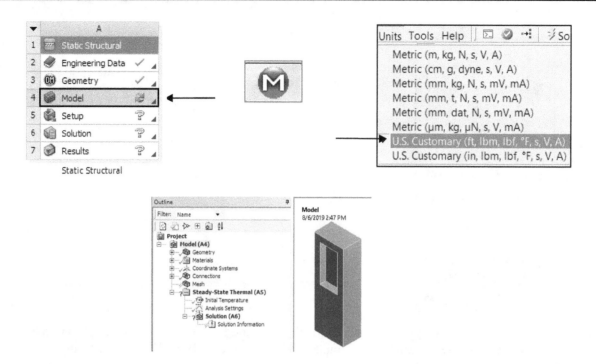

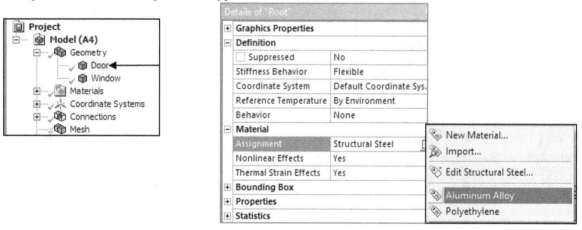

From the Project tree, expand Geometry and the three solid models (Root, Cement and Glass) are listed. Highlight **Door**. In the Details of "Door" window, Structural Steel is shown in Assignment. Right-click to pick **Aluminum Alloy** material type.

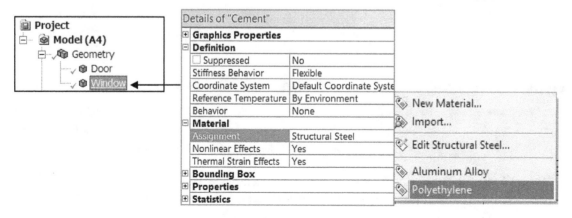

Now let us highlight **Window** listed in the Project tree. In the Details of "Window" window, Structural Steel is shown in Assignment. Right-click to pick Polyethylene material type.

From the Project tree, expand **Connections** > **Contacts**. There is a contact region consisting 4 pairs of faces. Highlight the listed contact region. This region is between window and door: Contact Body is Door. Contact Target is Window. The type of contact region is **Bonded**, meaning no slide or separation between the contact faces is allowed.as shown below.

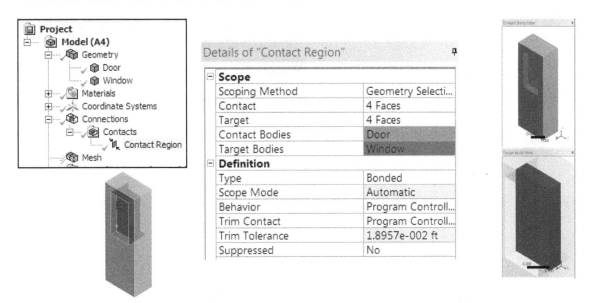

Step 6: Generate Mesh and Define the Boundary Condition.

From the Project tree, Highlight **Mesh.** Expand Size and set Resolution to 4, and right-click to pick **Generate Mesh**.

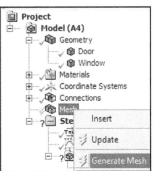

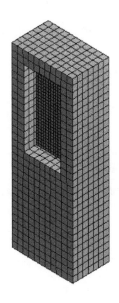

Details of "Mesh"

Display	
Display Style	Body Color
Defaults	
Physics Preference	Mechanical
Element Order	Program Controlled
☐ Element Size	Default
Sizing	
Use Adaptive Sizing	Yes
Resolution	4
Mesh Defeaturing	Yes
☐ Defeature Size	Default
Transition	Fast
Span Angle Center	Coarse
Initial Size Seed	Assembly
Bounding Box Diagonal	602.41 mm
Average Surface Area	27862 mm²
Minimum Edge Length	20.0 mm

To define the boundary conditions, first highlight **Initial Temperature** listed under **Steady-State Thermal (A5)**, and specify 50 degree as the initial temperature setting.

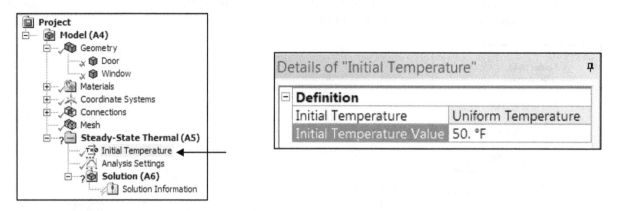

Highlight **Steady-State Thermal (A5)**. Pick **Insert** and **Convection**. Pick the 6 surfaces at the left side of this assembly, as shown. Click **Apply**. Specify 50 as the environment temperature setting, and specify 10.5 BTU/s/ft²/°F as the coefficient of convection.

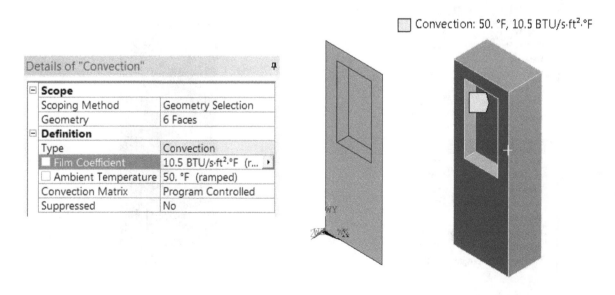

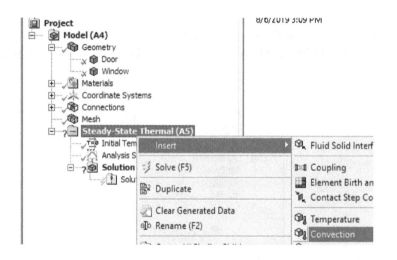

Highlight **Steady-State Thermal (A5)**. Pick **Insert** and **Convection**. Pick the 6 surfaces at the right side of this assembly, as shown. Click **Apply**. Specify 50 as the environment temperature setting, and specify 1.46 BTU/s/ft²/°F as the coefficient of convection.

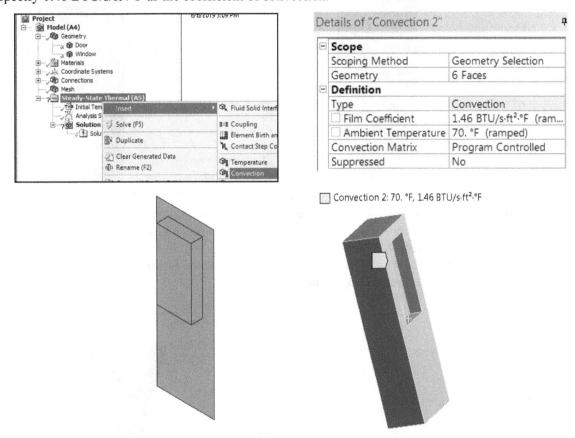

Convection 2: 70. °F, 1.46 BTU/s·ft².°F

Step 6: Run FEA program. Highlight **Solution (A6)** and pick **Solve**.

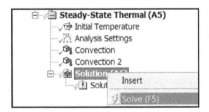

To examine the obtained results, highlight **Solution (A6)**. Right-click to pick **Insert** > **Thermal** > **Temperature** as well as **Heat Flux**. Right-click to pick **Evaluation All Results**. The maximum and minimum values of temperature are 71.3 °F and 49.8 °F, respectively.

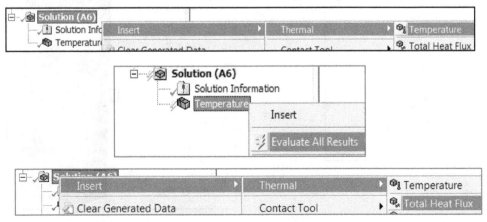

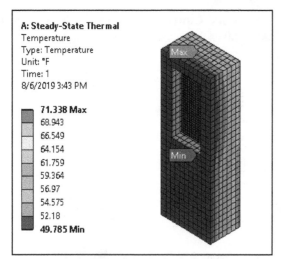

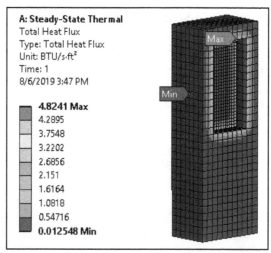

Users are able to study the stress distribution with an individual component, say the temperature distribution associated with the window component. In the Project tree, highlight **Window** listed under Geometry, right-click to pick **Hide All Other Bodies**. Afterwards, highlight Temperature listed under **Solution (A6)**. User may click the icon of **Max** and/or **Probe** to view the location and numerical value of temperature. As indicated, the maximum temperature is 72.89 °F.

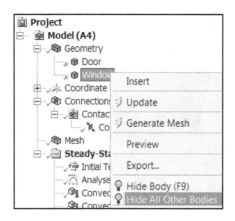

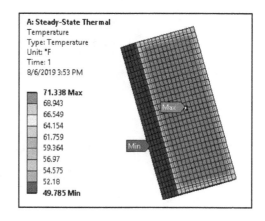

Repeat the above process to show the temperature distribution of the door component. Users may need to click Show All Body to recover the assembly first.

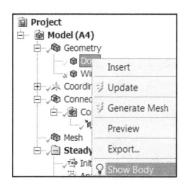

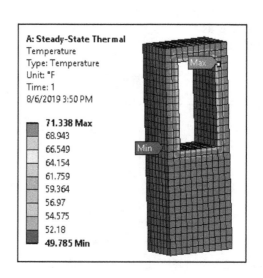

Click **File** and **Close Mechanical** to go back to the **Project Schematic** screen. The displayed Steady-State Thermal panel shows all check marks, indicating that the FEA run was successful. From the top menu, click the icon of **Save Project** to complete this study.

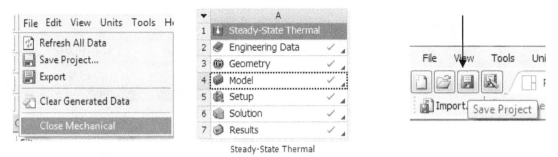

Steady-State Thermal

4.8 Thermal Analysis for an Assembly with 3 Material Types

The displayed drawing represents an assembly: The main body, the two small blocks on the right side to the main body, and the two small blocks on the left side to the main body. There are three (3) types of material: structural steel for the main body, aluminum alloy for the 2 small blocks on right, and magnesium alloy for the 2 small blocks on left. Assume that the environmental temperature setting is 10 °C. The left side of the assembly is subjected to a pre-defined temperature condition. The temperature setting is 20 °C. The surface on the right side is subjected to convection condition. The temperature setting is 10 °C, and coefficient is 4 mW/mm²°C.

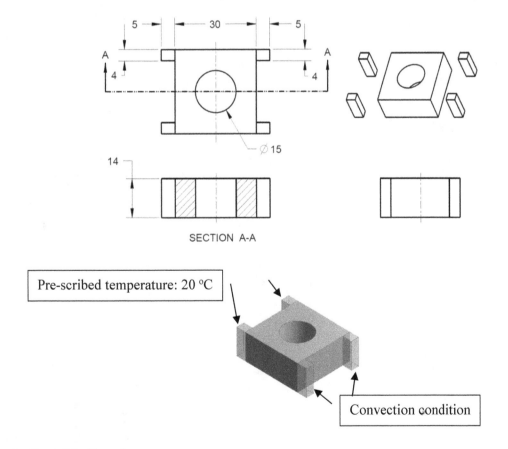

Step 1: Open Workbench.

From the start menu, click **ANSYS 19** > **Workbench 19**. Users may directly click Workbench 19 when the symbol of Workbench 19 is on display.

Workbench 19.2

Step 2: From **Toolbox**, highlight **Steady-State Thermal** and drag it to the Project Schematic screen.

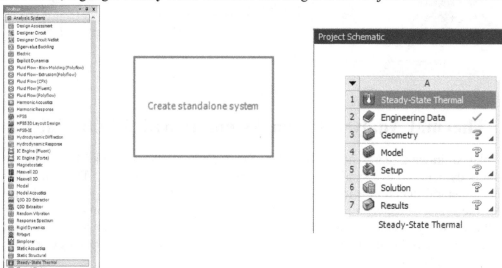

Step 3: In Design Modeler, Create the Assembly Model.

Right-click the Geometry cell and pick New **DesignModeler Geometry** to start **Design Modeler**. Users may double click **Geometry** to directly enter **DesignModeler** if **New SpaceClaim Geometry** is not presented. Click the Units drop down on the top menu, and select **Millimeter**.

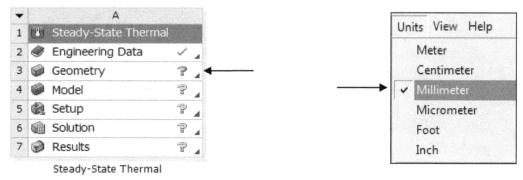

Select ZXPlane as the sketching plane. Click Look At to orient the sketching plane.

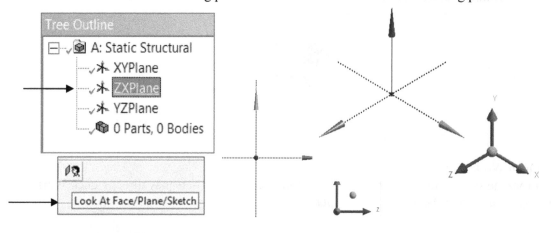

Click **Sketching**, and click **Draw**. In the Draw window, click **Circle**. Sketch a circle shown below.

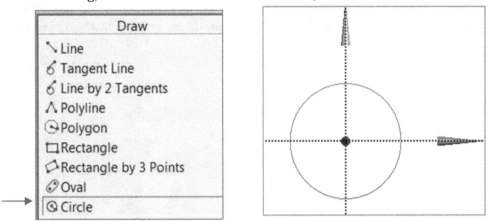

Click **Dimensions** and from the **Dimensions** panel, select **General**. Pick the circle and make a left click to position the diameter dimension. Specify 15 as its value.

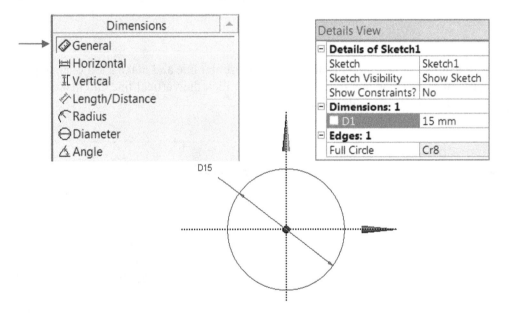

From **Draw** to select **Rectangle**. Sketch a rectangle as shown.

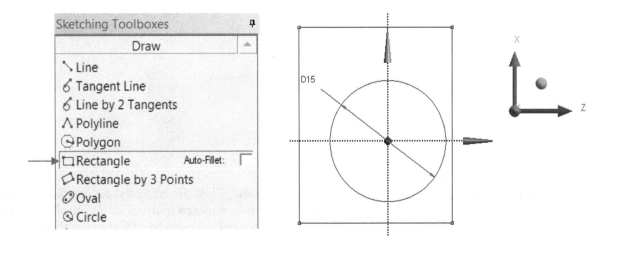

To add two symmetrical constraints, click **Constraints** and select **Symmetry**. Using Right-click, pick the vertical axis as the centerline to be symmetric about, and pick the two vertical sides of the rectangle. Using Right-click, pick the horizontal axis as the centerline to be symmetric about, and pick the two horizontal sides of the rectangle.

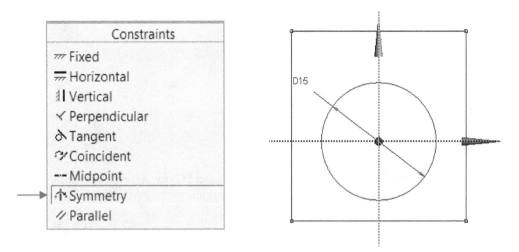

Click **Dimensions** and select **General**. Pick the horizontal line and make a left click to position the dimension. Specify 30 as its value and repeat this process. Pick the vertical line and make a left click to position the dimension. Specify 30 as its value.

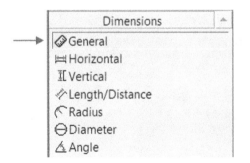

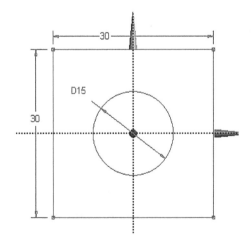

Upon completing the sketch, click the icon of **Extrude**. Click the created sketch or sketch 1 from the Tree Outline, and click **Apply**. For Direction, select **Both-Symmetric** and specify 7 as the height (both = 14). Click the icon of **Generate** to obtain the 3D solid model of the cylindrical feature, as shown.

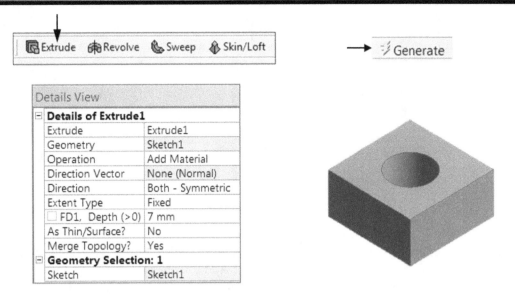

Now let us create the second feature: first sketch a rectangle. The 2 size dimensions are 5 and 4, respectively. In the Tree Outline, select ZXPlane as the sketching plane, click **New Sketch**. Click the icon of **Look At** to orient the sketching plane. From **Draw** to select **Rectangle**. Sketch a rectangle as shown.

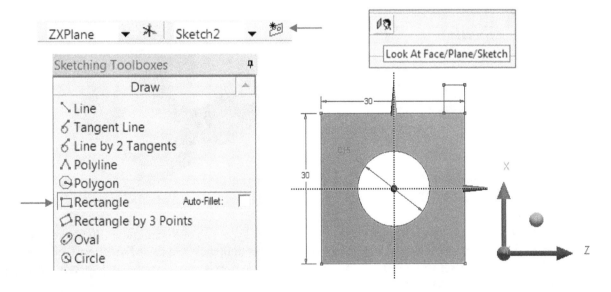

Click **Dimensions** and click **General**. Pick the horizontal line and make a left click to position the dimension. Specify 5 as its value and repeat this process. Pick the vertical line and make a left click to position the dimension. Specify 4 as its value.

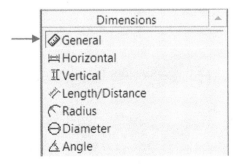

Details View	
Details of Sketch2	
Sketch	Sketch2
Sketch Visibility	Show Sketch
Show Constraints?	No
Dimensions: 2	
☐ H4	5 mm
☐ V5	4 mm

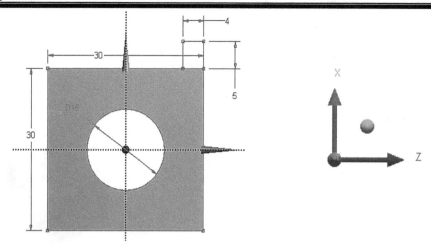

Upon completing the sketch, click the icon of **Extrude**. Click the created sketch or sketch 2 from the Tree Outline, and click **Apply**. In Operation, select Add Frozen so that a new Body is created. For Direction, select **Both-Symmetric** and specify 7 as the extrusion distance (both = 14). Click the icon of **Generate** to obtain a small block feature, as shown.

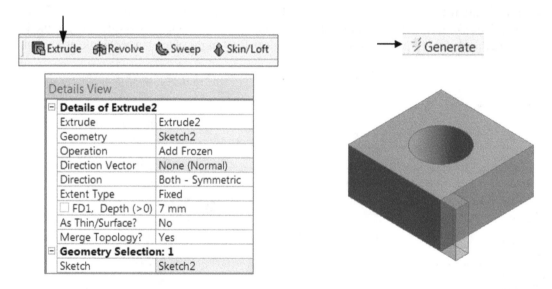

From the top menu, click Create > Pattern. In Pattern Type, select Rectangular. In Geometry, pick the displayed small rectangular block and click Apply. In Direction, pick X-axis as Direction 1, click **Apply**, and specify -35 as the offset value (30+5). In Direction 2, pick Z-axis, click Apply, and specify -26 as the offset value (30- 4). Click **Generate**.

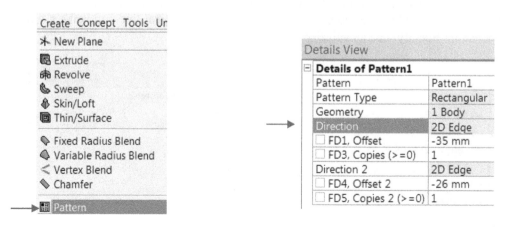

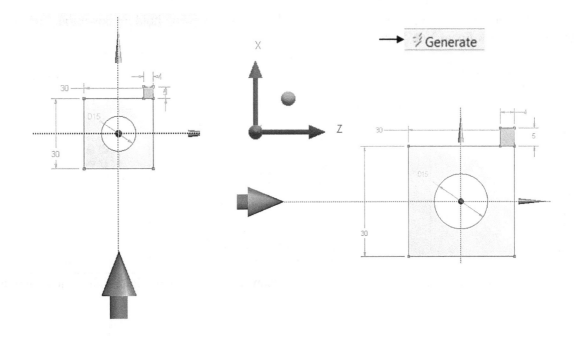

As shown below, a set of 3 additional small rectangular blocks are generated. In Tree Outline, there 5 Parts, 5 Bodies are listed.

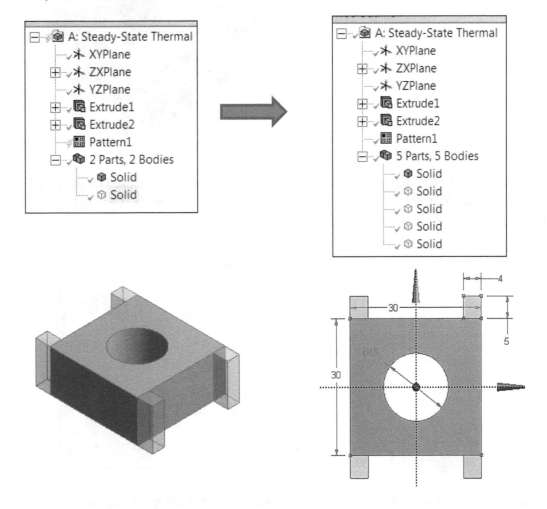

To combine the 5 parts to a single part, pick the 5 solids while holding down the **Ctrl** key, and right-click to pick **Form New Part**.

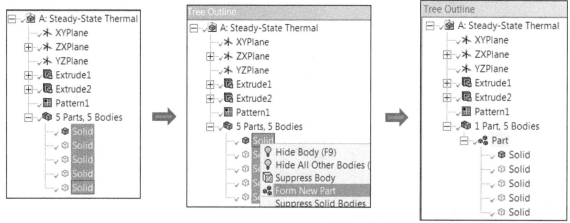

Now let us rename the 5 Solids listed to new names, namely, Steel, Magnesium, and Aluminum.

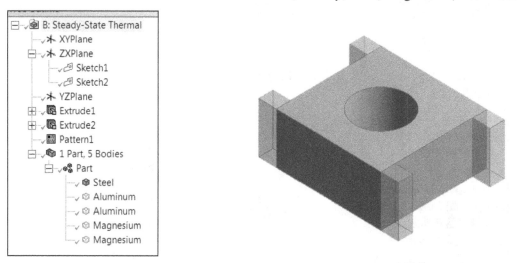

Upon completing the process of creating the geometry, we need to close Design Modeler. Click **File** > Close **Design Modeler**, and go back to **Project Schematic**. Specify **Thermal Three Types of Material** as the file name. Click **Save**.

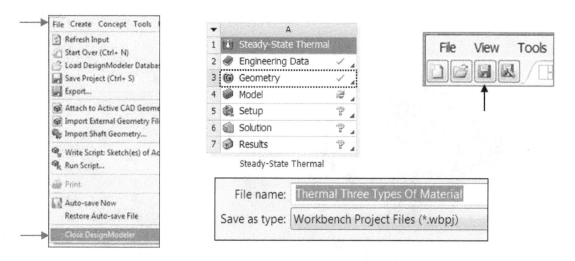

Step 3: Activate the Aluminum Alloy and the Magnesium Alloy in the Engineering Data Sources.

In the Steady-State Thermal panel, double click Engineering Data. We enter the Engineering Data Mode.

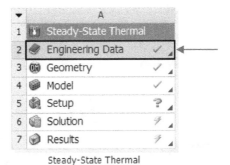

Steady-State Thermal

Click the icon of **Engineering Data Sources**. Click **General Materials**. We are able to locate Aluminum Alloy. Click the plus symbol nearby Aluminum Alloy to activate it. Also locate Magnesium Alloy and activate it, too. Afterwards, click the icon of Engineering Data Sources and click the icon of Project to go back to Project Schematic.

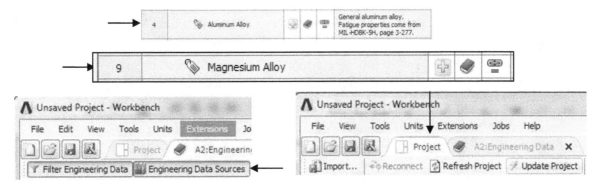

Step 4: Assign Three (3) Material Types of Material to the main body, and small blocks.
Double-click the icon of Model. We enter the Mechanical Mode.

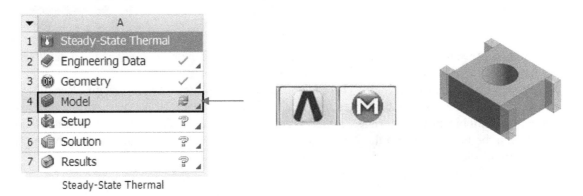

Steady-State Thermal

Check the unit system, and pick Metric (mm, t, N, s, mV, mA).

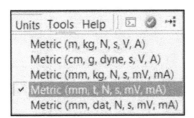

Now let us do the material assignment. From the Project tree, expand Geometry. While holding down the **Ctrl** key, highlight the 2 items marked as Aluminum. In Details of Multiple Selection, click material assignment, pick Aluminum Alloy to replace Structural Steel, which is the default material assignment.

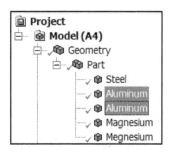
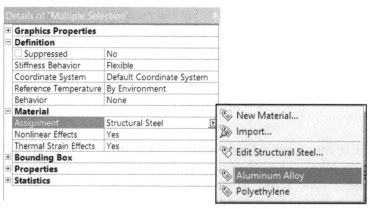

While holding down the Ctrl key, highlight the 2 items marked as Magnesium. In Details of Multiple Selection, click material assignment, pick Magnesium Alloy to replace Structural Steel, which is the default material assignment.

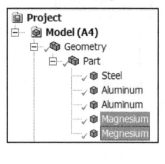
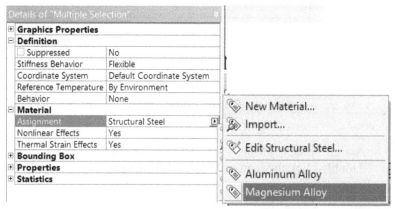

There is no need to make an assignment of Structural Steel to the main body as the default setting for assignment is Structural Steel.

In Project tree, click Connection. There is no response from the software system. Recall that we performed Form a New Part, which combined the 5 Solids into a new part. It is important to note that the main body and 4 small blocks have been assembled as a new part. The Bonded connection is automatically assumed between the main body and the 2 small aluminum blocks and between the main body and the 2 small magnesium blocks.

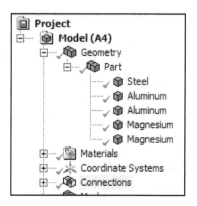

From the Project tree, Highlight **Mesh**, In Sizing, set Resolution to 6, and right-click to pick **Generate Mesh**.

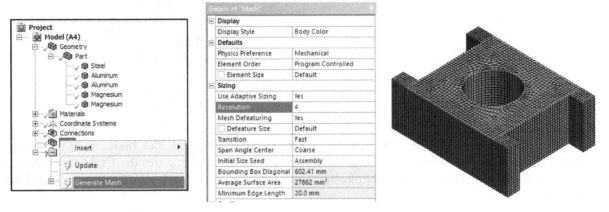

Step 5: Define the boundary condition.

Highlight **Initial Temperature** listed under **Steady-State Thermal (A5)**, and specify 10 degree as the initial temperature setting.

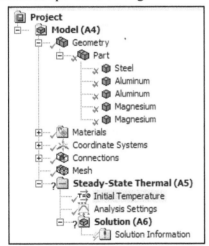

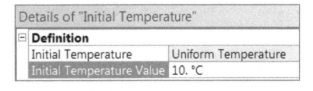

Highlight **Steady-State Thermal (A5)**. Pick **Insert** and **Temperature**. Pick the 2 surfaces on the left sides of the magnesium blocks while holding down the Ctrl key, as shown. Pick **Apply**. Specify 20 as the magnitude of pre-defined temperature.

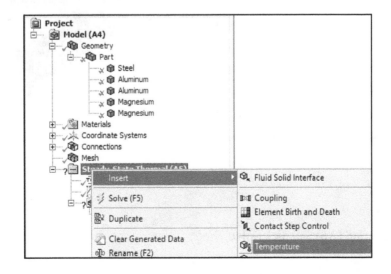

Details of "Temperature"	
Scope	
Scoping Method	Geometry Selection
Geometry	2 Faces
Definition	
Type	Temperature
☐ Magnitude	20. °C (ramped)
Suppressed	No

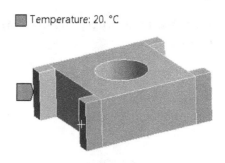

Temperature: 20. °C

To define the convection condition, highlight **Steady-State Thermal**. Pick **Insert** and **Convection**. Pick the 2 surfaces on the right sides of the aluminum blocks while holding down the **Ctrl** key, as shown. Specify 10 °C as the **Ambient** or environmental temperature, and 0.004 (W/°C/mm²) as the **Film Coefficient**.

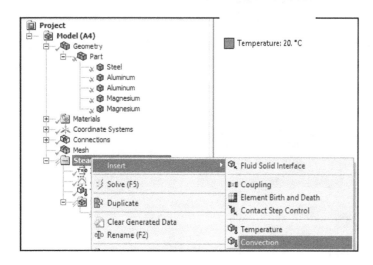

Temperature: 20. °C

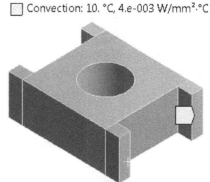

Convection: 10. °C, 4.e-003 W/mm².°C

Details of "Convection"	
Scope	
Scoping Method	Geometry Selection
Geometry	2 Faces
Definition	
Type	Convection
☐ Film Coefficient	4.e-003 W/mm².°C (ra..
☐ Ambient Temperature	10. °C (ramped)
Convection Matrix	Program Controlled
Suppressed	No

Highlight **Solution (A6)** and pick **Solve**.

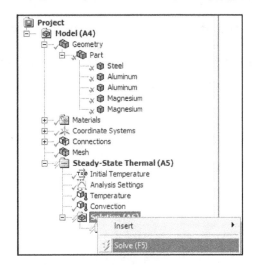

Step 6: Displaying Results

Highlight **Solution (A6)**. Right-click to pick **Insert** > **Thermal** > **Temperature**.

Highlight **Temperature** under **Solution (A6)**. Right-click to pick **Evaluation All Results**. The maximum and minimum values of temperature are 20°C and 14.34 °C, respectively.

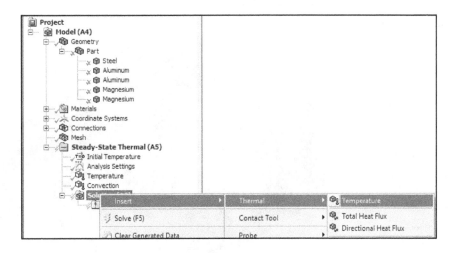

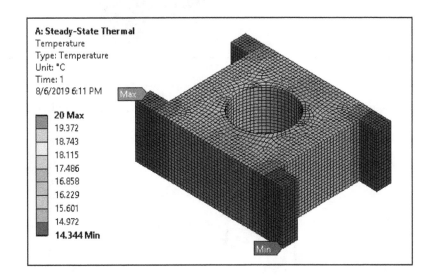

Highlight **Solution (A6)**. Right-click to pick **Insert > Thermal > Total Heat Flex**.

Highlight **Total Heat Flux** under **Solution (A6)**. Right-click to pick **Evaluation All Results**. The maximum and minimum values of total hear flux are 0.058 W/mm² and 0.69 mW/m², respectively.

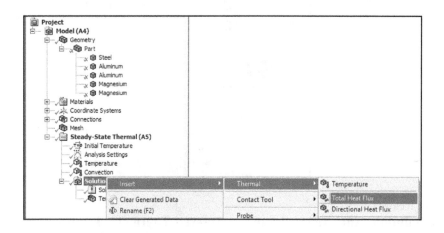

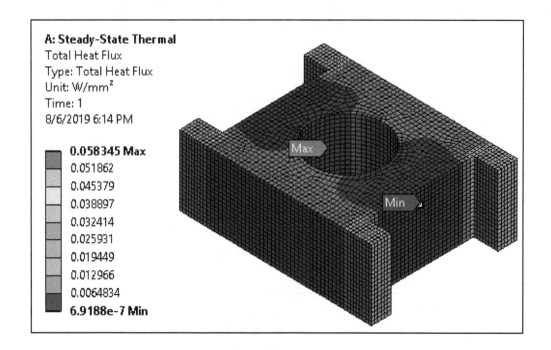

Click **File** and **Close Mechanical** to go back to **Project Schematic** and click **Save**.

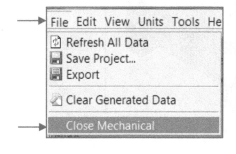

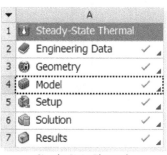

4.9   References

1. ANSYS Parametric Design Language Guide, Release 15.0, Nov. 2013, ANSTS, Inc.
2. ANSYS Workbench User's Guide, Release 15.0, Nov. 2013, ANSYS, Inc.
3. X. L. Chen and Y. J. Liu, Finite Element Modeling and Simulation with ANSYS Workbench, 1st edition, Barnes & Noble, January 2004.
4. J. W. Dally and R. J. Bonnenberger, Problems: Statics and Mechanics of Materials, College House Enterprises, LLC, 2010.
5. J. W. Dally and R. J. Bonnenberger, Mechanics II Mechanics of Materials, College House Enterprises, LLC, 2010.
6. X. S. Ding and G. L. Lin, ANSYS Workbench 14.5 Case Studies (in Chinese), Tsinghua University Publisher, Feb. 2014.
7. G. L. Lin, ANSYS Workbench 15.0 Case Studies (in Chinese), Tsinghua University Publisher, October 2014.
8. K. L. Lawrence. ANSYS Workbench Tutorial Release 13, SDC Publications, 2011.
9. K. L. Lawrence. ANSYS Workbench Tutorial Release 14, SDC Publications, 2012
10. Huei-Huang Lee, Finite Element Simulations with ANSYS Workbench 14, Theory, Applications, Case Studies, SDC Publications, 2012.
11. Huei-Huang Lee, Finite Element Simulations with ANSYS Workbench 16, Theory, Applications, Case Studies, SDC Publications, 2015.
12. Jack Zecher, ANSYS Workbench Software Tutorial with Multimedia CD Release 12, Barnes & Noble, 2009.
13. G. M. Zhang, Engineering Design and Creo Parametric 3.0, College House Enterprises, LLC, 2014.
14. G. M. Zhang, Engineering Analysis with Pro/Mechanica and ANSYS, College House Enterprises, LLC, 2011.

4.10   Exercises

1.   A component called bearing housing is shown below. The unit used in the drawing is millimeter. Assume that the material type is Structural Steel. The frictional action in the contact area of the inner cylinder representing the rotating shaft and the cylindrical surface of the hole is modeled as a heat flow load. The numerical value of the heat flow is 10 W. The convection condition for the two surfaces (the outer cylindrical surface and the bottom surface) is defined by film coefficient equal to 0.0007 cal/s/cm^2 and the bulk ambient temperature equal to 20 °C.

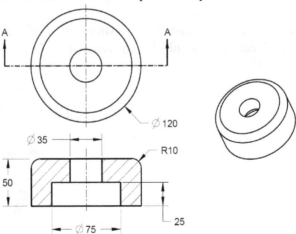

        The following steps are suggested for creating a 3D solid model of this bearing housing component in DesignModeler.

Step 1: select Millimeter unit. Select ZXPlane as the sketching plane to create a plate feature. Afterwards create the Imprint Faces, as shown. It is important that you click Freeze from Tools listed on the top menu.

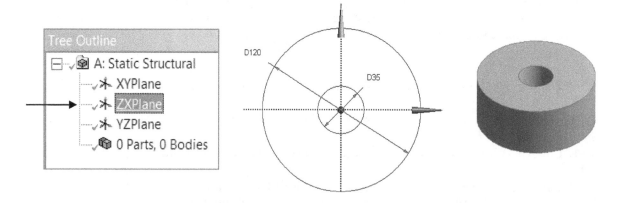

Step 2: select ZXPlane as the sketching plane to create the cut feature.

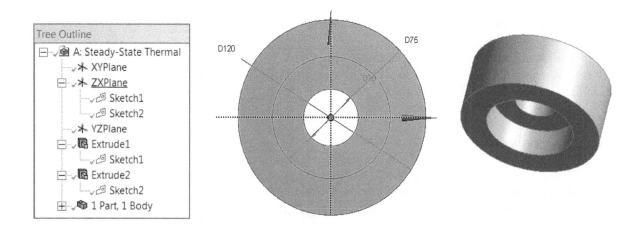

Step 3: Add a fixed radius blend feature. Afterwards perform FEA. Note that the film coefficient equal to 0.0007 cal/s/cm^2 should be converted to 0.00029 W/mm^2 at the time to define the convection condition.

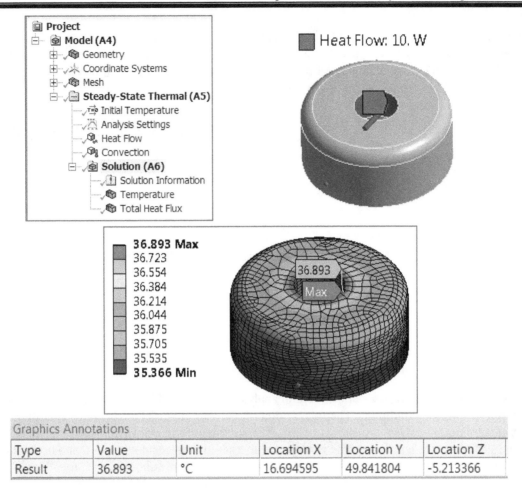

Graphics Annotations					
Type	Value	Unit	Location X	Location Y	Location Z
Result	36.893	°C	16.694595	49.841804	-5.213366

2.   The engineering drawing of a heat sink component is shown below. You are asked to perform structural analysis, steady-state thermal and combined structural and thermal analysis.

Step 1: you are asked to create a 3D model for the plate component using Geometry, as shown below. The unit is mm. The material type is structural steel.

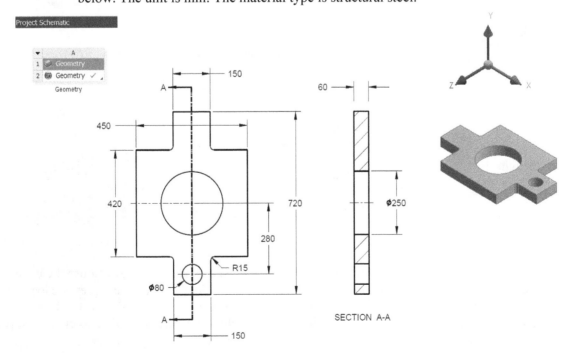

Step 2: You are asked to perform a steady-state thermal analysis. The initial temperature or environment temperature is 5 degree Celsius. The convection condition is the surface on the **left** side with film coefficient 0.0025 W/(°C mm²) and the ambient temperature 5 °C. The pre-scribed temperature condition is the surface on the **right** side with 30 degree Celsius, as shown below. Mesh size requirements: Resolution 4 and click Generate Mesh.

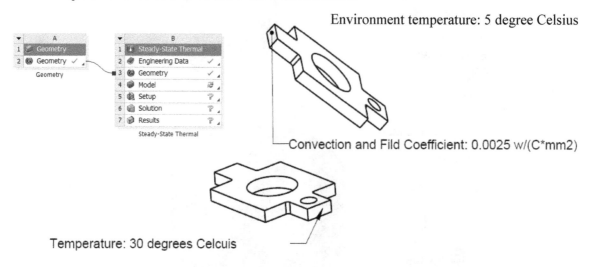

Environment temperature: 5 degree Celsius

Convection and Fild Coefficient: 0.0025 w/(C*mm2)

Temperature: 30 degrees Celcuis

Step 3: You are asked to perform a static structural analysis, using the obtained temperature distribution in step 2 as the load condition. There is no other load condition. The constraint conditions for the static structural analysis are shown below. The surface on the right (front) side is fixed support. The surface on the left (back) side is subjected to a set of displacements: x = 0 and both y and z are set free. It is important to note that the environment temperature should be set to 5 degree Celsius.

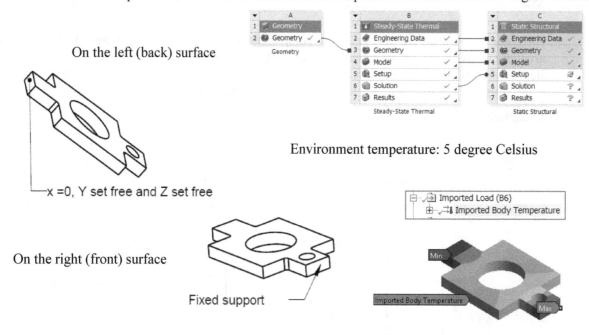

On the left (back) surface

x =0, Y set free and Z set free

Environment temperature: 5 degree Celsius

On the right (front) surface

Fixed support

Step 4: You are asked to perform a new static structural analysis. Make sure that you use the geometric model and connect it with Static Structural, as shown. The load condition is a pressure load acting on the cylindrical surface of the large hole, as shown. The constraint conditions for the new static structural analysis are shown below. The surface on the right side is fixed support. The surface on the left side is subjected to a set of displacements: x = 0 and both y and z are set free.

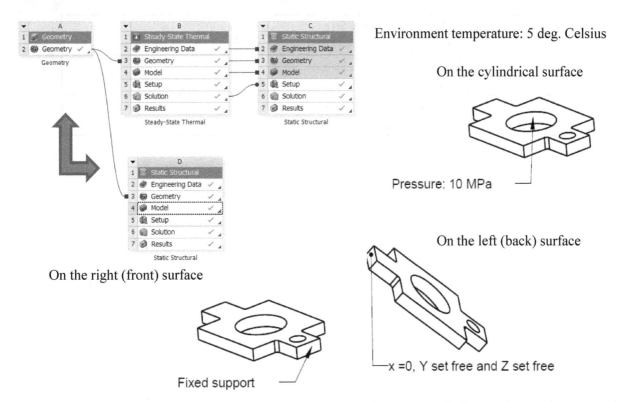

Environment temperature: 5 deg. Celsius

On the cylindrical surface

Pressure: 10 MPa

On the left (back) surface

x =0, Y set free and Z set free

On the right (front) surface

Fixed support

Step 5: You are asked to perform a combined mechanical and thermal analysis. Such a static structural analysis has two load conditions. The first load condition is the pressure load of 10 MPa. The second load condition is the imported load, as shown. The constraint conditions for the combined mechanical and thermal analysis are shown below. The surface on the right (front) side is fixed support. The surface on the left (back) side is subjected to a set of displacements: x = 0 and both y and z are set free.

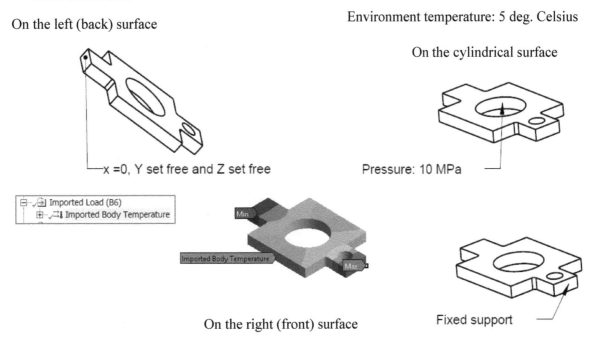

On the left (back) surface

Environment temperature: 5 deg. Celsius

On the cylindrical surface

x =0, Y set free and Z set free

Pressure: 10 MPa

Imported Load (B6)
Imported Body Temperature

On the right (front) surface

Fixed support

List the three (3) results of Maximum values of von Mises stress from your previous studies. Fill in the numerical values in the following table with the units.

Max Value of von Mises Stress due to Thermal Load only (MPa)	
Max Value of von Mises Stress due to Pressure Load only (MPa)	
Max Value of von Mises Stress Obtained from the Combined Study (MPa)	

CHAPTER 5

Frequency and Vibration Analysis

5.1   Introduction

In Chapter 2 and Chapter 3, we discussed static structural analysis with components and assemblies. All case studies in those 2 chapters required a fixed support constraint and the objects, either a component or an assembly, under study are in a status of static equilibrium condition. The case studies presented in Chapter 4 focused on the temperature distribution under a set of thermal constraints and load conditions. In this chapter, we present 6 case studies where our main interest is in the frequency domain. We focus on the dynamic response of the object under investigation. Section 5.2 presents a music fork or a tuning fork widely used in tuning musical instruments. The objective of this case study is to determine its fundamental frequency. Section 5.3 presents a diving board commonly used in diving sports. Section 5.4 and Section 5.5 present two case studies in buckling. The object under study in Section 5.4 is a steel stool. We determine that the magnitude of the critical load leading to the occurrence of a buckling failure mode. The object under study in Section 5.5 is a slender column, on which a compressive stress is built up to lead to a sudden sideway deflection of the column. Section 5.6 is a case study dealing with vibration of a string commonly used in musical instruments. Section 5.7 presents a cam-follower system. Our focus is to determine the time-varying displacement, velocity and acceleration of the follower component as the cam rotates continuously.

5.2   Modal Analysis of a Tuning Fork

The following figure illustrates a tuning fork used to tune a piano. A tuning fork is a simple two-pronged metal instrument with a handle and tines that form a U shape. Tuning forks will vibrate at a set frequency to produce a musical tone when struck and therefore is used for many applications such as piano tuning as well as assessing a person's ability to hear. Tuning forks work by holding the handle and striking the tines with a force great enough for the tines to vibrate at a frequency, resulting in a sound. The geometry of a music fork under study is shown below. Assume that the material is structural steel. Use FEA to determine the frequencies of the first 4 modes and the mode shapes of the first 4 modes.

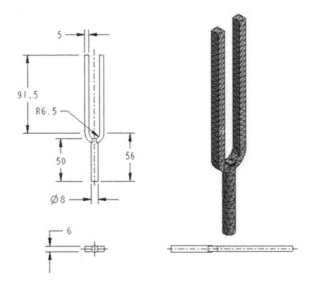

Step 1: Open Workbench.
From the start menu, click **ANSYS 19** > **Workbench 19**.

Step 2: From **Toolbox**, highlight **Modal** and drag it to the **Project Schematic** area.

Step 3: Start Design Modeler and create a 2D sketch.
Right-click the **Geometry** cell and pick **New DesignModeler Geometry** to start **Design Modeler**. Users may double click **Geometry** to directly enter **Design Modeler** if **New SpaceClaim Geometry** is not present. Click the **Units** drop down on the top menu, and select **Millimeter**.

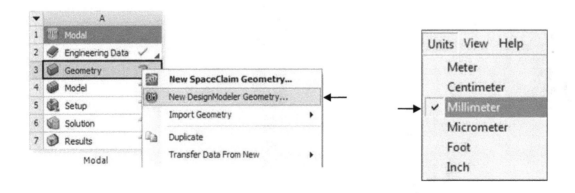

In the Tree Outline, select **ZXPlane** as the sketching plane.

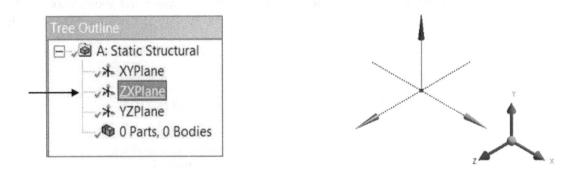

To orient the sketching plane, from the top menu, click the icon of **Look At** so that the selected sketching plane is oriented to be parallel to the screen and ready for making a sketch.

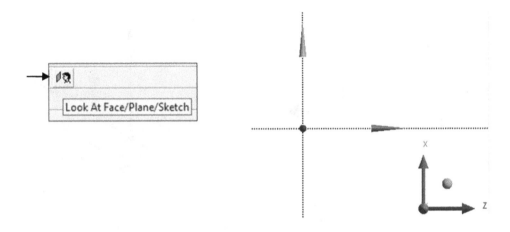

At the bottom of **Tree Outline**, click the icon of **Sketching**. The Sketching Toolboxes window appears. Click **Draw**. Select **Rectangle**, and sketch a rectangle.

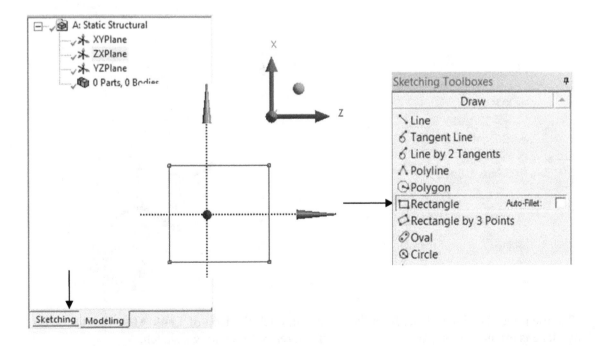

Add two symmetric constraints so that the sketched rectangle is symmetric about X-axis and Y-axis. Afterwards, click Dimensions > General. Click the top edge and specify 5. Click the edge on the left or right side and specify 6.

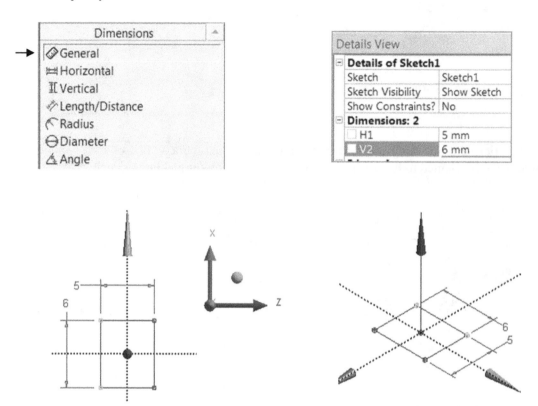

Upon completing this sketch (profile), click **Modeling** for creating the second sketch (path). In the **Tree Outline**, select **ZYPlane** as the sketching plane. Click New Sketch.

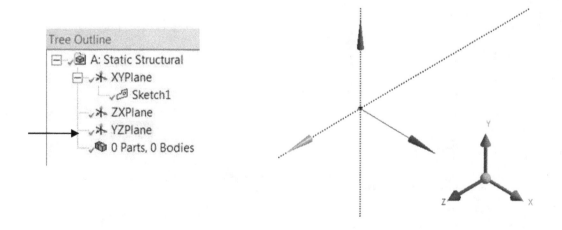

To orient the sketching plane, from the top menu, click the icon of **Look At** so that the selected sketching plane is oriented to be parallel to the screen and ready for making a sketch.

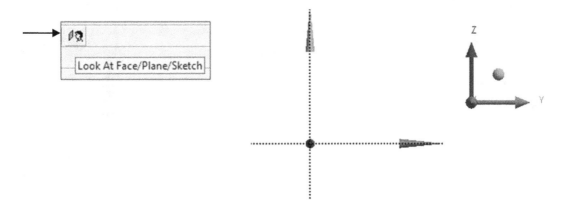

At the bottom of **Tree Outline**, click the icon of **Sketching**. Click **Draw**. Select **Line.** Sketch 1 horizontal line, as shown.

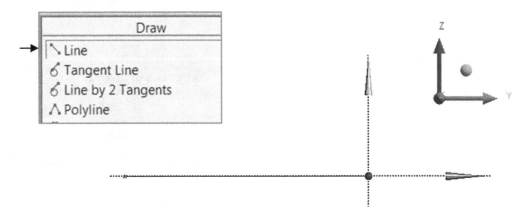

Click **Arc by 3 Points** and sketch an arc, as shown. Make sure "T" is shown, indicating the sketched arc is tangent to the sketched line.

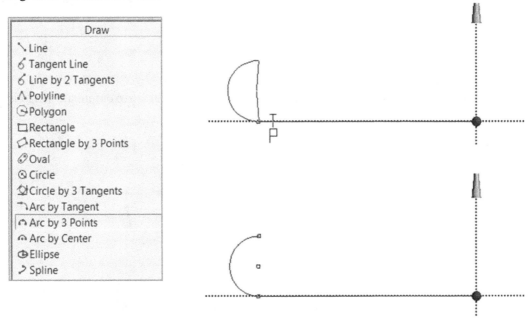

Click **Line** and sketch a line. This line is tangent to the arc and parallel to the sketched line. This line stops at the Z-axis.

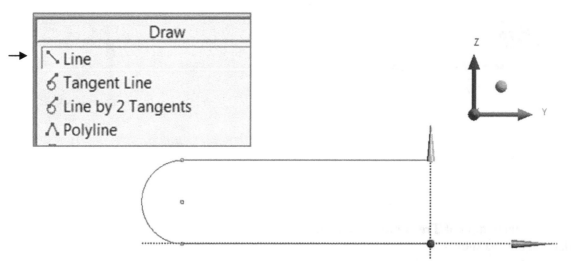

Click **Dimensions,** and select **General**. Specify 91.5 as the length value and specify 9 as the radius value, as shown.

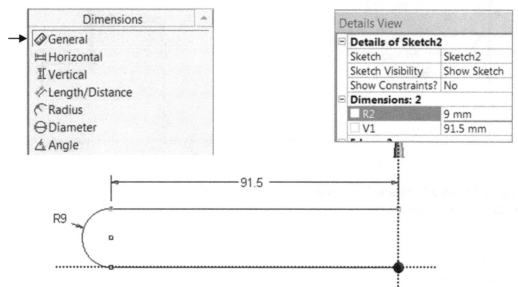

Upon completing this sketch (path), click **Modeling.** Click **Sweep**. Click Sketch1 (profile) first and click **Apply.** Afterwards, click Sketch2 (path) and **Apply.** Click **Generate** to obtain the 3D solid model of the wrench model, as shown.

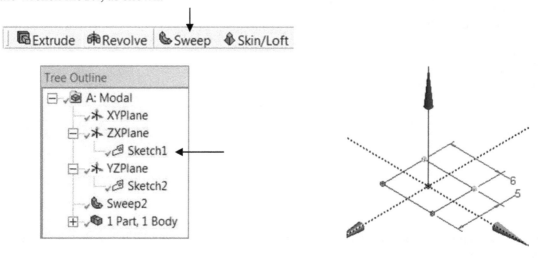

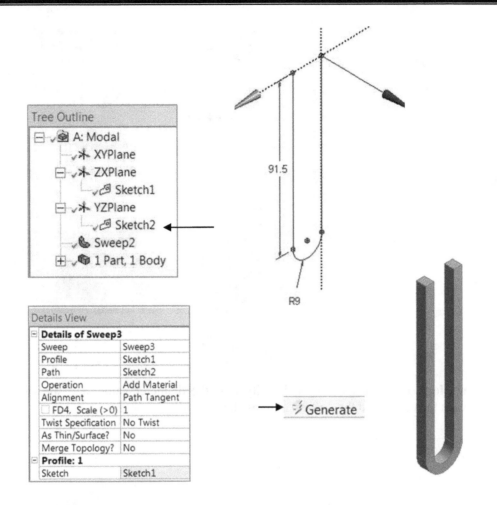

Step 4: Create the handle.

We first create a datum plane. Select ZX plane first. Afterwards, click the icon of **New Plane**. In Details View, select **Offset Z** in **Transform**. Specify -147.5 as the offset value. Click **Generate**. Plane4 is created.

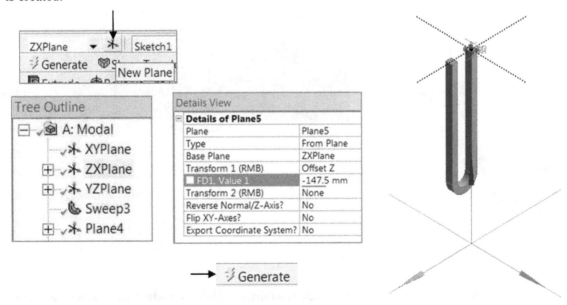

Click **New Sketch**. Afterwards, click **Look At**. From **Draw**, select **Circle** and sketch a circle, as shown below:

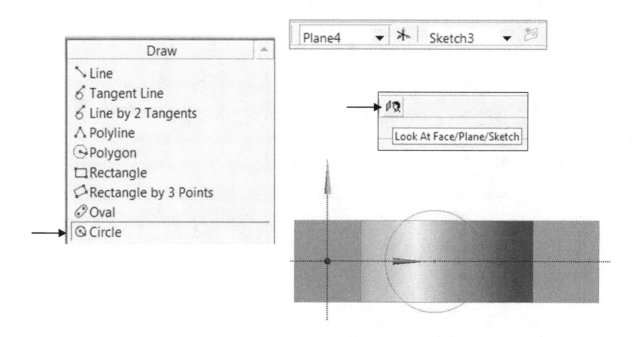

Click **Dimensions,** and select **General**. Specify 8 as the diameter value, and specify 9 as the distance value between the center and the vertical axis, as shown.

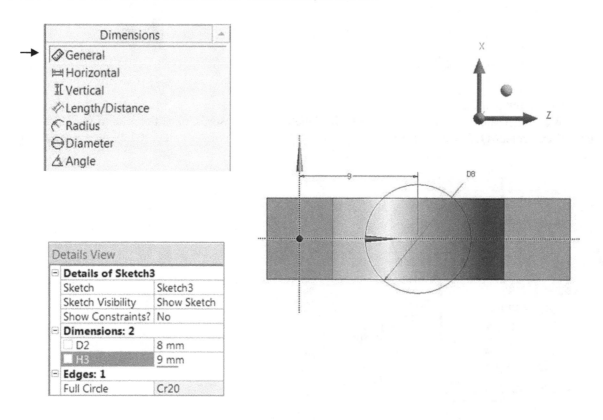

Upon completing this sketch, or Sketch3, click **Modeling.** Click **Extrude**. Click Sketch3 and click **Apply**. Specify 49 as the extrusion distance value. Click **Generate** to obtain the 3D solid model of the handle feature, as shown.

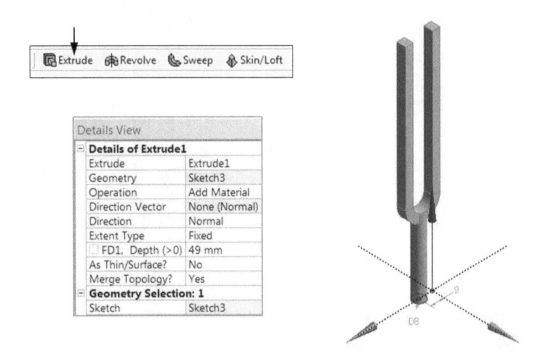

Upon completing the process of creating the geometry, we need to close Design Modeler. Click **File** > Close **Design Modeler**, and go back to **Project Schematic**. Specify **Music Fork** as the file name. Click **Save**.

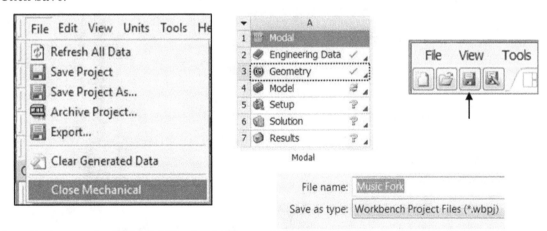

Step 5: Accept the Default Material Assignment of Structural Steel and Perform FEA

Double-click the icon of **Model**. We enter the **Mechanical Mode**. Check the unit system, and pick Metric (mm, t, N, s, mV, mA).

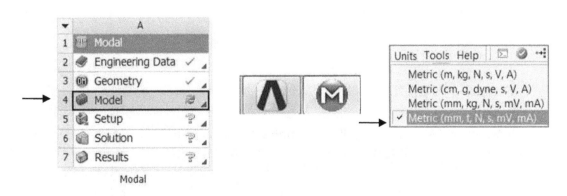

Highlight **Mesh**. In Details of "Mesh", expand **Sizing** and set Resolution to 4. Right-click **Mesh** to pick **Generate Mesh**.

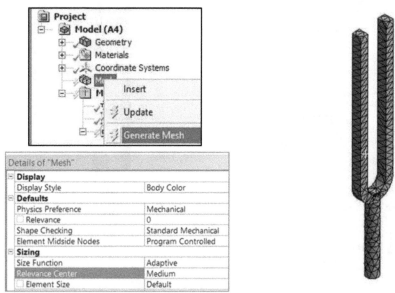

Highlight **Modal (A5)** and right-click to pick **Insert**. Afterwards, select **Fixed Support**. From the screen, pick the cylindrical surface from the handle, as shown. Click **Apply**.

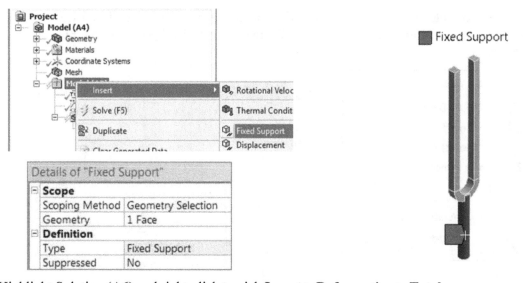

Highlight Solution (**A6**) and right-click to pick **Insert > Deformation > Total.**

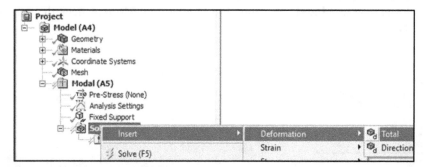

Highlight Solution (**A6**) and right-click to pick **Solve.** Workbench system is running FEA. The first six (6) modes are listed in a table and shown in a graph. The frequency of the first mode is 383 Hz. The frequency of the second mode is 384 Hz. The frequency of the third mode is 446 Hz.

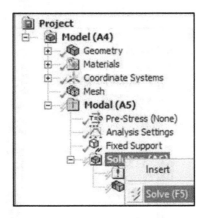

	Mode	✔ Frequency [Hz]
1	1.	383.13
2	2.	384.32
3	3.	445.65
4	4.	446.2
5	5.	2390.3
6	6.	2398.9

Tabular Data

To plot the first mode shape. Highlight Total Deformation. Mode 1 is listed in Details of "Total Deformation".

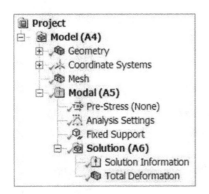

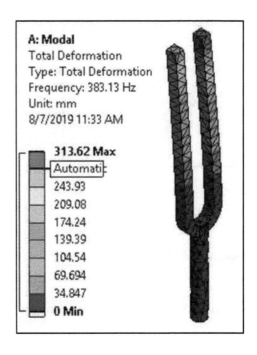

To plot the second mode shape. Highlight Total Deformation. Change Mode 1 to Mode 2 in Details of "Total Deformation". Afterwards, highlight Total Deformation and right-click to pick Evaluation All Results.

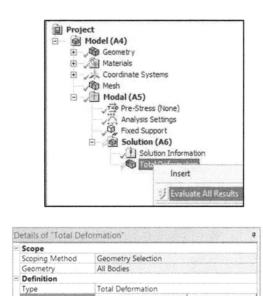

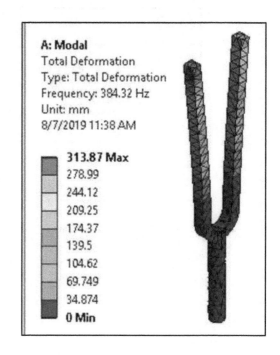

To plot the third mode shape. Highlight Total Deformation. Change Mode 2 to Mode 3 in Details of "Total Deformation". Afterwards, highlight Total Deformation and right-click to pick Evaluation All Results.

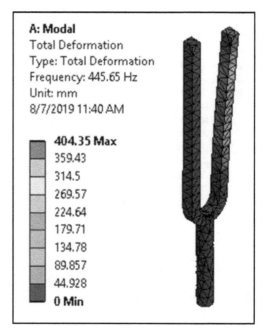

To animate the motion of a mode shape, just click the icon of **Play**.

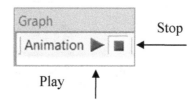

From the top menu, click **File** > **Close Mechanical**. In this way, we return to the main page of Workbench, or Project Schematic. From the top menu, click the icon of **Save Project.** From the top menu, click **File** > **Exit**.

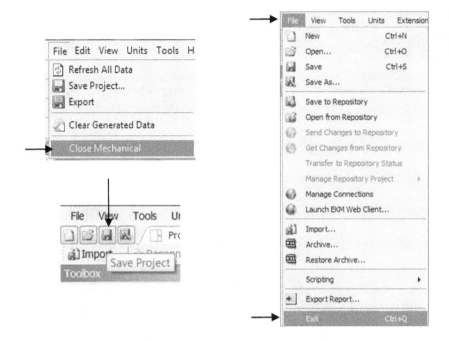

5.3   Modal Analysis of a Pre-Stressed Diving Board

A diving board is shown below. The three dimensions of this diving board are 4000 x 750 X 75 mm. There is a surface region on the top surface of the board, serving as a loading area on which a diver stands. Assuming the load magnitude is 400 Newton. Perform a static structural analysis first and import the results to a modal analysis panel to perform a modal analysis to determine the first 4 modes of this diving board.

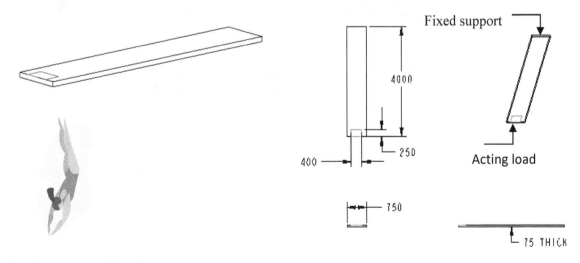

Step 1: Open Workbench.

From the start menu, click **ANSYS 19** > **Workbench 19**. Users may directly click Workbench 19 when the symbol is on display.

Step 2: From **Toolbox > Analysis System**, highlight **Model** and drag it to the main screen.

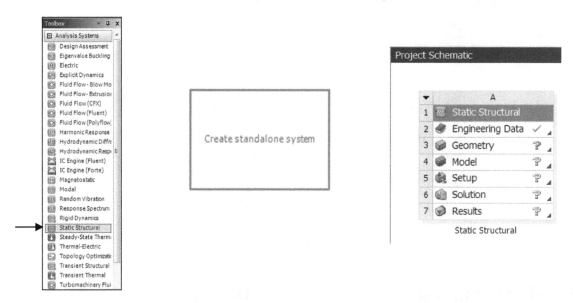

From Analysis System, highlight Modal and, using the mouse, drag it to Solution of the Static Structural panel and make sure **Share A2 to A4** is on display. Afterwards, release the mouse. In the Project Schematic page, both Panel A and Panel B are on display.

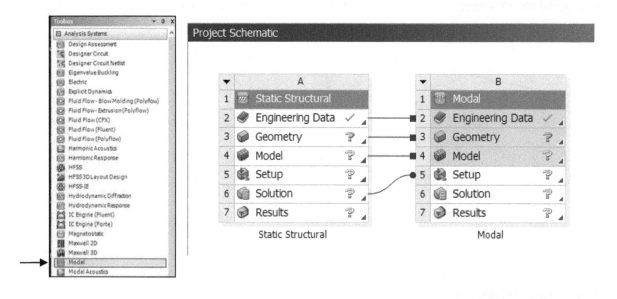

Right-click the **Geometry** cell and pick **New DesignModeler Geometry** to start **Design Modeler**. Users may double click **Geometry** to directly enter **Design Modeler** if **New SpaceClaim Geometry** is not present. Click the **Units** drop down on the top menu, and select **Millimeter**.

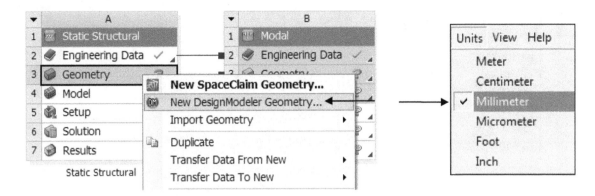

To create the first sketch, select **ZXPlane** as the sketching plane. Click the icon of **Look At** so that the sketching plane is parallel to the screen and ready for preparing a sketch.

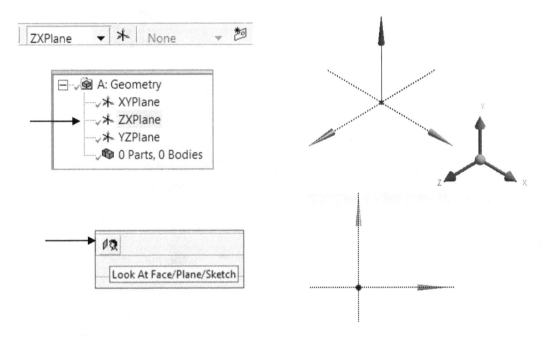

Click **Sketching**, and click **Draw**. In the Draw window, click **Rectangle**. Sketch a rectangle and add a symmetrical constraint so that the sketched rectangle is symmetric about the Z- axis.

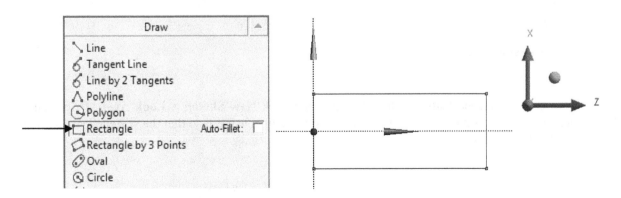

Click **Dimensions** and click **General**. Specify the 2 size dimensions. They are 4000 and 750, respectively.

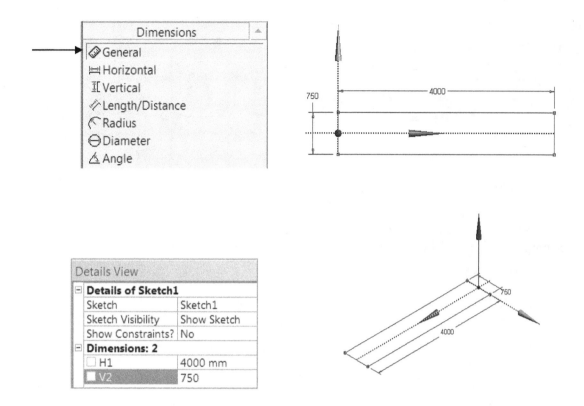

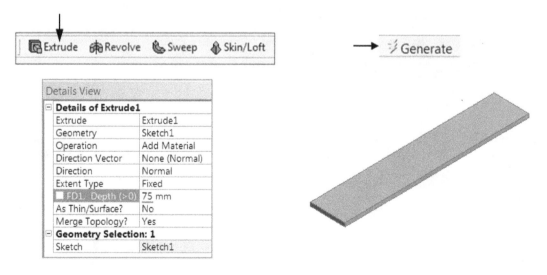

Upon completing the sketch, click the icon of **Extrude**. Click the created sketch or sketch 1 from the Tree Outline, and click **Apply**. In Operation, select Add Material. Specify 75 as the thickness value. Click **Generate** to obtain the 3D solid model of the plate feature, as shown.

To create the second feature, select ZXPlane and click **New Sketch > Look At**. In **Draw**, click **Rectangle**, and sketch a rectangle. Afterwards, add a symmetry constraint so that the sketched rectangle is symmetric about the Z-axis.

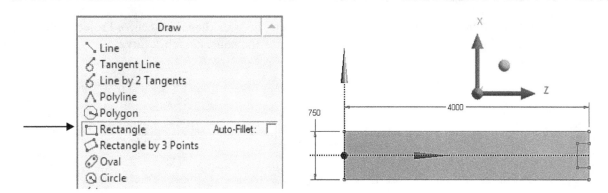

Specify the 2 size dimensions: 400 and 250, as shown below.

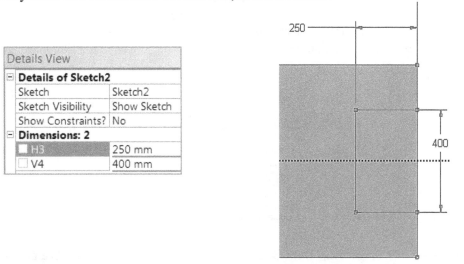

Upon completing the sketch, click the icon of **Extrude**. Click the created sketch or sketch 2 from the Tree Outline, and click **Apply**. In Operation, select **Imprint Faces**. In Extent Type, select To Faces. Pick the top surface of the plate feature, as shown. Click **Generate** to obtain the surface region feature for applying the load condition.

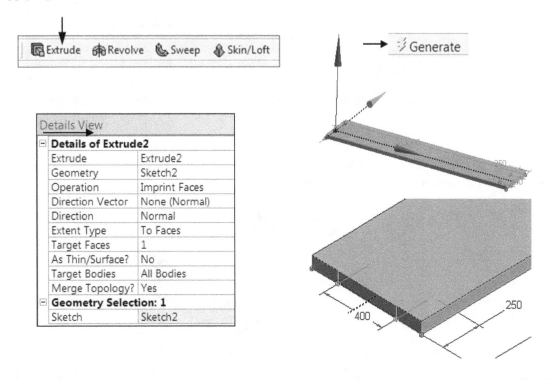

Click **File** and **Close Design Modeler**. We go back to the **Workbench** screen, or Project Schematic. Click **Save Project**. Specify **Diving Board Combined** as the file name and click **Save**.

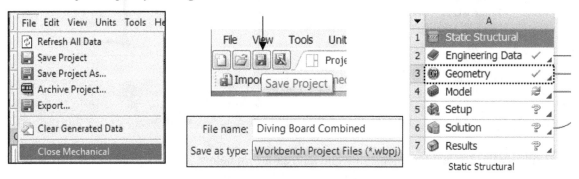

Step 3: Activate Polyethylene from the material library, or Engineering Data Sources.
In the Static Structural panel, double click Engineering Data. We enter the Engineering Data Mode.

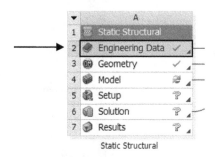

Click the icon of **Engineering Data Sources**. Click **General Materials**. We are able to locate Polyethylene and activate it. Afterwards, click the icon of Engineering Data Sources and click the icon of Project to go back to Project Schematic.

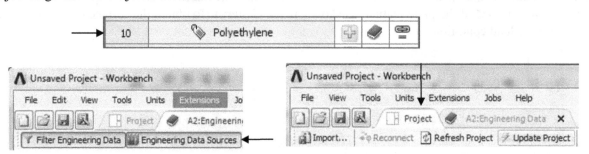

Step 4: Assign Polyethylene to the 3D solid Model
Double-click the icon of Model. We enter the Mechanical Mode.

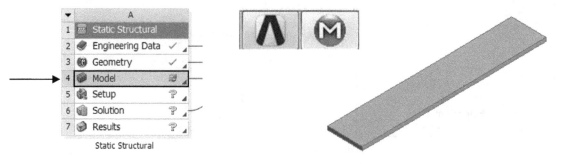

Now let us do the material assignment. From the Project tree, expand Geometry. Highlight Solid. In Details of "Solid", click material assignment, pick Polyethylene to replace Structural Steel, which is the default material assignment.

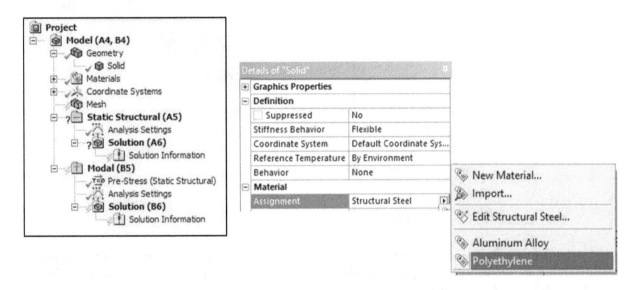

From the Project tree, Highlight **Mesh**. In the Detailed of "Mesh" window, click **Sizing**. Set Resolution to 4. Right-click **Mesh** again to pick **Generate Mesh**.

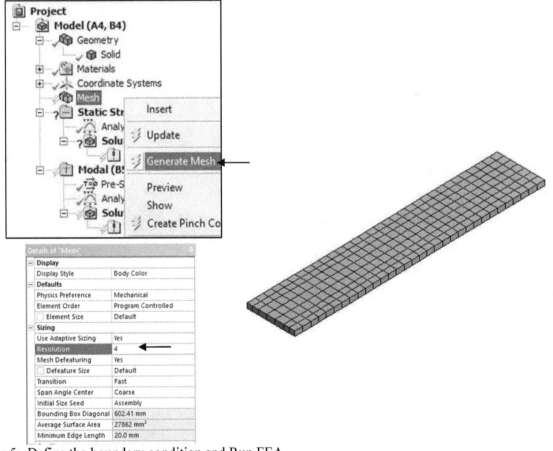

Step 5: Define the boundary condition and Run FEA.

Highlight **Static Structural (A5)**. Pick Insert and **Fixed Support**. Pick the surface at the end of the plate feature, as shown. Pick **Apply**.

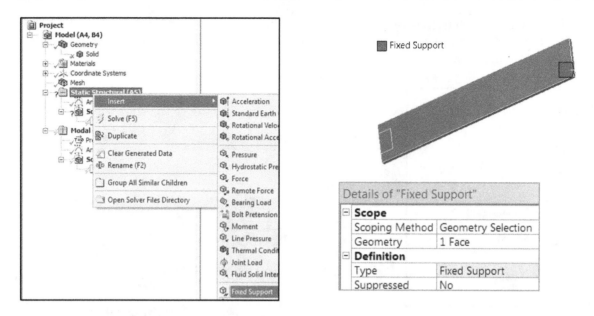

Highlight **Static Structural (A5)**. Pick Insert and **Force**. Pick the defined surfaces region or imprint surface, as shown. Pick **Apply**. Specify -400 as the value for Y Component, and set 0 for the values for both X and Z Components.

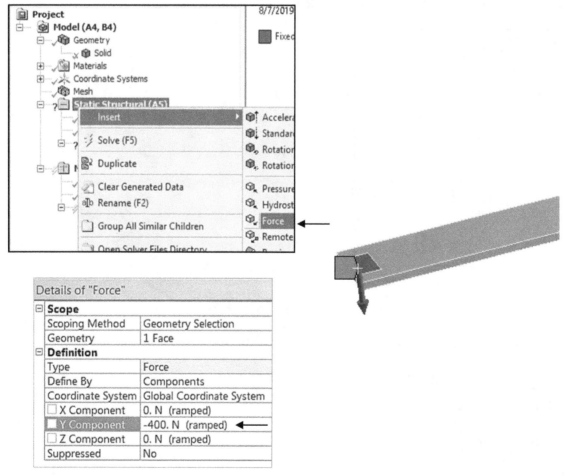

From the project tree, highlight **Solution (A6)** and right-click to pick **Solve**. Workbench system is running FEA.

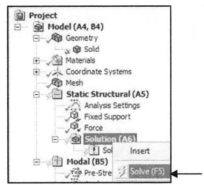

At this moment, we have completed the stress analysis. Now we work on the modal analysis. Highlight Analysis Settings under **Modal (B5).** In Details of "Analysis Settings", specify 4 as the **Max Modes to Find**.

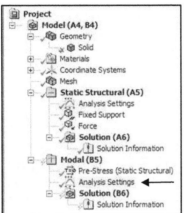

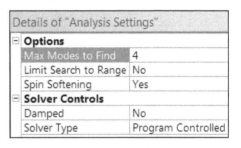

Highlight **Solution (B6)** and right-click to pick **Solve**. The natural frequencies of the first 4 modes are listed.

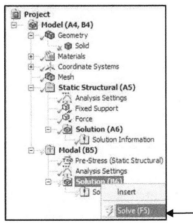

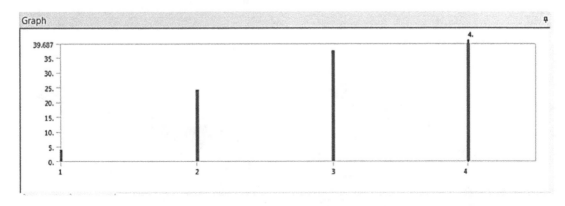

To visualize the shape of each of the first 4 modes, highlight **Solution (B6),** and right-click to pick Insert > Deformation > Total. Highlight Total Deformation and right-click to pick Evaluate All Results.

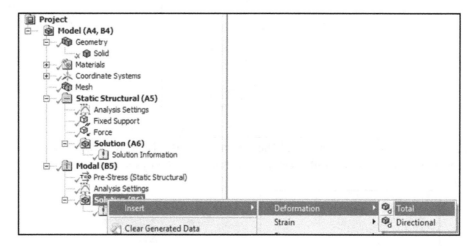

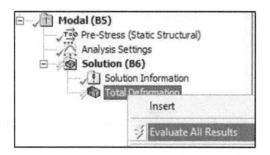

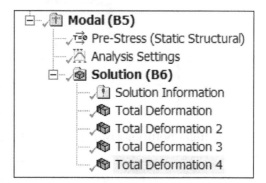

The 4 mode shapes are shown below.

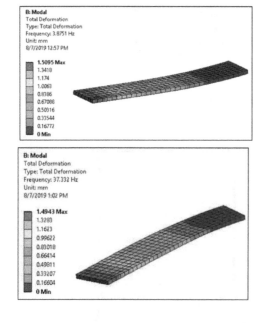

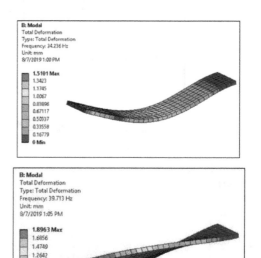

Users may plot the above 4 graphs on a single graph. To do so, go to the top menu, look for the icon of **Viewports**. Expand this icon to pick Four Viewports.

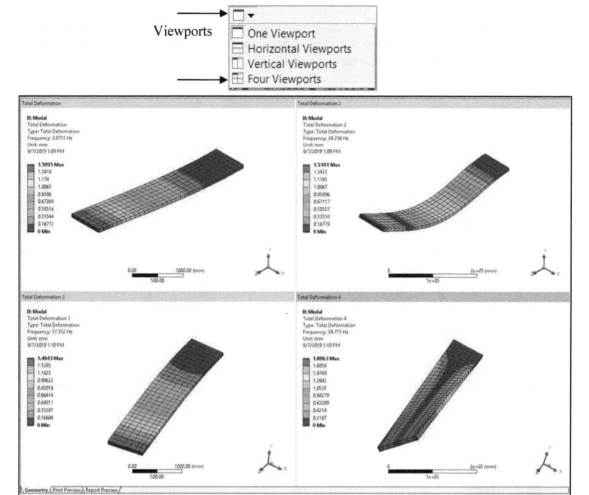

From the top menu, click **File** > **Close Mechanical**. In this way, we return to the main page of Workbench.

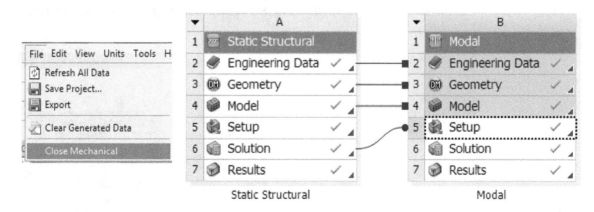

From the top menu, click **File** > **Save Project**. Afterwards, click **File** > **Exit**.

5.4 Eigenvalue Buckling Analysis of a Steel Stool

A stool is shown below. The seat is made of stainless steel. The 4 legs are made of steel. Assume that a weight of 1000 N is acting on the top surface, simulating a person sits on the stool. Determine the maximum deformation and the maximum value of von Mises stress developed on the stool. Afterwards, perform an eigenvalue buckling analysis to determine the magnitude of the critical load that may lead to the performance failure of the stool under the critical load.

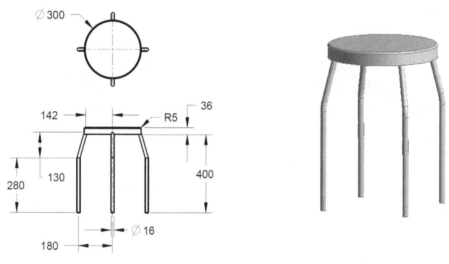

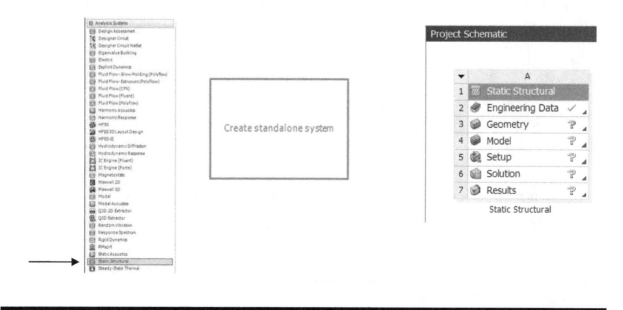

Step 1: Open Workbench.

From the start menu, click **ANSYS 19 > Workbench 19**. Users may directly click Workbench 19 when the symbol is on display.

Step 2: From **Toolbox > Analysis System**, highlight **Static Structural** and drag it to the main screen.

Right-click the **Geometry** cell and pick **New DesignModeler Geometry** to start **Design Modeler**. Users may double click **Geometry** to directly enter **Design Modeler** if **New SpaceClaim Geometry** is not present. Click the **Units** drop down on the top menu, and select **Millimeter**.

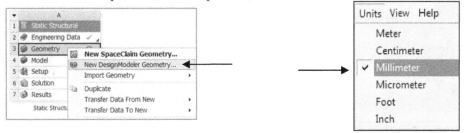

We first create a datum plane, which is parallel to **ZXPlane** with an offset value equal to 400. To do so, highlight **ZX** Plane. Click the icon of New Plane. Specify 464 as the offset Z value. Click **Generate**. Plane4 is created.

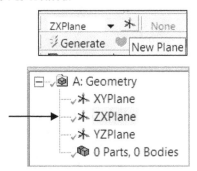

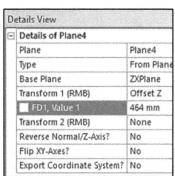

Highlight **Plane4**. Click **New Sketch**. To orient the sketching plane, from the top menu, click the icon of **Look At** so that the sketching plane is parallel to the screen and ready for preparing a sketch.

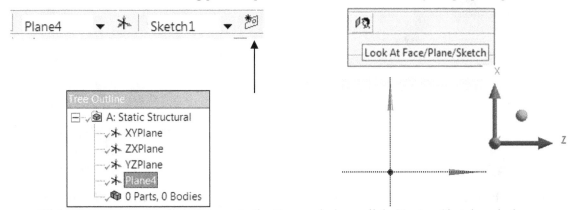

Click **Sketching**, and click **Draw**. In the Draw window, click **Circle**. Sketch a circle.

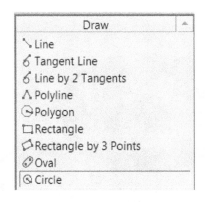

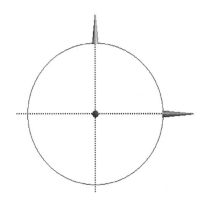

Specify the diameter dimension of 320.

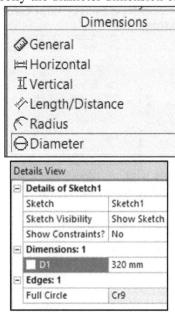

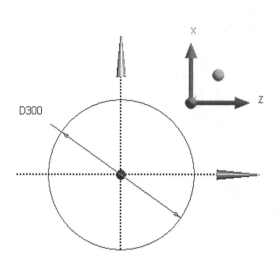

Upon completing the sketch, click the icon of **Extrude**. Click the created sketch or sketch 1 from the Tree Outline, and click **Apply**. In Operation, select Add Material. Select Both Asymmetric and specify 5 and 35 as the 2 values. Click **Generate** to obtain the cylindrical plate feature, as shown.

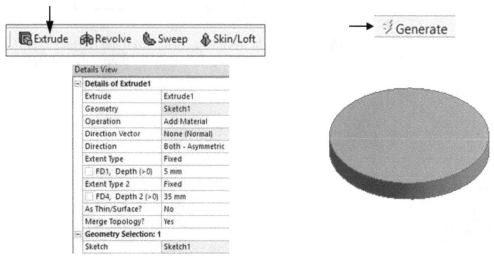

Click **Blend** and select Fixed Radius. Specify 8 as the radius value. Pick the top edge. Click Generate.

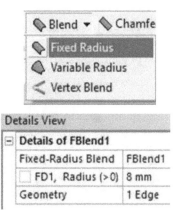

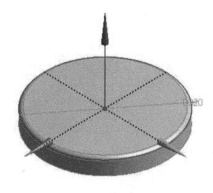

Click **Thin/Surface**. Select Face to Remove. Pick the bottom surface. Specify 5 as the thickness value and click Generate.

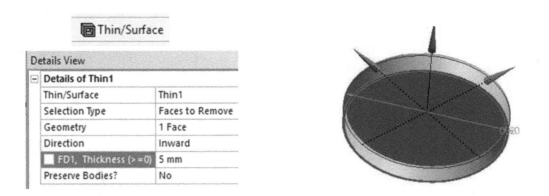

In Tree Outline, highlight **XY Plane**, and click the icon of **New Sketch > Look At**. Click Draw and select Arc by Center and sketch a half circle. Click Line to sketch 3 lines. Afterwards, specify 3 dimensions: radius 9, vertical dimension 442 with respect to ZXPlane and line length 14, as shown.

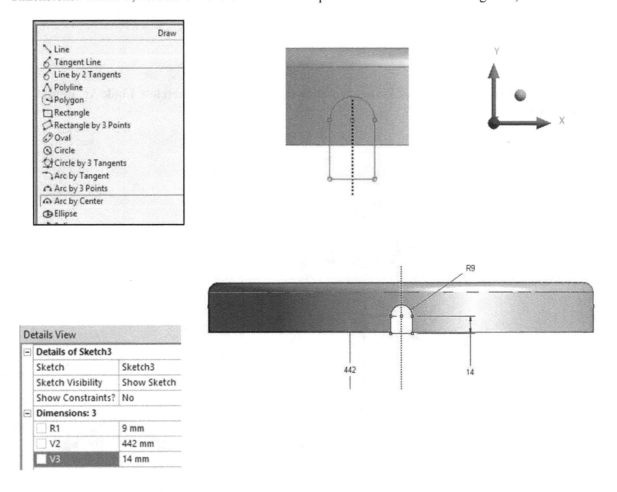

Upon completing the sketch, click the icon of **Extrude**. Click the created sketch or sketch 1 from the Tree Outline, and click **Apply**. In Operation, select Add Material. Select Both Asymmetric and specify 5 and 35 as the 2 values. Click **Generate** to obtain the cylindrical plate feature, as shown.

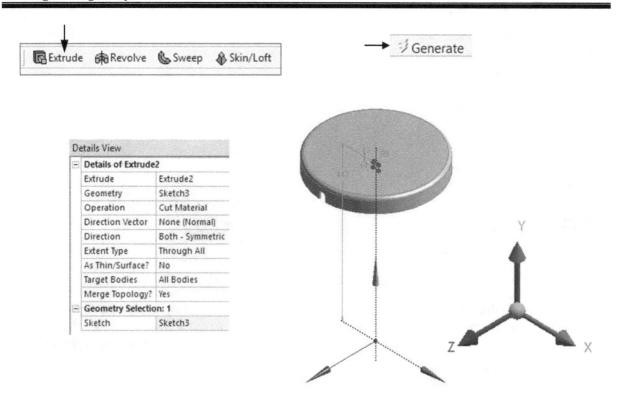

In Tree Outline, highlight **YZPlane**, and click the icon of **New Sketch > Look At**. Repeat this process to cut the small areas.

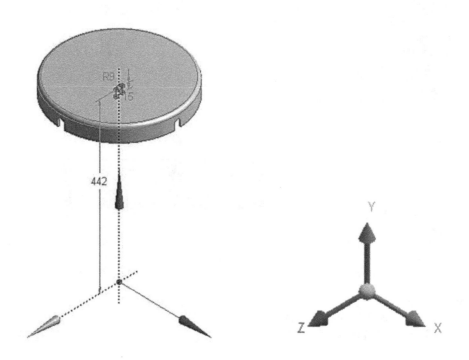

In the model tree, expand 1 Part 1 Body, rename the Solid to Seat. Afterwards. From the top menu, select **Freeze**.

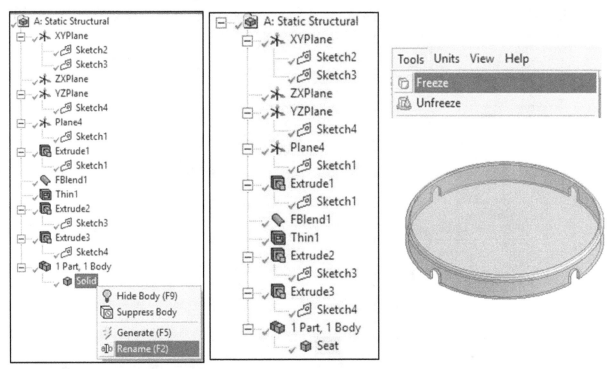

To create the second feature of a leg, select ZXPlane and click **New Sketch > Look At**. In **Draw**, click **Circle**, and sketch a circle. Add a diameter dimension of 16, and a position dimension of 180.

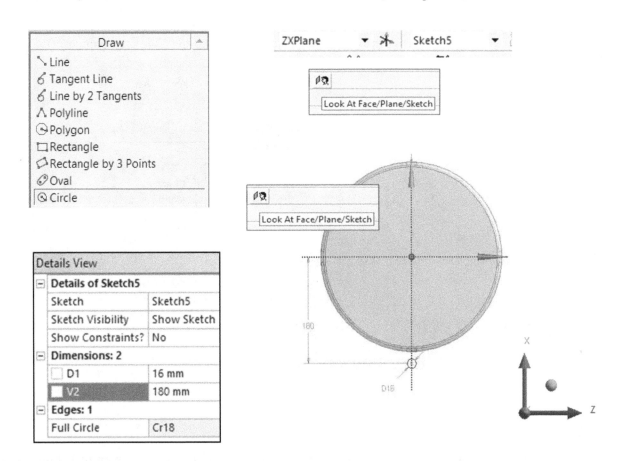

In Tree Outline, highlight ZX Plane, and click the icon of **New Sketch > Look At**. Sketch a circle, and specify the 2 dimensions: 180 and 280, as shown below.

In Tree Outline, highlight XY Plane, and click the icon of **New Sketch > Look At**. Sketch a vertical line, and specify the 2 dimensions: 180 and 280, as shown below.

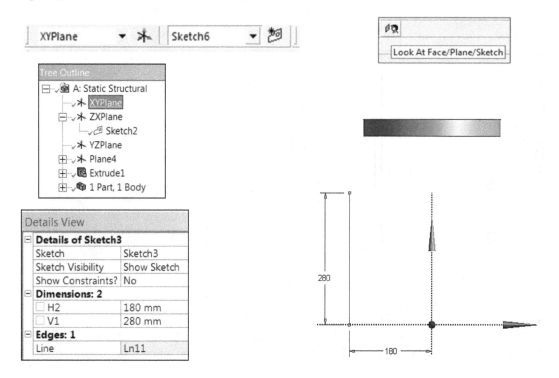

Now sketch one more line, as shown. Specify 2 dimensions of 180 and 284 (keep symmetry). Click **Modify**. Click **Copy**.

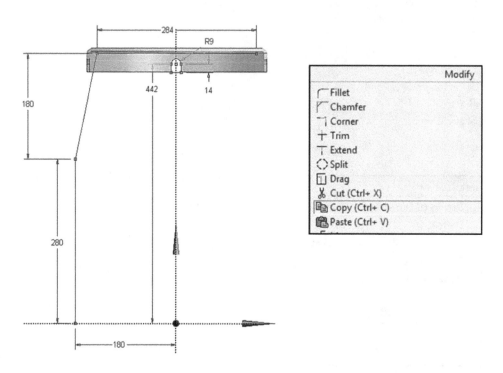

While holding down the Ctrl key, pick the three lines. Right click on the screen pick **End/Use Place Origin as Handle**. Right click on the screen again and pick **Flip Horizontal**. Move your cursor to the origin and make a left click.

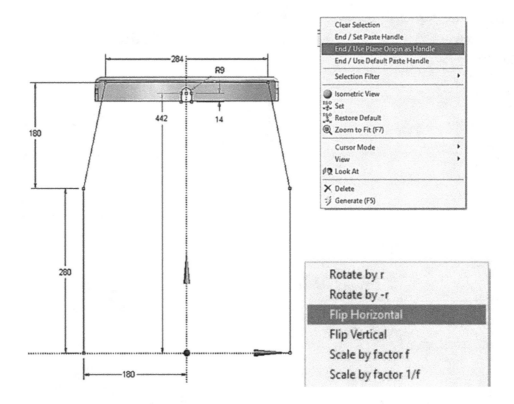

Upon completing the sketch, click the icon of **Sweep**. Pick Sketch5 (a circle) as Profile. Click **Apply**. Pick Sketch6 as Path. Click **Apply**. In **Operation**, select **Add Frozen**. Click **Generate** to create the leg feature. Note that **2 Parts, 2 Bodies** are listed in Tree Outline.

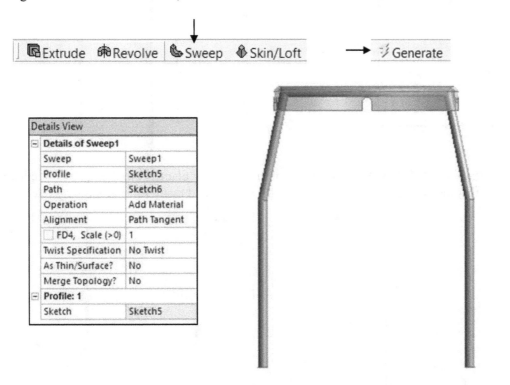

Click Create > Body Transformation > Rotate. Pick the created leg and Specify Yes to preserve the original body. Select Y axis and click **Apply**. Specify 90 degrees as the angle to rotate. Click **Generate**. The second leg is created, as shown.

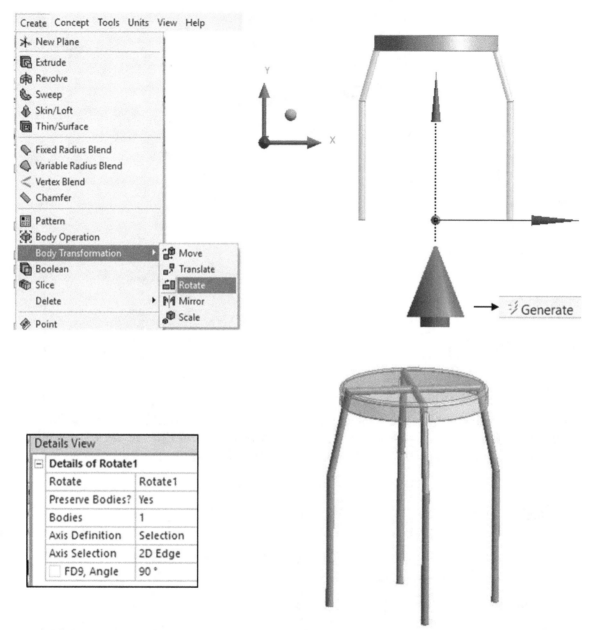

Highlight **Plane4**. Click **New Sketch**. To orient the sketching plane, from the top menu, click the icon of **Look At** so that the sketching plane is parallel to the screen and ready for preparing a sketch.

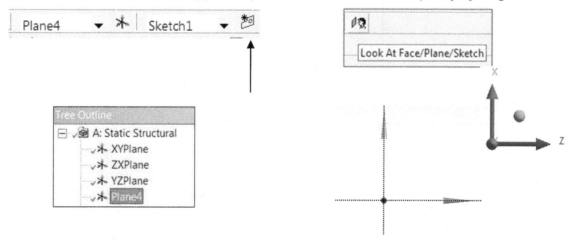

Click **Sketching**, and click **Draw**. Click **Circle**. Sketch a circle.

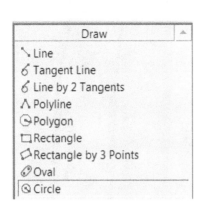

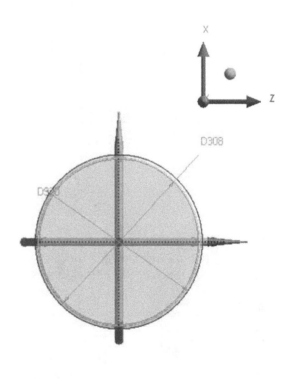

Specify the diameter dimension of 308.

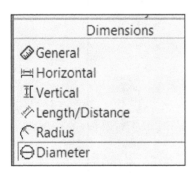

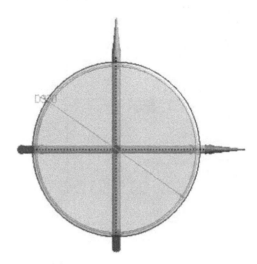

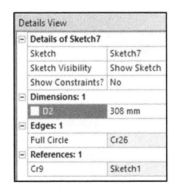

Upon completing the sketch, click the icon of **Extrude**. Click the created sketch or sketch 7 from the Tree Outline, and click **Apply**. In Operation, select Cut Material, and specify 5 as the cutting distance. Click **Generate** to obtain the flat surface on the top of the pair of legs, as shown.

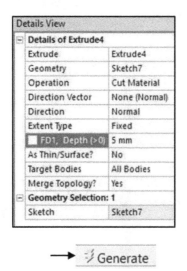

Details View	
Details of Extrude4	
Extrude	Extrude4
Geometry	Sketch7
Operation	Cut Material
Direction Vector	None (Normal)
Direction	Normal
Extent Type	Fixed
☐ FD1, Depth (>0)	5 mm
As Thin/Surface?	No
Target Bodies	All Bodies
Merge Topology?	Yes
Geometry Selection: 1	
Sketch	Sketch7

⟶ ⚡ Generate

In the model tree, rename the solid to Legs, as shown.

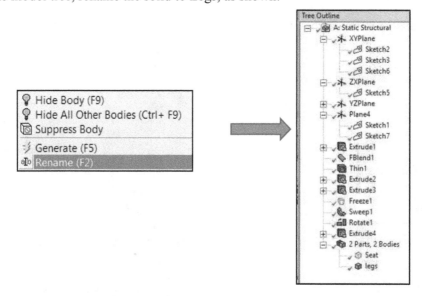

Click **File** and **Close Design Modeler**. We go back to the **Workbench** screen, or Project Schematic. Click **Save Project**. Specify **Stool Buckling** as the file name and click **Save**.

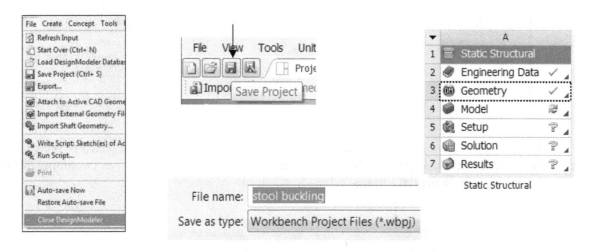

Step 3: Activate Stainless Steel from the material library, or Engineering Data Sources.
    In the Static Structural panel, double click Engineering Data. We enter the Engineering Data Mode.

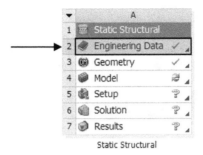

Click the icon of **Engineering Data Sources**. In Outline of Favorites, Structural Steel is listed and activated with the activated symbol listed in column C Location by the system default. Click the icon of **Engineering Data Sources**. Click **General Materials**. We are able to locate Stainless Steel and activate it. Afterwards, click the icon of Engineering Data Sources and click the icon of Project to go back to Project Schematic.

Step 4: Assign Stainless Steel to the 3D solid Model
    Double-click the icon of Model. We enter the Mechanical Mode. Check the unit system.

Now let us do the material assignment. From the Project tree, expand Geometry. Highlight Seat. In Details of "Seat", click material assignment, pick Stainless Steel to replace Structural Steel, which is the default material assignment.

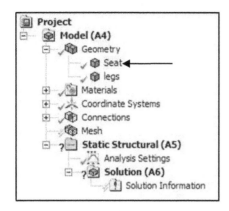

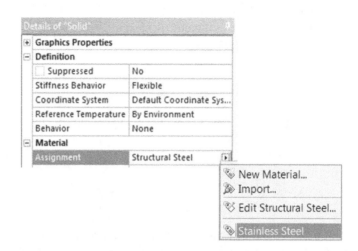

For the material type of Legs is Structural Steel. There is no need to do the new assignment.

From the Project tree, Highlight **Mesh**. In the Detailed of "Mesh" window, click **Sizing**. Set Resolution to 4. Right-click **Mesh** again to pick **Generate Mesh**.

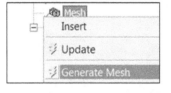

Step 5: Define the boundary condition and Run FEA.

Highlight **Static Structural (A5)**. Pick Insert and **Fixed Support**. Pick the 4 bottom surfaces of the 4 legs while holding down the Ctrl key. Pick **Apply**.

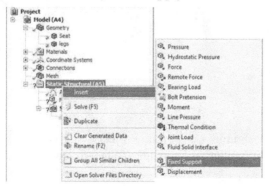

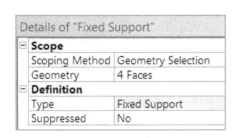

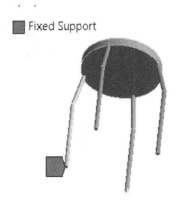

Highlight **Static Structural (A5)**. Pick Insert and **Force**. Pick the top surface of the seat, as shown. Pick **Apply**. Specify -1000 as the value for Y Component, and set 0 for the values for both X and Z Components.

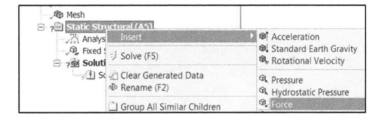

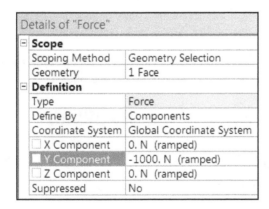

From the project tree, highlight **Solution (A6)** and right-click to pick **Solve**. Workbench system is running FEA.

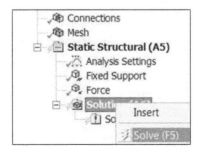

At this moment, we have completed the static stress analysis. We may plot the deformation distribution and the von Mises stress distribution.

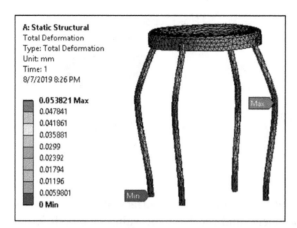

Now let us close the Mechanical Analysis. From the top menu, click **File** > **Close Mechanical**. In this way, we return to the main page of Workbench. From the top menu, click **File** > **Save Project**.

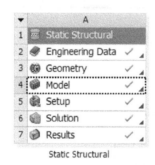

Step 6: Eigenvalue Buckling Analysis

To get on the buckling analysis, from **Toolbox** > **Analysis System**, highlight **Eigenvalue Buckling** and drag it to A6 Solution. Upon seeing Share A2:A4 Transfer A6, release the mouse button.

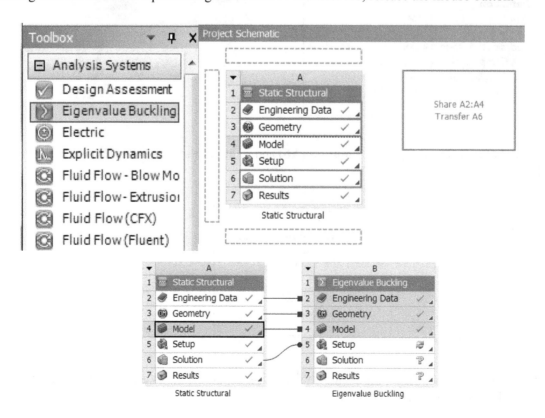

Highlight B5 Setup of Eigenvalue Buckling and right-click to pick Edit, or just double click Setup to start the buckling analysis. In the Project tree, highlight Solution (A6) and right-click to pick **Solve**.

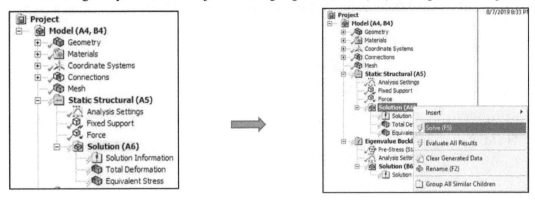

Highlight Analysis Settings listed under Eigenvalue Buckling (B5), specify 4 as Max Modes to Find.

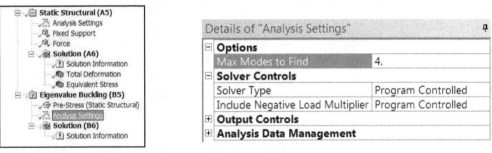

Highlight **Solution (B6)** and right-click to pick **Solve**. The system starts to run the buckling analysis.

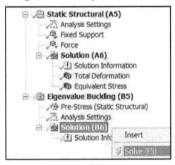

In the graph area, four load multiplier values are listed. For mode 1, the load multiplier value is 132.81.

For mode 1, the load multiplier value is 132.81. The critical load, which leads to the occurrence of the first model buckling, is given by:

$$P_{critical\_load} = P_{applied\_load} * buckling\_load\_multiplier = (1000)*(132.81) = 132810 \ (N)$$

In order to visualize the mode shape for each of the 4 buckling modes, right-click in the Tabular Data area to pick Select All. Afterwards, right-click again to pick Create Mode Shape Results. In Project tree, Total Deformation, Total Deformation 2, Total Deformation 3 and Total Deformation 4 are created.

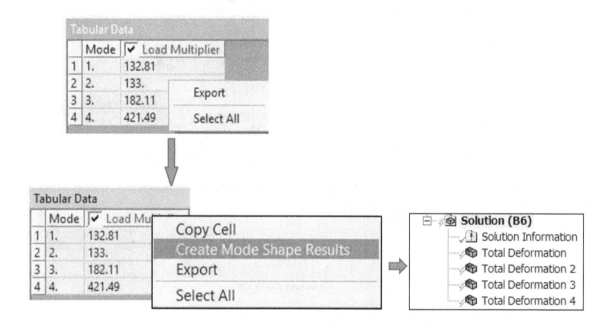

Highlight Total Deformation and right-click to pick Evaluate All Results. The first mode shape is on display.

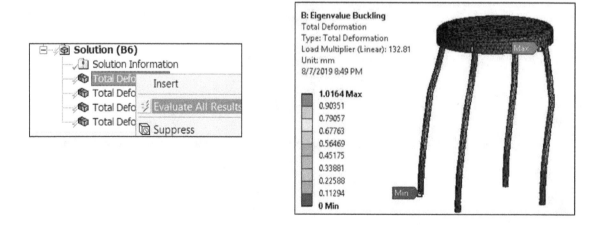

In order to display the 4 mode shapes together, go to the top menu, look for the icon of **Viewports.** Expand this icon to pick Four Viewports.

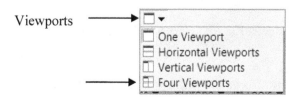

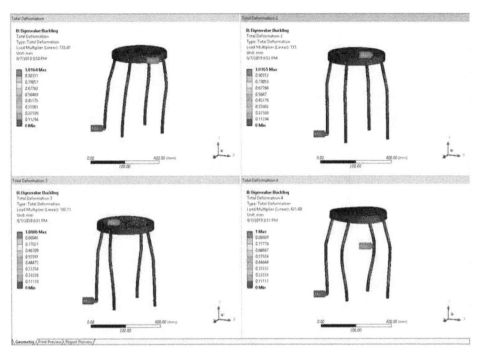

From the top menu, click **File** > **Close Mechanical**. In this way, we return to the main page of Workbench.

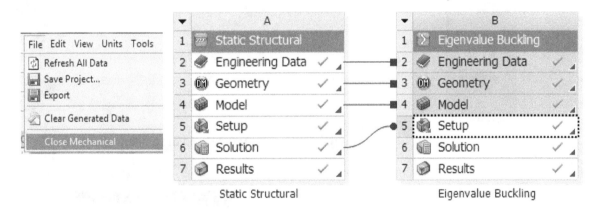

From the top menu, click **File** > **Save Project**. Afterwards, click **File** > **Exit**.

5.5   Eigenvalue Buckling Analysis of a Slender Vertical Column

A slender column is positioned vertically. Assume that the material type is structural steel. The bottom surface is fixed to the ground. As a load, a vertical force is acting on the cylindrical surface of the hole with 1000 N as its magnitude. The direction of this load is downward. Determine the maximum deformation and the maximum value of von Mises stress developed on this column. Afterwards, perform an eigenvalue buckling analysis to determine the magnitude of the critical load that may lead to the performance failure of this column under the critical load.

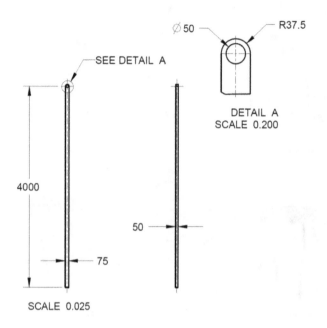

Step 1: Open Workbench.

From the start menu, click **ANSYS 19** > **Workbench 19**. Users may directly click Workbench 19 when the symbol is on display.

Step 2: From **Toolbox** > **Analysis System**, highlight **Static Structural** and drag it to the main screen.

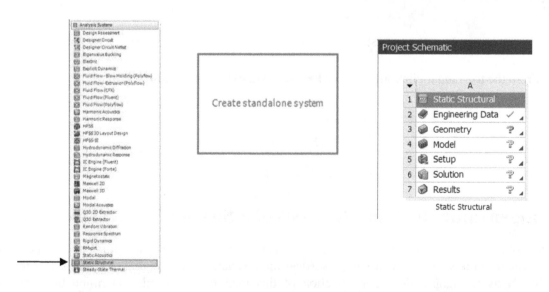

Right-click the **Geometry** cell and pick **New DesignModeler Geometry** to start **Design Modeler**. Users may double click **Geometry** to directly enter **Design Modeler** if **New SpaceClaim Geometry** is not present. Click the icon of **Units** listed on the top menu, and select **Millimeter**.

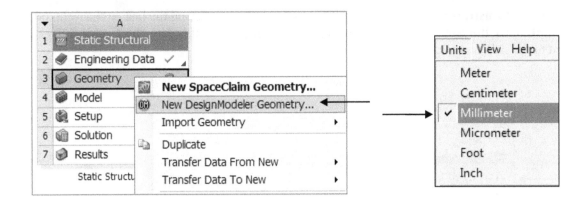

In Tree Outline, highlight **XYPlane**, and click **Look At** to orient the sketching plane.

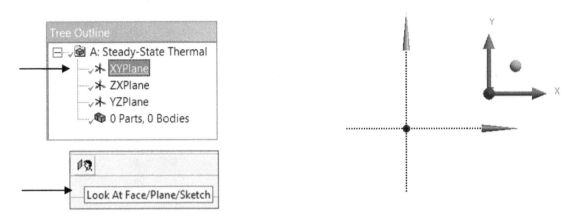

Click **Sketching**, and click **Draw**. In the Draw window, click **Circle**. Sketch a circle.

Afterwards, click **Line**. Sketch 3 lines (2 vertical lines and a horizontal line) to form a closed sketch, as shown.

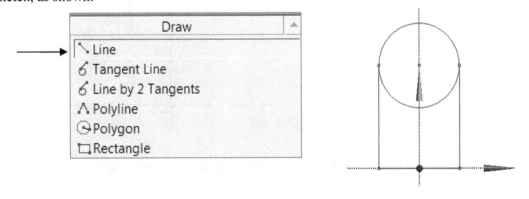

Click **Constraints**, select **Tangent**. Click the sketched circle and the line on the right-side. Afterwards, click the sketched circle and the line on the left-side. In this way, the 2 vertical lines are tangent to the sketched circle.

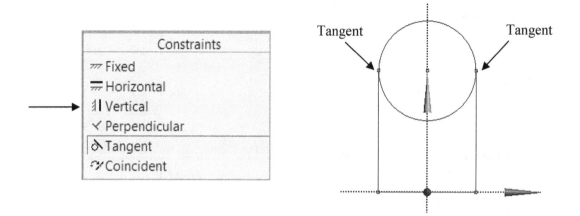

Click **Modify**, select **Trim**. Delete the half-circle, as shown.

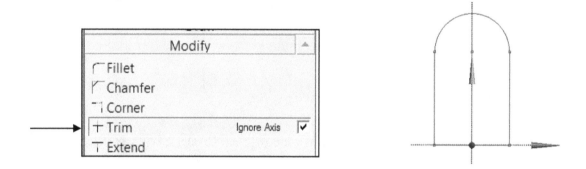

Now click **Dimensions**, click **Radius** first. Specify 37.5 as the radius value.

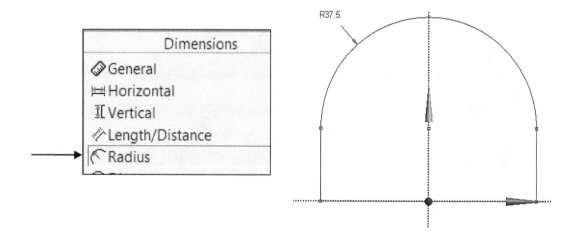

Now click **Dimensions**, click **Vertical**. Specify 4000 as the distance value between the arc center and the horizontal axis.

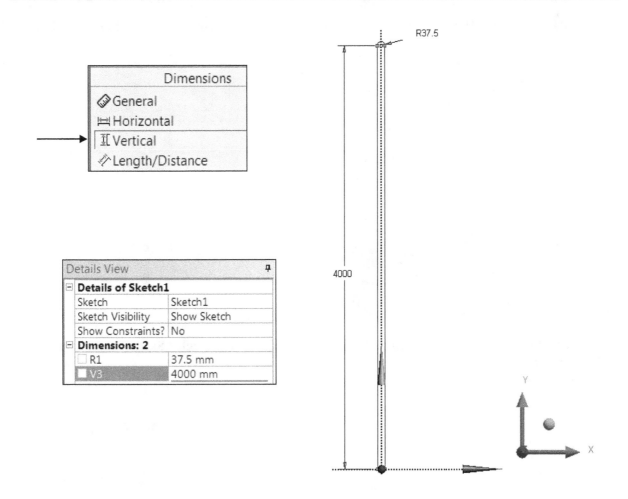

Upon completing the sketch, click the icon of **Extrude**. Click the created sketch or sketch 1 from the Tree Outline, and click **Apply**. Select Add Material. Select Both – Symmetric and specify 25 as the half-thickness value (the total is 50). Click **Generate** to obtain the 3D solid model of the vertical column, as shown.

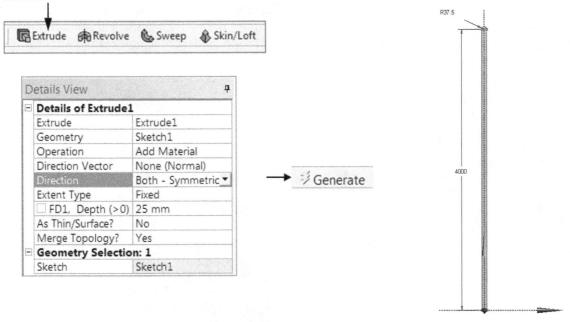

To create a hole feature, highlight **XYPlane** and click **New Sketch**. Afterwards click **Look At**.

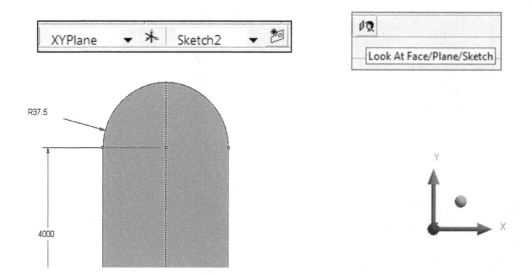

Click **Sketching**, and click **Draw**. In the Draw window, click **Circle**. Sketch a circle. Specify 50 as its diameter value.

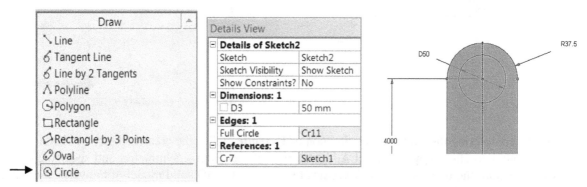

Upon completing the sketch, click the icon of **Extrude**. Click the created sketch or sketch 2 from the Tree Outline, and click **Apply**. Select Cut Material. Select Both – Symmetric and select Through All as the depth of cut choice. Click **Generate** to obtain the 3D solid model of the hole feature, as shown.

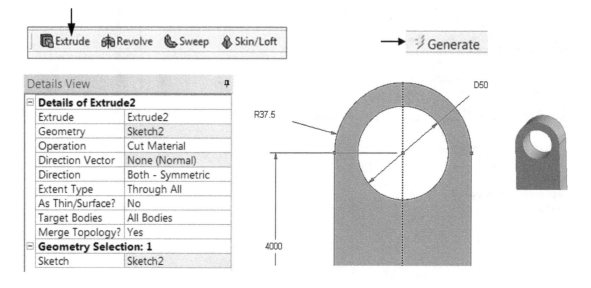

Click **File** and **Close Design Modeler**. We go back to the **Workbench** screen, or Project Schematic. Click **Save Project**. Specify **Slender Bar Buckling** as the file name and click **Save**.

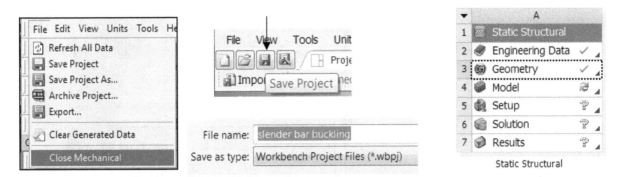

Step 3: Accept the Default Material Assignment of Structural Steel and Mesh Generation
Double-click the icon of **Model**. We enter the Mechanical Mode. Check the unit system.

For material assignment, Structural is the default material assignment. From the Project tree, Highlight **Mesh**. In the Detailed of "Mesh" window, click **Sizing**. Specify 25 as the Element Size. Right-click **Mesh** to pick **Generate Mesh**.

Details of "Mesh"	
Display	
Display Style	Body Color
Defaults	
Physics Preference	Mechanical
Element Order	Program Controlled
☐ Element Size	25.0 mm
Sizing	
Use Adaptive Sizing	Yes
Resolution	Default (2)
Mesh Defeaturing	Yes
☐ Defeature Size	Default
Transition	Fast
Span Angle Center	Coarse
Initial Size Seed	Assembly
Bounding Box Diagonal	4038.5 mm
Average Surface Area	1.4542e+005 mm²
Minimum Edge Length	50.0 mm

Step 4: Define the boundary condition and Run FEA.

Highlight **Static Structural (A5)**. Pick Insert and **Fixed Support**. Pick the bottom surface of the column bar. Pick **Apply**.

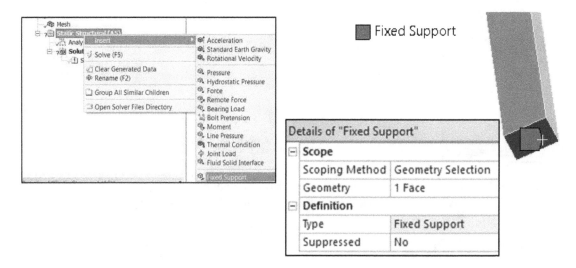

Highlight **Static Structural (A5)**. Pick Insert and **Force**. Pick the cylindrical surface of the hole feature, as shown. Pick **Apply**. Specify -1000 as the value for Y Component, and set 0 for the values for both X and Z Components.

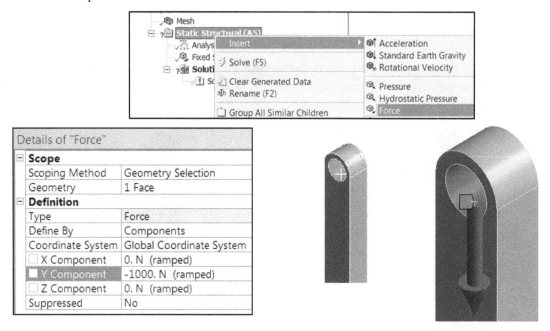

From the project tree, highlight **Solution (A6)** and right-click to pick **Solve**. Workbench system is running FEA.

At this moment, we have completed the static stress analysis. We may plot the deformation distribution and the von Mises stress distribution.

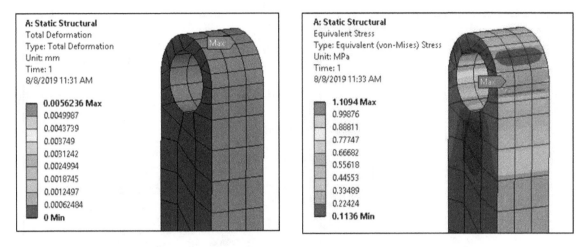

Now let us close the Mechanical Analysis. From the top menu, click **File** > **Close Mechanical**. In this way, we return to the main page of Workbench. From the top menu, click **File** > **Save Project**.

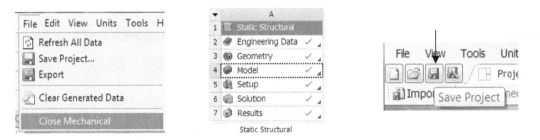

Step 5: Eigenvalue Buckling Analysis

To get on the buckling analysis, from **Toolbox** > **Analysis System**, highlight **Eigenvalue Buckling** and drag it to A6 Solution. Upon seeing Share A2:A4 Transfer A6, release the mouse button.

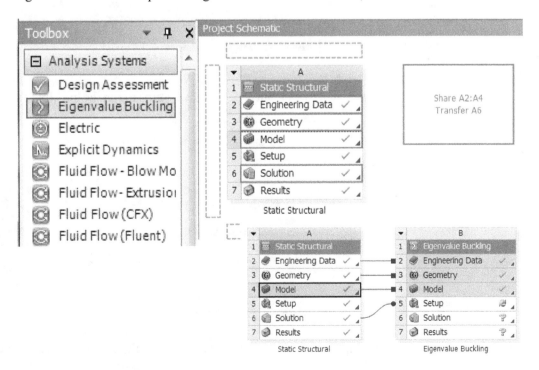

Highlight B5 Setup of Eigenvalue Buckling and right-click to pick Edit, or just double click Setup to start the buckling analysis. In the Project tree, highlight Solution (A6) and right-click to pick **Solve**.

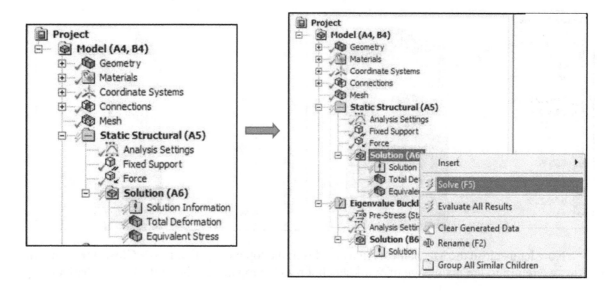

Highlight Analysis Settings listed under Eigenvalue Buckling (B5), specify 4 as Max Modes to Find.

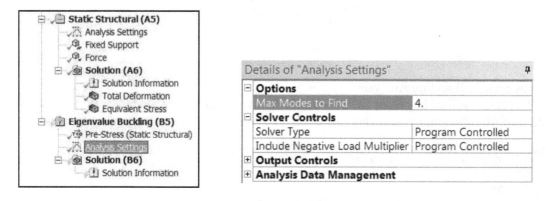

Highlight **Solution (B6)** and right-click to pick **Solve**. The system starts to run the buckling analysis.

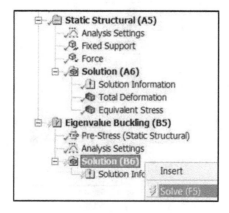

In the graph area, four load multiplier values are listed. For mode 1, the load multiplier value is 24.117.

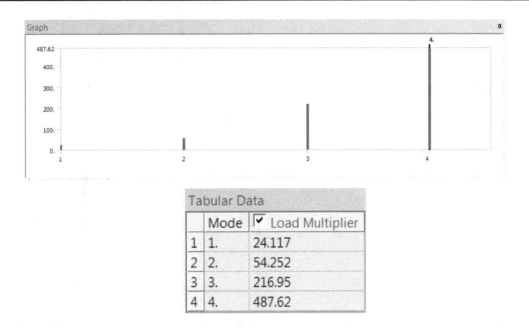

For mode 1, the load multiplier value is 24.118. The critical load, which leads to the occurrence of the first model buckling, is given by:

$$P_{critical\_load} = P_{applied\_load} * buckling\_load\_multiplier = (1000)*(24.118) = 24118 \ (N)$$

In order to visualize the mode shape for each of the 4 buckling modes, right-click in the Tabular Data area to pick Select All. Afterwards, right-click again to pick Create Mode Shape Results. In Project tree, Total Deformation, Total Deformation 2, Total Deformation 3 and Total Deformation 4 are created.

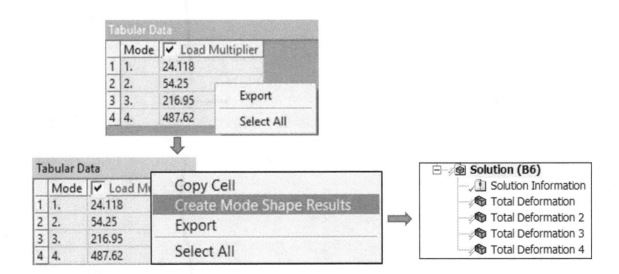

Highlight Total Deformation and right-click to pick **Evaluate All Results**. The first mode shape is on display.

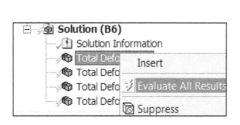

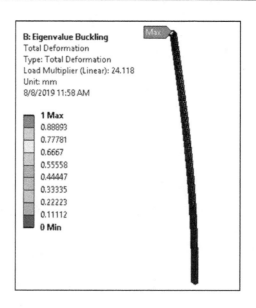

In order to display the 4 mode shapes together, go to the top menu, look for the icon of **Viewports**. Expand this icon to pick Four Viewports.

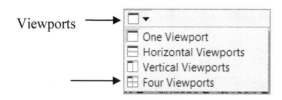

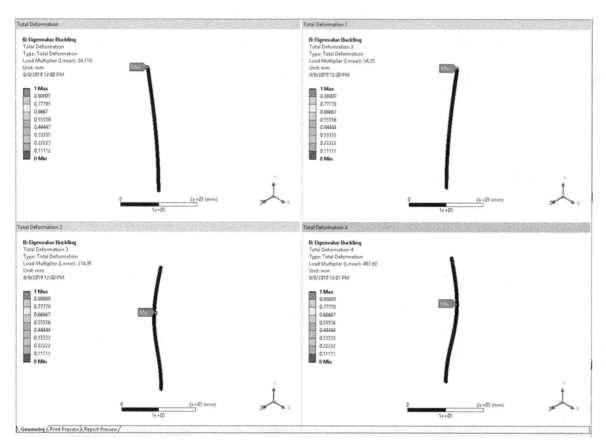

From the top menu, click **File** > **Close Mechanical**. In this way, we return to the main page of Workbench.

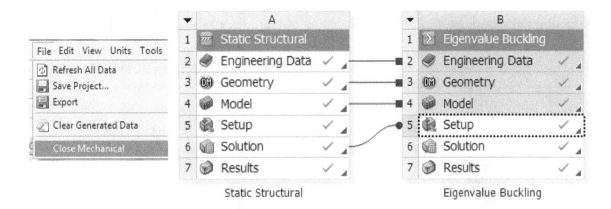

Static Structural   Eigenvalue Buckling

From the top menu, click **File** > **Save Project**. Afterwards, click **File** > **Exit**.

5.6   Vibration Analysis of a String Subjected to Excitation

Strings are widely used in music instruments, such as guitars, violins, harps, etc. In the vibration study, we model a music string as a simple support object. Assume that the string length is 1000 mm. The cross-section is a square with the side dimension equal to 1 mm. The magnitude of the tension force is 2000 N. The Young's modules value of the material type is 2e11 N/m2, and Poisson's ration is 0.30. The density of the material type is 7800 kg/m3. At the top surface of the string, create a rectangular area (0.2 mm x 1 mm) with a distance value of 200 to be used to define a vertical load of 20 N.

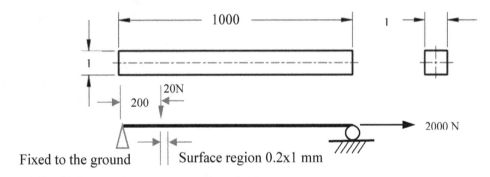

Base on the vibration theory, the natural frequencies of such a string can be determined by the following formulas. First, the linear density value is given by

$$\gamma = \rho A = (7800)(10^{-6}) = 7.8 \times 10^{-3} \quad (\text{kg/m})$$

The wave velocity value is given by

$$\alpha = \sqrt{\frac{T}{\gamma}} = \sqrt{\frac{2000}{7.8 \times 10^{-3}}} = 506.4 \quad (\text{m/s})$$

The natural frequencies are governed by the following formula:

$$f_i = \frac{i\alpha}{2L} = \frac{i \times (506.4)}{2(1)} = 253.2i \quad (\text{Hz}) \quad \text{for } i = 1,2,...$$

Calculate the first 8 mode shape frequencies, using the above formula, and list the calculated values in the table:

Mode No	1	2	3	4	5	6	7	8
Frequency (Hz) calculated	253.2	253.2	506.4	506.4	759.6	759.6	1013	1013

Step 1: Open Workbench.

From the start menu, click **ANSYS 19** > **Workbench 19**. Users may directly click Workbench 19 when the symbol is on display.

Step 2: From **Toolbox > Analysis System**, highlight **Static Structural** and drag it to the main screen.

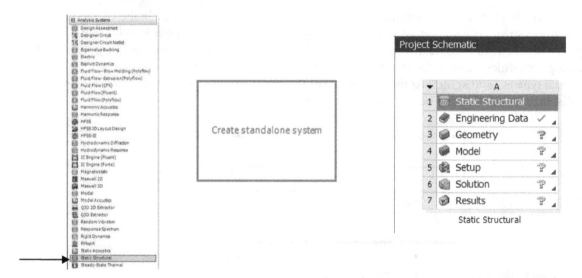

Right-click the **Geometry** cell and pick **New DesignModeler Geometry** to start **Design Modeler**. Users may double click **Geometry** to directly enter **Design Modeler** if **New SpaceClaim Geometry** is not present. Click the **Units** drop down on the top menu, and select **Millimeter**.

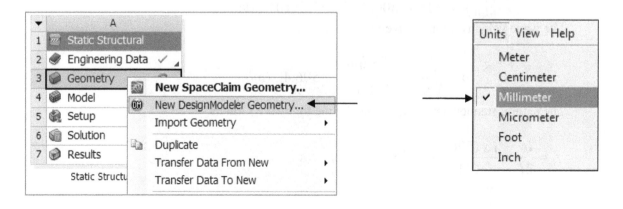

To create the first sketch, select **XYPlane** as the sketching plane.

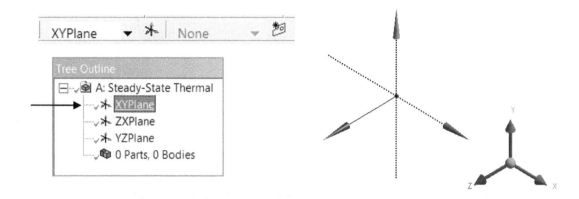

To orient the sketching plane, from the top menu, click the icon of **Look At** so that the sketching plane is parallel to the screen and ready for preparing a sketch.

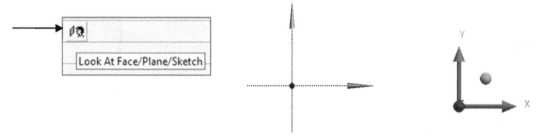

Click **Sketching**, and click **Draw**. In the Draw window, click **Rectangle**. Sketch a rectangle and add 2 symmetric constraints so that the sketched rectangle is symmetric about both X- axis and Y-axis.

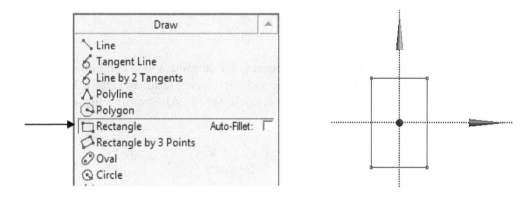

Specify the 2 size dimensions. They are 1 and 1, respectively.

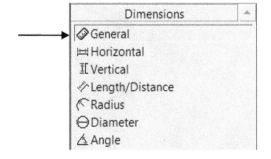

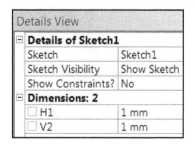

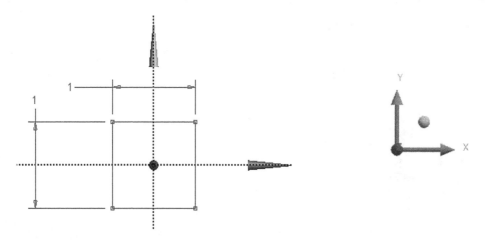

Upon completing the sketch, click the icon of **Extrude**. Click the created sketch or sketch 1 from the Tree Outline, and click **Apply**. In Operation, select Add Material. Specify 1000 as the length value. Click **Generate** to obtain the 3D solid model of the string, as shown.

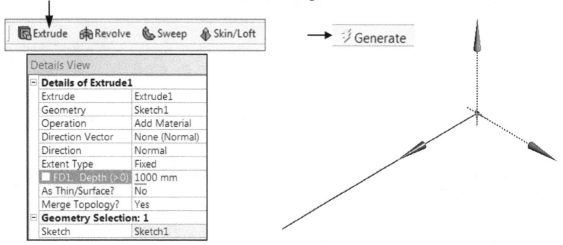

We need to create a surface region (imprint surface) for defining a load condition. First we need to create a new plane. In Tree Outline, highlight ZXPlane and click **New Plane**. In Details View, select From Face. Pick the top surface of the rectangular string and click Apply. Afterwards, click Generate. Plane4 is created.

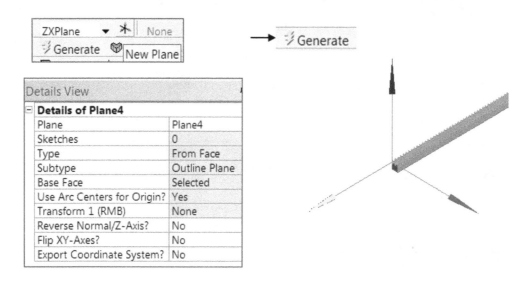

Now Plane4 is listed. **Click New Sketch > Look At**. In **Draw**, click **Rectangle**, and sketch a rectangle. Afterwards, add a position dimension of 200, and a size dimension of 0.2.

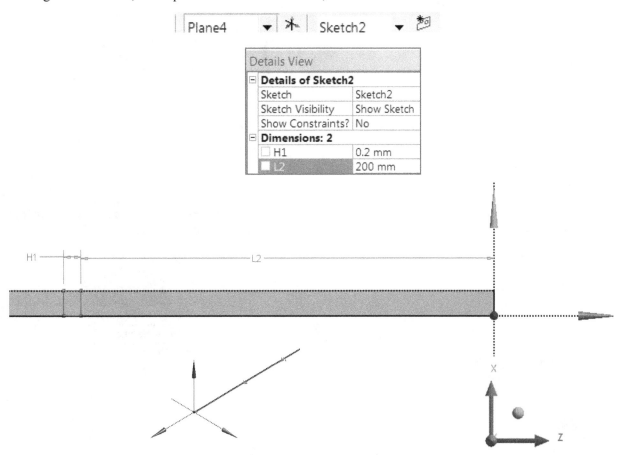

Upon completing the sketch, click the icon of **Extrude**. Click the created sketch or sketch 2 from the Tree Outline, and click **Apply**. In Operation, select **Imprint Faces**. Click **Generate** to create the imprint face to be used as a surface region for defining a load condition.

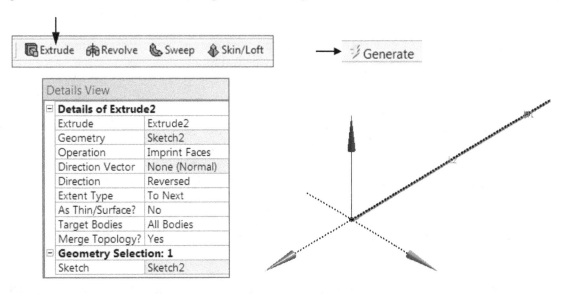

Now we have completed the creation of a rectangular string with an imprint face.

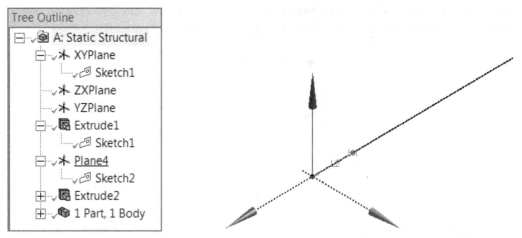

Click **File** and **Close Design Modeler**. We go back to the **Workbench** screen, or Project Schematic. Click **Save Project**. Specify **harmonic response 1** as the file name and click **Save**.

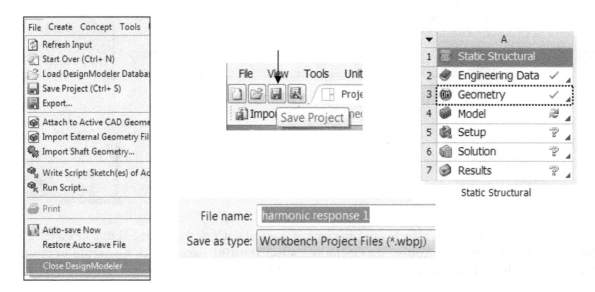

Step 3: Assign material properties, generate mesh and define the force and constraint conditions.
Double-click the icon of Model. We enter the Mechanical Mode.

Note the default assignment of material properties is Structural Steel. When the user expands **Model** > **Geometry,** and click **Solid**, Structural Steel is listed in Default of "Solid". There is no need to change the assignment of material properties.

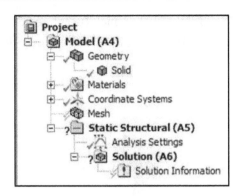

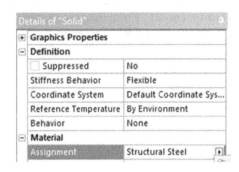

From the Project tree, Highlight **Mesh**. In the Detailed of "Mesh" window, click **Sizing**. Specify 0.75 as **Element Size**. Right-click **Mesh** again to pick **Generate Mesh**.

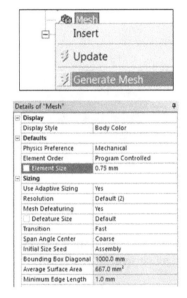

From the project tree, highlight **Static Structural (A5)** and right-click to pick **Insert**. Afterwards, select **Fixed Support**. From the screen, pick the back surface of the string, as shown. Click **Apply**. As a result, this 3D solid model is fixed to the ground through the back surface.

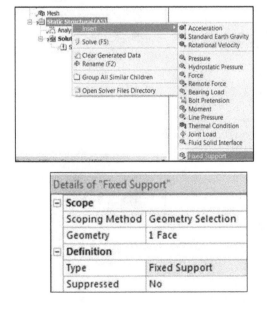

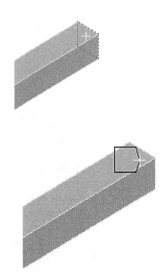

From the project tree, highlight **Static Structural (A5)** and right-click to pick **Insert**. Afterwards, select **Displacement**. From the screen, pick the front surface of the string, as shown. Click Apply. Specify 0 for X component and Y component, and set Z component free.

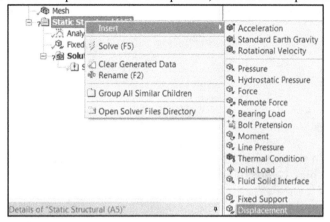

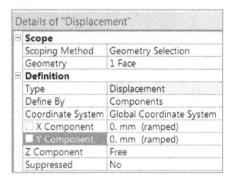

From the project tree, highlight **Static Structural (A5)** and right-click to pick **Insert**. Afterwards, select **Force**. From the screen, pick the front surface of the string, as shown. Click **Apply**. Specify 2000 N as Z Component, and set 0 for both X and Y components.

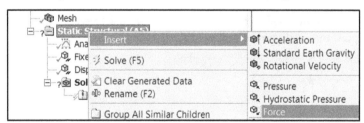

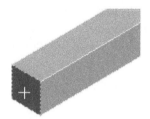

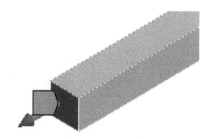

Step 4: Run FEA.

From the project tree, highlight **Solution (A6)** and right-click to pick **Solve**. Workbench system is running FEA.

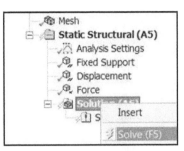

To plot the von Mises stress distribution, highlight **Solution (A6)** and pick **Insert**. Afterwards, highlight **Equivalent Stress** and pick **Equivalent (von Mises).**

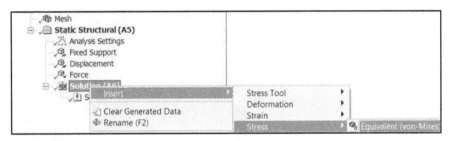

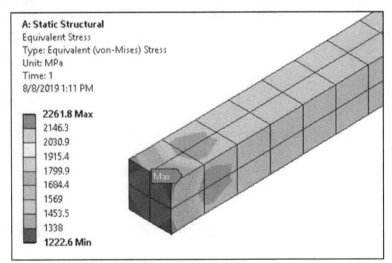

To plot the displacement distribution, highlight **Solution (A6)** and pick **Insert**. Afterwards, highlight **Deformation** and pick **Total.** Afterwards, highlight **Equivalent Stress** and pick **Equivalent (von Mises).**

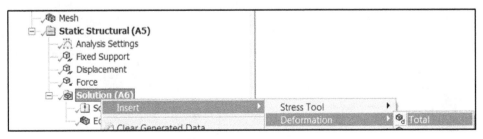

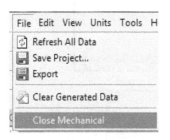

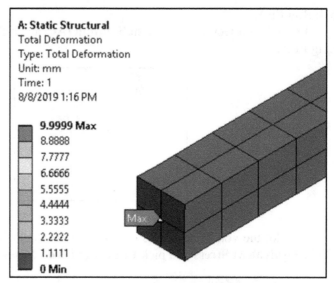

A: Static Structural
Total Deformation
Type: Total Deformation
Unit: mm
Time: 1
8/8/2019 1:16 PM

9.9999 Max
8.8888
7.7777
6.6666
5.5555
4.4444
3.3333
2.2222
1.1111
0 Min

From the top menu, click **File** > **Close Mechanical**. In this way, we return to the main page of Workbench. , click **File** > **Save.Project**.

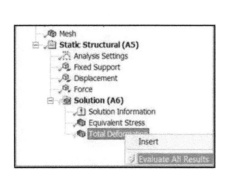

From Analysis System, highlight Modal and, using the mouse, drag it to Solution of the Static Structural panel and make sure **Share A2 to A4** is on display. Afterwards, release the mouse button. In the Project Schematic page, both Panel A and Panel B are on display.

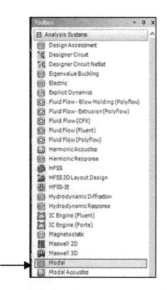

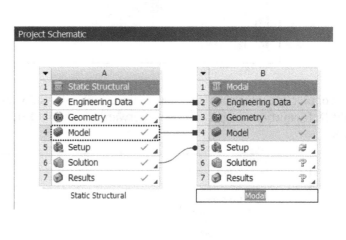

At this moment, we have completed the stress analysis. Now we work on the modal analysis. Highlight **Setup (B5)** and right-click to pick **Edit** or double click **Setup (B5)**.

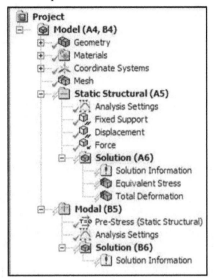

Highlight **Solution (A6)** and right-click to pick **Solve** to update the data base.

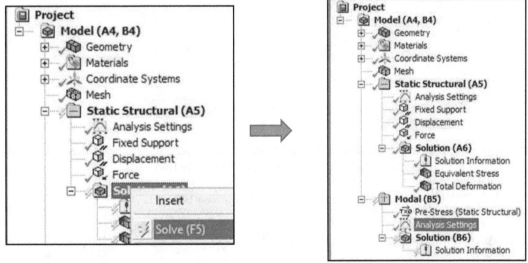

Now let us work with the modal analysis. Highlight **Analysis Settings** under **Modal (B5).** In Details of "**Analysis Settings**", specify 8 as the **Max Modes to Find**.

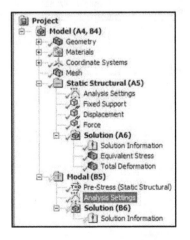

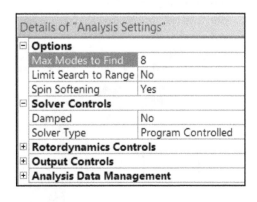

Details of "Analysis Settings"	
Options	
Max Modes to Find	8
Limit Search to Range	No
Spin Softening	Yes
Solver Controls	
Damped	No
Solver Type	Program Controlled
Rotordynamics Controls	
Output Controls	
Analysis Data Management	

Highlight **Solution (B6)** and right-click to pick **Solve**. The natural frequencies of the first 8 modes are listed.

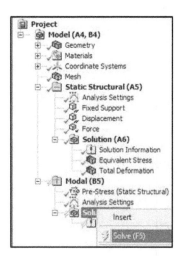

	Mode	☑ Frequency [Hz]
1	1.	253.12
2	2.	253.12
3	3.	506.31
4	4.	506.31
5	5.	759.61
6	6.	759.61
7	7.	1013.1
8	8.	1013.1

In the Tabular Area, right-click to pick Select All. Afterwards, right-click to pick Create Mode Shape Results.

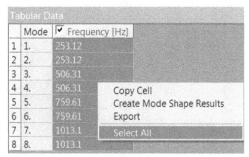
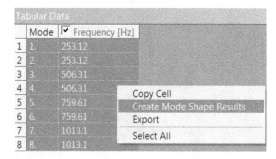

In the project tree, there are 8 Total Deformation items are listed. Select one of them and right-click to pick Evaluation All Results. Click Graph to show the 8 mode listing.

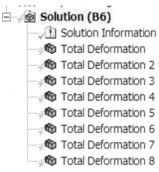
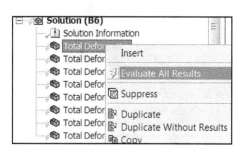

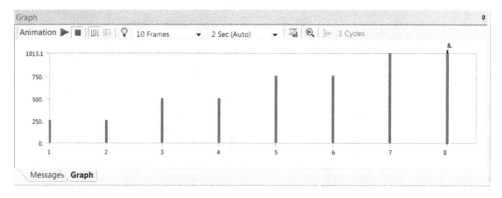

Users may plot 4 mode shapes in a single graph. To do so, go to the top menu, look for the icon of **Viewpoints**. Expand this icon to pick Four Viewports.

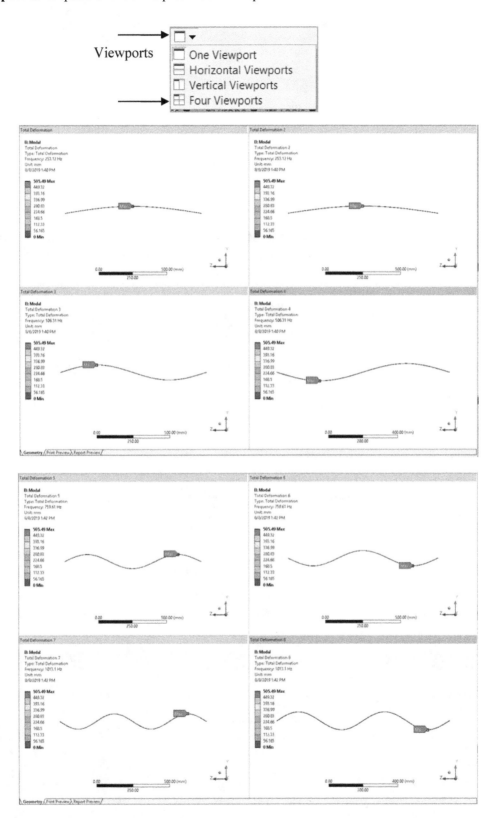

From the top menu, click **File** > **Close Mechanical**. In this way, we return to the main page of Workbench. From the top menu, click **File** > **Save Project**.

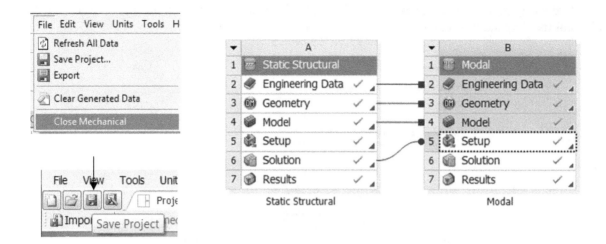

From Analysis System, highlight **Harmonic Response** and, using the mouse, drag it to Solution of the Modal panel and make sure **Share B2 to B4** is on display. Afterwards, release the mouse button. In the Project Schematic page, Panel A, Panel B and Pane C are all on display.

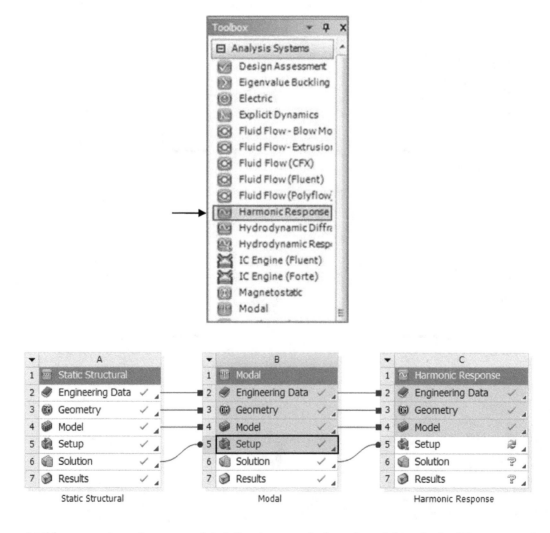

At this moment, we have completed the stress analysis and modal analysis. Now we work on the harmonic response analysis. Highlight **Setup (C5)** and right-click to pick **Edit** or double click **Setup (C5)**.

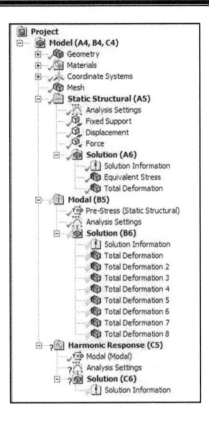

Highlight **Modal (B5)** and right-click to pick **Solve** to update the data base.

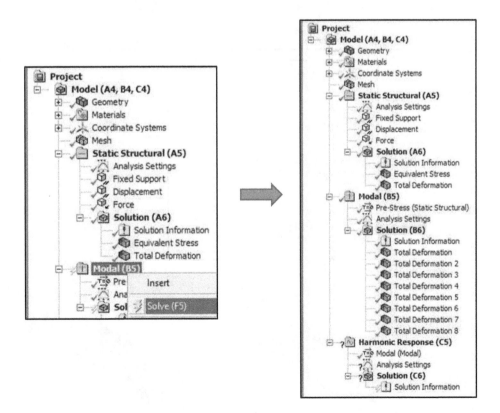

Now let us work with the harmonic response analysis. Highlight **Analysis Settings** under Harmonic Response **(C5).** In Details of "Analysis Settings", specify 0 as Range Minimum, 600 as Range Maximum and 50 as Solution Interval.

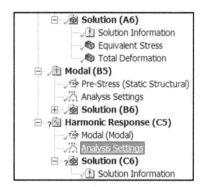

Details of "Analysis Settings"	
Options	
Frequency Spacing	Linear
☐ Range Minimum	0. Hz
☐ Range Maximum	600. Hz
☐ Solution Intervals	50
User Defined Frequencies	Off
Solution Method	Mode Superposition
Include Residual Vector	No
Cluster Results	No
Store Results At All Frequencies	Yes

Highlight Harmonic Response **(C5)** and right-click to pick **Insert > Force**. Pick the surface region or imprint face. Click **Apply**. Specify 20 N as Y Component and 30 as Y Phase Angle.

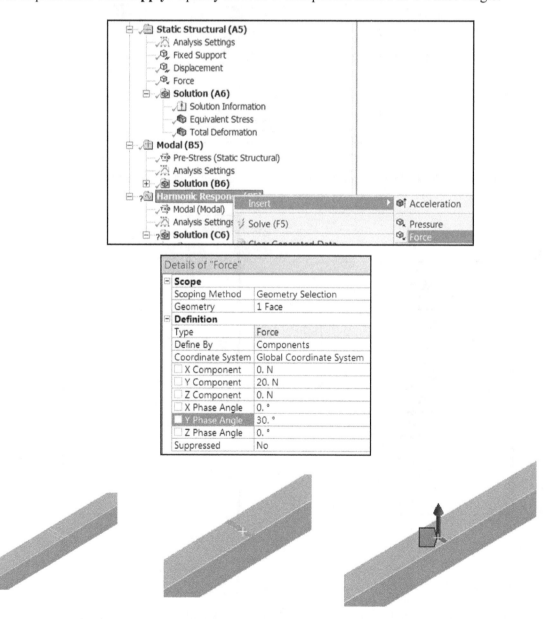

Highlight **Solution (C6)** and right-click to pick **Solve.**

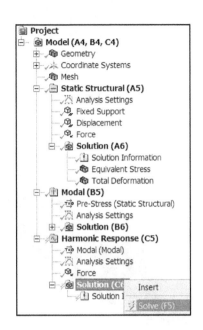

Tabular Data

	Set	☑ Frequency [Hz]
1	1.	12.
2	2.	24.
3	3.	36.
4	4.	48.
5	5.	60.
6	6.	72.
7	7.	84.
8	8.	96.
9	9.	108.
10	10.	120.
11	11.	132.
12	12.	144.
13	13.	156.
14	14.	168.
15	15.	180.
16	16.	192.
17	17.	204.
18	18.	216.
19	19.	228.
20	20.	240.
21	21.	252.
22	22.	264.
23	23.	276.
24	24.	288.
25	25.	300.
26	26.	312.

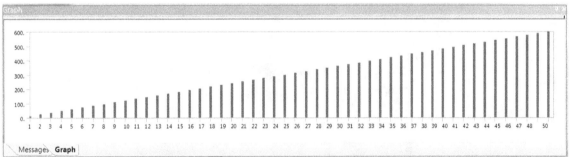

Highlight **Solution (C6)** and pick **Insert**. Afterwards, highlight **Deformation** and pick **Total.** Afterwards, click **Evaluate All Results**.

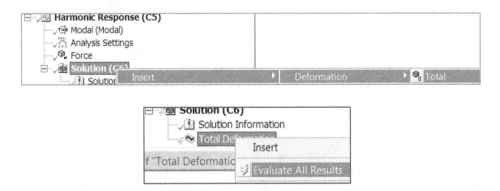

The deformation distribution on display represents the deformation under the excitation of 600 Hz and 0 degrees of phase angle. The maximum deformation is 1.26 mm, and located around 800 mm with respect to the left end.

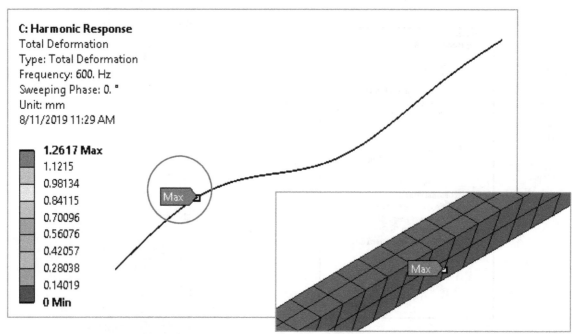

Highlight **Solution (C6)** and pick **Insert > Frequency Response > Deformation**. Pick the string and click Apply. In Orientation, pick Y-Axis. Afterwards, highlight **Deformation** and pick **Total**. Afterwards, click **Evaluate All Results**.

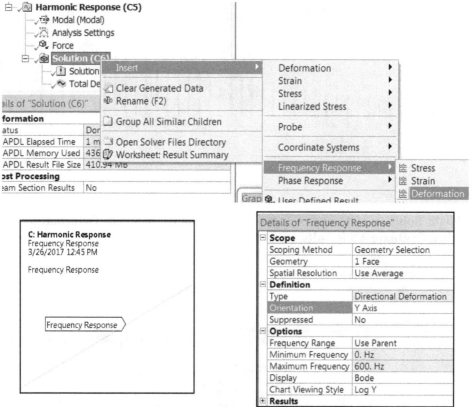

There are 2 graphs on display. The first graph presents the frequency response curve with the frequency range from 0 Hz to 600 Hz. The frequency interval is 12 Hz. Note there are 2 peaks located at 252 Hz and 504 Hz. These two frequency values match the natural frequency of the first mode and the natural frequency of the second mode of this string under the pre-stress condition. The second graph presents the phase angle vs the frequency of excitation. Note that there are two (2) phase shift angles of 180° from 30° to -150°, located at 252 Hz and 504 Hz where system resonances occur under excitation.

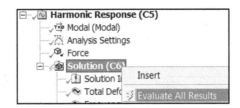

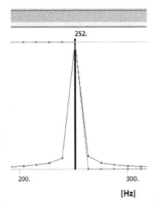

	Frequency [Hz]	☑ Amplitude [mm]	☑ Phase Angle [°
17	204.	2.8117	30.
18	216.	3.4067	30.
19	228.	4.5621	30.
20	240.	7.8101	30.
21	252.	80.214	30.
22	264.	7.0902	-150.
23	276.	2.7846	-150.

Frequency Response Curve

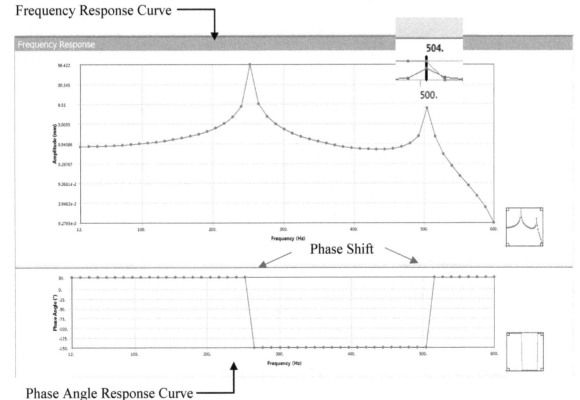

Phase Shift

Phase Angle Response Curve

In order to display the stress distribution, say von Mises stress distribution at the frequency setting of 252 Hz, highlight Solution (C6) and right-click to pick Insert > Stress > Equivalent (von Moses). Afterward, click **Evaluate All Results**. Highlight Equivalent Stress under Solution (C6). In the Graph area, select No 21, which corresponds to 252 Hz. Right-click to pick Retrieve This Results. The von Mises stress distribution under excitation at 252 Hz is shown. The maximum value is 10786 MPa.

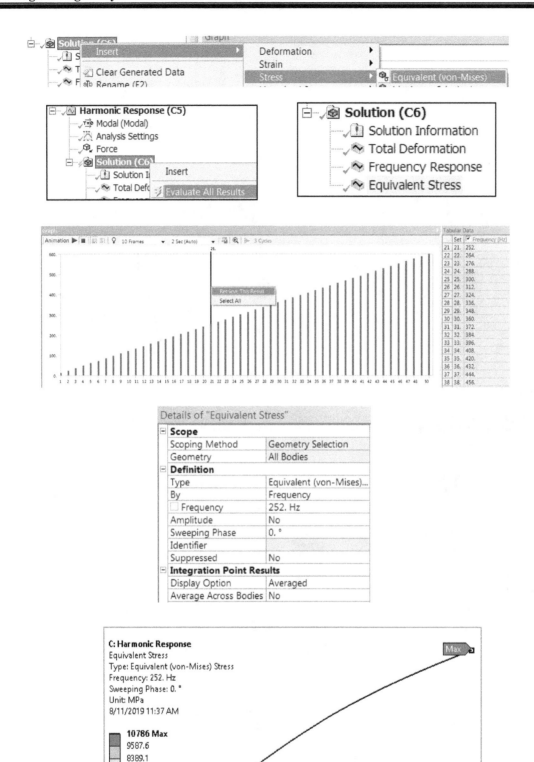

Now let us choose No. 10, which corresponds to 120 Hz. Right click to pick Retrieve This Results. The von Mises stress distribution under excitation at 120 Hz is shown. The maximum value is 279 MPa.

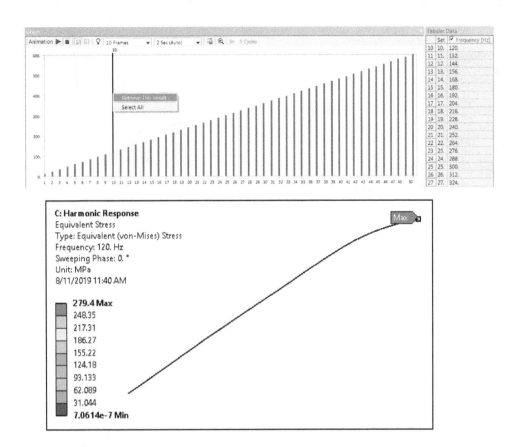

There is a significant difference between the 2 values of von Mises stress with a static load of 20 N and under the excitation with the magnitude equal to 20 N. When the excitation frequency approaches to 252 Hz, the natural frequency of the first mode of this music string under the current pre-stress condition, the resonance occurs, magnifying the values of von Mises stress. Therefore, engineers should avoid the working condition where the excitation frequency is close to 252 Hz.

It is important to note that the setting of mode numbers when performing the Modal Analysis was 8. As a result, the upper limit of the frequency on display was 1013.1 Hz. If dividing 1013.1 Hz by 1.5, a constant, we have 675.4 Hz. This is the reference number we have used when defining the upper limit of frequency in the process of setting up the upper limit of the frequency range. It is the common sense that the response frequency range should be less than the natural frequency of the upper limit of the frequency range obtained during the modal analysis. The constant of 1.5 is a general rule based on the system default settings.

5.7   Study of Transient Response of a Cam-Follower Mechanism

The following Cam-Follow mechanism is taken from a textbook [2]. The rotating cam drives the follower to perform a translation motion along the horizontal axis. Assume the rotation speed is 60 RPM, or one revolution per minute (or 60 seconds). The displacement trajectory as a function of rotation time is shown below with the maximum displacement equal to 25 mm. The force acting at the tip of the follower is equal to 1000 N. In the design process, engineers first design a displacement trajectory of the cam component. Engineers position the cam and the follower positions relative to each other, for example, the two centerlines are aligned horizontally. Afterwards, engineers want to know the velocity and acceleration trajectories as a function of time. More important is engineers want to know the transient and steady-state stages in operation. The objective of this study is to obtain the information related to the dynamic nature of motion during the design stage. The Transient Structural Analysis provides a unique platform for design engineers to gain a deep understanding in the design process.

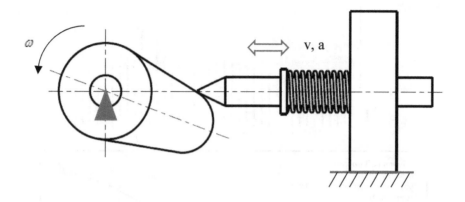

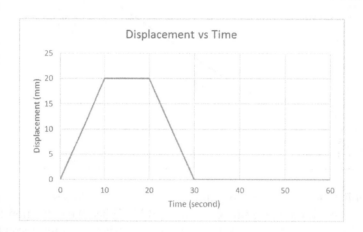

Step 1: Open Workbench.
From the start menu, click **ANSYS 19 > Workbench 19**. Users may directly click Workbench 19 when the symbol is on display.

Step 2: Start Transient Structural Analysis.
From **Toolbox > Analysis System**, highlight **Transient Structural** and drag it to the main screen.

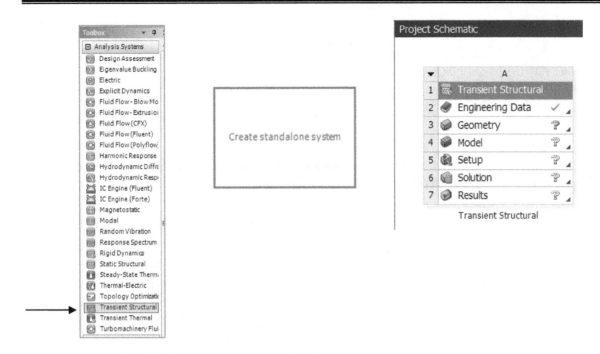

Right-click the **Geometry** cell and pick **New DesignModeler Geometry** to start **Design Modeler**. Users may double click **Geometry** to directly enter **Design Modeler** if **New SpaceClaim Geometry** is not present. Click the **Units** drop down on the top menu, and select **Millimeter**.

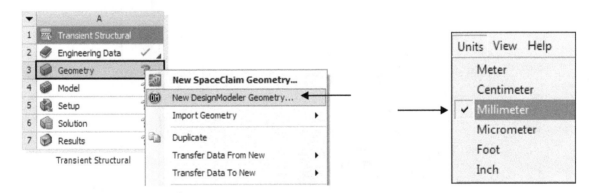

To create the first sketch, select **XYPlane** as the sketching plane.

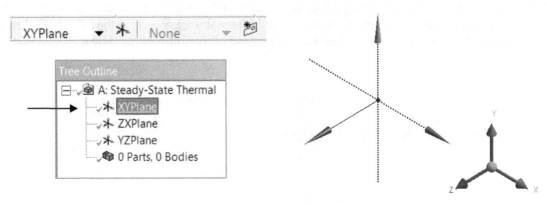

To orient the sketching plane, from the top menu, click the icon of **Look At** so that the sketching plane is parallel to the screen and ready for preparing a sketch.

Click **Sketching**, and click **Draw**. In the Draw window, click **Polyline**. Sketch four lines to form a pin shape, as shown. Right-click on the screen to pick **Closed End**.

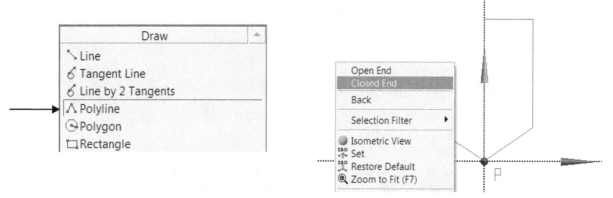

Specify the 3 size dimensions. They are 50, 10 and 10, respectively.

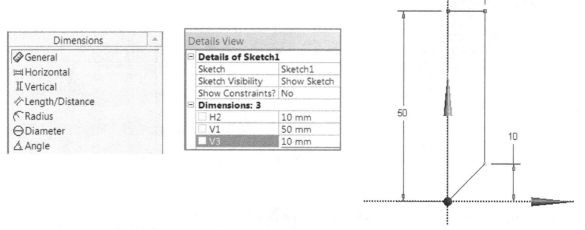

Upon completing the sketch, click the icon of **Revolve**. Click the created sketch or sketch 1 from the Tree Outline, and click **Apply**. In Operation, select Add Material. Pick Y-axis as the axis of revolution. Click **Generate** to obtain the 3D solid model of the contact head of the follower, as shown.

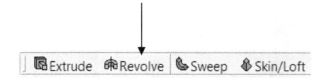

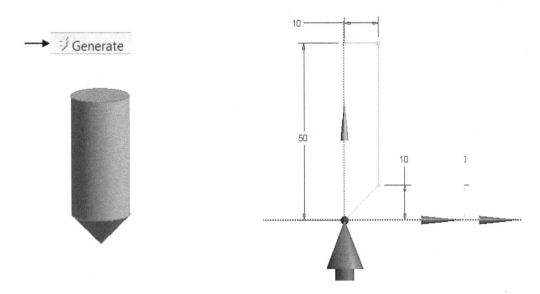

Click **File** and **Close Design Modeler**. We go back to the **Workbench** screen, or Project Schematic. Click **Save Project**. Specify **Transient Dynamics** as the file name and click **Save**.

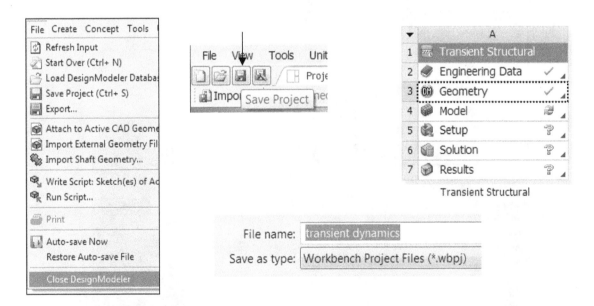

Step 3: Assign Structural Steel to the 3D solid Model
Double-click the icon of Model. We enter the Mechanical Mode. Check the unit system.

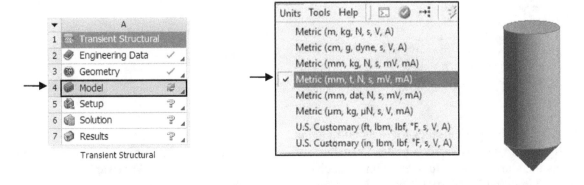

Now let us do the material assignment. From the Project tree, expand Geometry. Highlight Solid. In Details of "Solid", Structure Steel is listed in **Material Assignment**, which is the default material assignment.

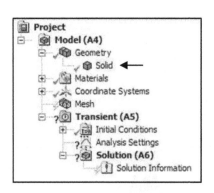

Details of "Solid"	
⊞ **Graphics Properties**	
⊟ **Definition**	
☐ Suppressed	No
Stiffness Behavior	Flexible
Coordinate System	Default Coordinate System
Reference Temperature	By Environment
Behavior	None
⊟ **Material**	
Assignment	Structural Steel
Nonlinear Effects	Yes
Thermal Strain Effects	Yes

From the Project tree, Highlight **Mesh**. In the Detailed of "Mesh" window, click **Sizing**. Set Resolution to 4. Right-click **Mesh** again to pick **Generate Mesh**.

Details of "Mesh"	
⊟ **Display**	
Display Style	Body Color
⊟ **Defaults**	
Physics Preference	Mechanical
Element Order	Program Controlled
☐ Element Size	Default
⊟ **Sizing**	
Use Adaptive Sizing	Yes
Resolution	4
Mesh Defeaturing	Yes
☐ Defeature Size	Default
Transition	Fast
Span Angle Center	Coarse
Initial Size Seed	Assembly
Bounding Box Diagonal	602.41 mm
Average Surface Area	27862 mm²
Minimum Edge Length	20.0 mm

Step 4: Define the integration steps for the load function.

Highlight **Analysis Settings** under **Transient (A5)**. First, specify 4 as the Number Of Steps. Second, specify 4 as the Current Step Number. Third, specify 60 as the Step End Time. Afterwards, specify 0.5 as the Initial Time Step. Specify 0.1 as the Minimum Time Step. Specify 1 as the Maximum Time Step.

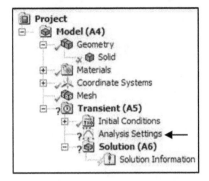

	Steps	End Time [s]
1	1	1.
2	2	2.
3	3	3.
4	4	60.
*		

Tabular Data

Details of "Analysis Settings"

Step Controls
Number Of Steps	4.
Current Step Number	4.
Step End Time	60. s
Auto Time Stepping	On
Define By	Time
Carry Over Time Step	Off
Initial Time Step	0.5 s
Minimum Time Step	0.1 s
Maximum Time Step	1. s
Time Integration	On

Solver Controls
Solver Type	Program Controlled
Weak Springs	Off
Large Deflection	On

Highlight **Analysis Settings** under **Transient (A5)**. First, specify 4 as the Number Of Steps. Second, specify 3 as the Current Step Number. Third, specify 30 as the Step End Time. Afterwards, specify 0.5 as the Initial Time Step. Specify 0.1 as the Minimum Time Step. Specify 1 as the Maximum Time Step.

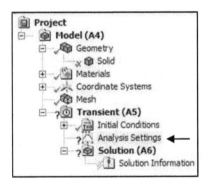

Tabular Data

	Steps	End Time [s]
1	1	1.
2	2	2.
3	3	30.
4	4	60.
*		

Details of "Analysis Settings"

Step Controls
Number Of Steps	4.
Current Step Number	3.
Step End Time	30. s
Auto Time Stepping	On
Define By	Time
Carry Over Time Step	Off
Initial Time Step	0.5 s
Minimum Time Step	0.1 s
Maximum Time Step	1. s
Time Integration	On

Solver Controls
Solver Type	Program Controlled
Weak Springs	Off
Large Deflection	On

Highlight **Analysis Settings** under **Transient (A5)**. First, specify 4 as the Number Of Steps. Second, specify 2 as the Current Step Number. Third, specify 20 as the Step End Time. Afterwards, specify 0.5 as the Initial Time Step. Specify 0.1 as the Minimum Time Step. Specify 1 as the Maximum Time Step.

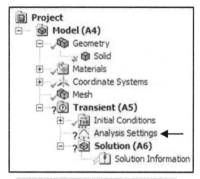

	Steps	End Time [s]
1	1	1.
2	2	20.
3	3	30.
4	4	60.
*		

Tabular Data

Details of "Analysis Settings"

Step Controls	
Number Of Steps	4.
Current Step Number	2.
Step End Time	20. s
Auto Time Stepping	On
Define By	Time
Carry Over Time Step	Off
Initial Time Step	0.5 s
Minimum Time Step	0.1 s
Maximum Time Step	1. s
Time Integration	On
Solver Controls	
Solver Type	Program Controlled
Weak Springs	Off
Large Deflection	On

Highlight **Analysis Settings** under **Transient (A5)**. First, specify 4 as the Number Of Steps. Second, specify 1 as the Current Step Number. Third, specify 10 as the Step End Time. Afterwards, specify 0.5 as the Initial Time Step. Specify 0.1 as the Minimum Time Step. Specify 1 as the Maximum Time Step.

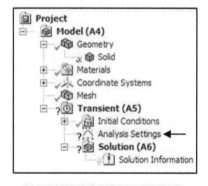

	Steps	End Time [s]
1	1	10.
2	2	20.
3	3	30.
4	4	60.
*		

Tabular Data

Details of "Analysis Settings"

Step Controls	
Number Of Steps	4.
Current Step Number	1.
Step End Time	10. s
Auto Time Stepping	On
Define By	Time
Initial Time Step	0.5 s
Minimum Time Step	0.1 s
Maximum Time Step	1. s
Time Integration	On
Solver Controls	
Solver Type	Program Controlled
Weak Springs	Off
Large Deflection	On

Step 5: Define the load condition.

Highlight **Transient (A5)**, and right-click to pick **Insert > Force**. Pick the top surface from the follow head. Click **Apply**. Specify -1000 as the magnitude along the Y Direction.

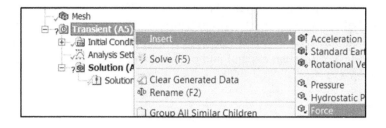

Details of "Force"

Scope	
Scoping Method	Geometry Selection
Geometry	1 Face
Definition	
Type	Force
Define By	Components
Coordinate System	Global Coordinate System
☐ X Component	0. N (step applied)
■ Y Component	-1000. N (step applied)
☐ Z Component	0. N (step applied)
Suppressed	No

Tabular Data

	Steps	Time [s]	☑ X [N]	☑ Y [N]	☑ Z [N]
1	1	0.	= 0.	= -1000.	= 0.
2	1	10.	0.	-1000.	0.
3	2	20.	= 0.	= -1000.	= 0.
4	3	30.	= 0.	= -1000.	= 0.
5	4	60.	= 0.	= -1000.	= 0.
*					

Step 6: Define the constraint for the motion of the follower head.

Highlight **Transient (A5)**, and right-click to pick **Insert > Displacement**. Pick the cylindrical surface from the follow head. Click **Apply**. Constrain the motion in both the X and Z directions, and set the motion in the Y direction free.

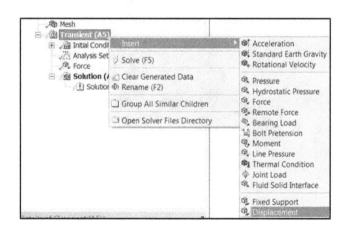

Details of "Displacement"

Scope	
Scoping Method	Geometry Selection
Geometry	1 Face
Definition	
Type	Displacement
Base Excitation	No
Define By	Components
Coordinate System	Global Coordinate System
☐ X Component	0. m (step applied)
Y Component	Free
■ Z Component	0. m (step applied)
Suppressed	No

Tabular Data

	Steps	Time [s]	☑ X [m]	☑ Z [m]
1	1	0.	= 0.	= 0.
2	1	10.	0.	0.
3	2	20.	= 0.	= 0.
4	3	30.	= 0.	= 0.
5	4	60.	= 0.	= 0.
*				

Step 6: Define the displacement trajectory of the tip of the follower head.

Highlight **Transient (A5)**, and right-click to pick **Insert > Displacement**. Pick the point or the tip of the follower head. Users may need to check the icon of Vertex before picking the point. After picking the point, click Apply. For the displacement in Y direction, expand and pick Tabular Data. In the Tabular Data window, specify 0.02 in Step 2 and Step 3.

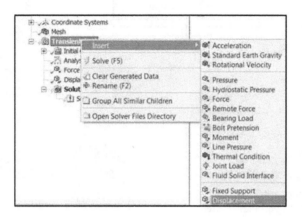

Details of "Displacement 2"

Scope	
Scoping Method	Geometry Selection
Geometry	1 Vertex
Definition	
Type	Displacement
Base Excitation	No
Define By	Components
Coordinate System	Global Coordinate System
X Component	Free
Y Component	Tabular Data
Z Component	Free
Suppressed	No

Tabular Data

	Steps	Time [s]	✔ Y [m]
1	1	0.	0.
2	1	10.	2.e-002
3	2	20.	2.e-002
4	3	30.	0.
5	4	60.	= 0.
*			

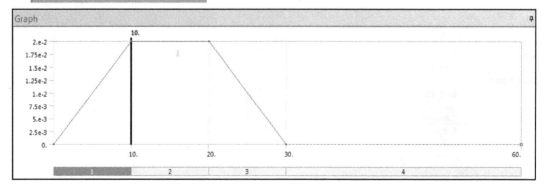

Step 7: Define the system outputs for recording the data obtain from the FEA run.

Highlight **Solution (A6)**, and right-click to pick **Insert > Deformation > Directional**. Pick the point or the tip of the follower head. Click **Apply**. In Orientation, select Y Axis.

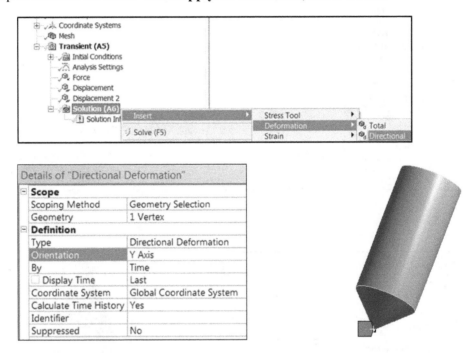

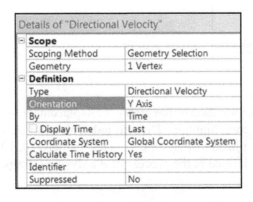

Highlight **Solution (A6)**, and right-click to pick **Insert > Deformation > Directional Velocity**. Pick the point or the tip of the follower head. Click **Apply**. In Orientation, select Y Axis.

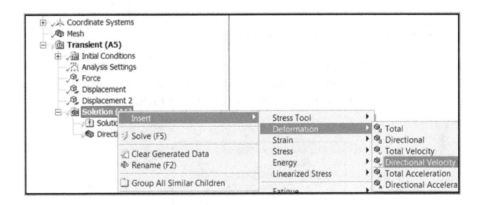

Highlight **Solution (A6)**, and right-click to pick **Insert > Deformation > Directional Acceleration**. Pick the point or the tip of the follower head. Click **Apply**. In Orientation, select Y Axis.

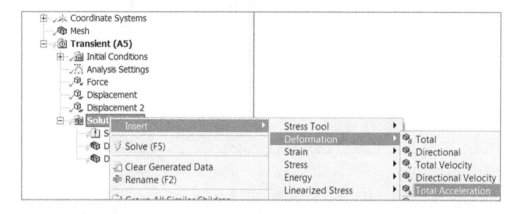

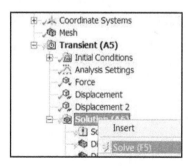

Step 7: Run FEA and review the results.

At this moment, we have completed the stress analysis. From the project tree, highlight **Solution (A6)** and right-click to pick **Solve**. Workbench system is running FEA.

Let us first to examine the graph of Directional Deformation or the graph of Displacement vs Time as defined in the textbook of Dynamics. In the Project tree, highlight **Directional Deformation** under **Solution (A6).** The displacement trajectory of the follower head displays the 4 stages, namely, [0, 10], [1, 20], [20, 30], and [30, 60]. This pattern matches what we defined in the process of **Analysis Settings.**

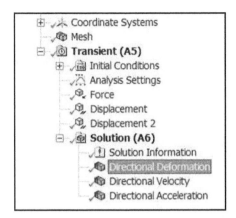

	Time [s]	☑ Minimum [mm]	☑ Maximum	▲
1	0.5	1.	1.	
2	0.75	1.5	1.5	
3	1.	2.	2.	
4	1.375	2.75	2.75	
5	1.9375	3.875	3.875	
6	2.7813	5.5625	5.5625	
7	3.7813	7.5625	7.5625	
8	4.7813	9.5625	9.5625	
9	5.7813	11.563	11.563	
10	6.7813	13.563	13.563	
11	7.7813	15.563	15.563	
12	8.7813	17.563	17.563	
13	9.3906	18.781	18.781	

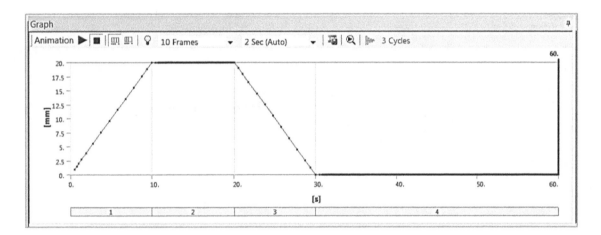

Let us first to examine the graph of Directional Velocity or the graph of Velocity vs Time as defined in the textbook of Dynamics. In the Project tree, highlight **Directional Velocity** under **Solution (A6).** The velocity trajectory of the follower head displays the 4 stages, namely, [0, 10], [10, 20], [20, 30], and [30, 60]. This pattern matches what we defined in the process of **Analysis Settings**. The oscillation of the magnitude of velocity clearly depicts the transient process of the follower head motion to its steady-state status.

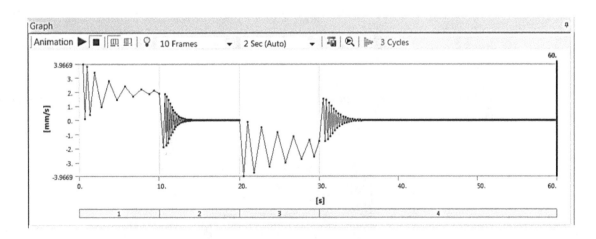

Tabular Data			
	Time [s]	☑ Minimum [mm/s]	☑ Maximun
1	0.5	3.9669	3.9669
2	0.75	9.289e-002	9.289e-002
3	1.	3.804	3.804
4	1.375	0.37406	0.37406
5	1.9375	3.3799	3.3799
6	2.7813	0.92408	0.92408
7	3.7813	2.7958	2.7958
8	4.7813	1.418	1.418
9	5.7813	2.4197	2.4197
10	6.7813	1.7029	1.7029
11	7.7813	2.2052	2.2052
12	8.7813	1.863	1.863
13	9.3906	2.1055	2.1055

The graph of Acceleration vs Time curve is shown below. The oscillation pattern clearly depicts that the transient and steady-state stages of the cam-follower system during the motion.

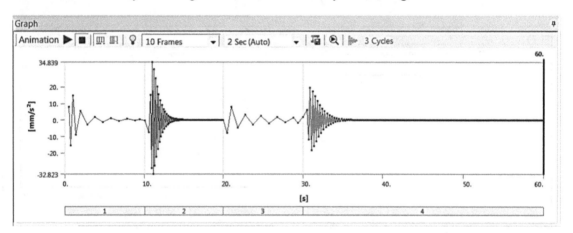

Tabular Data			
	Time [s]	☑ Minimum [mm/s²]	☑ Maximu
1	0.5	7.9339	7.9339
2	0.75	-15.496	-15.496
3	1.	14.844	14.844
4	1.375	-9.1465	-9.1465
5	1.9375	5.3438	5.3438
6	2.7813	-2.9107	-2.9107
7	3.7813	1.8718	1.8718
8	4.7813	-1.3779	-1.3779
9	5.7813	1.0017	1.0017
10	6.7813	-0.7168	-0.7168
11	7.7813	0.50237	0.50237
12	8.7813	-0.34222	-0.34222
13	9.3906	0.29784	0.29784

The FEA run also provides sets of numerical data, which can be used for the design process. To save those data sets, users may just right-click in the Tabular Data area to pick Export. Afterwards, users may save the data set in Excel File (*xls) format.

Tabular Data			
	Time [s]	☑ Minimum [mm/s²]	☑ Maximu
1	0.5	7.9339	7.9339
2	0.75	-15.496	Retrieve This Result
3	1.	14.844	Export
4	1.375	-9.1465	Select All
5	1.9375	5.3438	

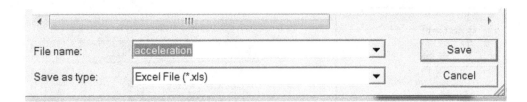

Let us modify the cam design. The original design is shown on the left side and the new design is shown in the right side. The 2 displacement vs time curves are also shown.

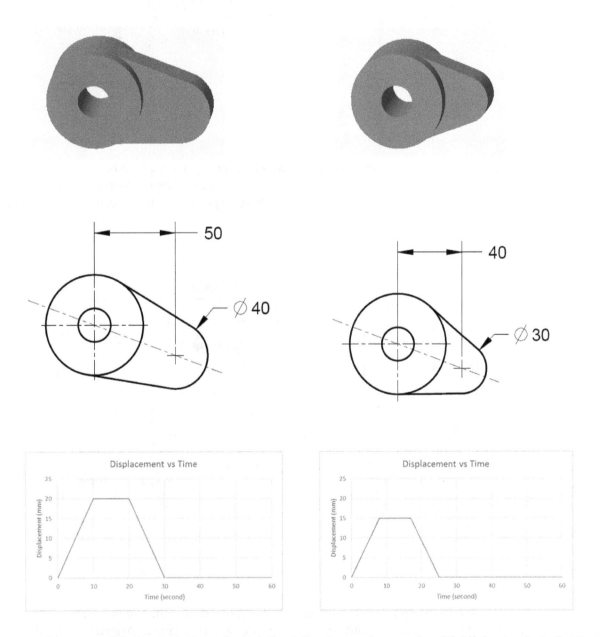

Let us modify the analysis settings defined for the previous study. Highlight **Analysis Settings** under **Transient (A5)**. First, specify 4 as the Number Of Steps. Second, specify 4 as the Current Step Number. Third, specify 60 as the Step End Time. Afterwards, specify 0.5 as the Initial Time Step. Specify 0.1 as the Minimum Time Step. Specify 1 as the Maximum Time Step.

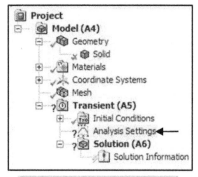

	Steps	End Time [s]
1	1	10.
2	2	20.
3	3	30.
4	4	60.
*		

Highlight **Analysis Settings** under **Transient (A5)**. First, specify 4 as the Number Of Steps. Second, specify 3 as the Current Step Number. Third, specify 25 as the Step End Time. Afterwards, specify 0.5 as the Initial Time Step. Specify 0.1 as the Minimum Time Step. Specify 1 as the Maximum Time Step.

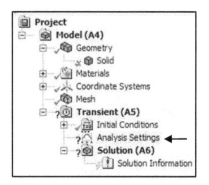

	Steps	End Time [s]
1	1	10.
2	2	20.
3	3	25.
4	4	60.
*		

Highlight **Analysis Settings** under **Transient (A5)**. First, specify 4 as the Number Of Steps. Second, specify 2 as the Current Step Number. Third, specify 17 as the Step End Time. Afterwards, specify 0.5 as the Initial Time Step. Specify 0.1 as the Minimum Time Step. Specify 1 as the Maximum Time Step.

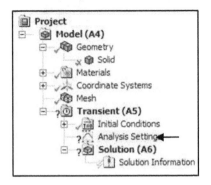

	Steps	End Time [s]
1	1	10.
2	2	17.
3	3	25.
4	4	60.
*		

Details of "Analysis Settings"

Step Controls	
Number Of Steps	4.
Current Step Number	2.
Step End Time	17. s
Auto Time Stepping	On
Define By	Time
Carry Over Time Step	Off
Initial Time Step	0.5 s
Minimum Time Step	0.1 s
Maximum Time Step	1. s
Time Integration	On
Solver Controls	
Solver Type	Program Controlled
Weak Springs	Off
Large Deflection	On

Highlight **Analysis Settings** under **Transient (A5)**. First, specify 4 as the Number Of Steps. Second, specify 2 as the Current Step Number. Third, specify 20 as the Step End Time. Afterwards, specify 0.5 as the Initial Time Step. Specify 0.1 as the Minimum Time Step. Specify 1 as the Maximum Time Step.

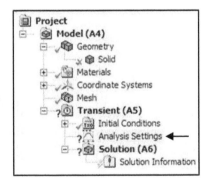

	Steps	End Time [s]
1	1	8.
2	2	17.
3	3	25.
4	4	60.
*		

Details of "Analysis Settings"

Step Controls	
Number Of Steps	4.
Current Step Number	1.
Step End Time	8. s
Auto Time Stepping	On
Define By	Time
Initial Time Step	0.5 s
Minimum Time Step	0.1 s
Maximum Time Step	1. s
Time Integration	On
Solver Controls	
Solver Type	Program Controlled
Weak Springs	Off
Large Deflection	On

After making the above modifications, highlight Displacement 2 listed under Transient (A5) modify the Tabular Data. Change 0.02 meter to 0.015 m for the time instants of 8 and 17, as shown below.

Tabular Data

	Steps	Time [s]	☑ Y [m]
1	1	0.	0.
2	1	8.	1.5e-002
3	2	17.	1.5e-002
4	3	25.	0.
5	4	60.	= 0.
*			

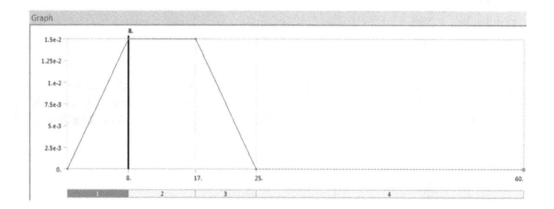

After those modification, highlight **Solution (A6)** and right-click to pick **Solve**. Now let us examine the graph of Directional Deformation or the graph of Displacement vs Time as defined in the textbook of Dynamics. In the Project tree, highlight **Directional Deformation** under **Solution (A6).** The displacement trajectory of the follower head displays the 4 stages, namely, [0, 8], [8, 17], [17, 25], and [25, 60]. This pattern matches what we defined in the process of **Analysis Settings.**

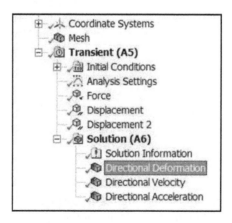

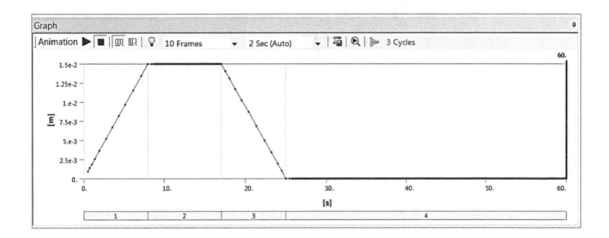

	Time [s]	☑ Minimum [m]	☑ Maximum [m]
1	0.5	9.375e-004	9.375e-004
2	0.75	1.4063e-003	1.4063e-003
3	1.	1.875e-003	1.875e-003
4	1.375	2.5781e-003	2.5781e-003
5	1.9375	3.6328e-003	3.6328e-003
6	2.7813	5.2148e-003	5.2148e-003
7	3.5812	6.7148e-003	6.7148e-003
8	4.3812	8.2148e-003	8.2148e-003
9	5.1812	9.7148e-003	9.7148e-003
10	6.1812	1.159e-002	1.159e-002
11	7.1812	1.3465e-002	1.3465e-002
12	8.	1.5e-002	1.5e-002
13	8.5	1.5e-002	1.5e-002

Let us first to examine the graph of Directional Velocity or the graph of Velocity vs Time as defined in the textbook of Dynamics. In the Project tree, highlight **Directional Velocity** under **Solution (A6).** The velocity trajectory of the follower head displays the 4 stages, namely, [0, 10], [1, 20], [20, 30], and [30, 60]. This pattern matches what we defined in the process of **Analysis Settings**. The oscillation of the magnitude of velocity clearly depicts the transient process of the follower head motion to its steady-state status.

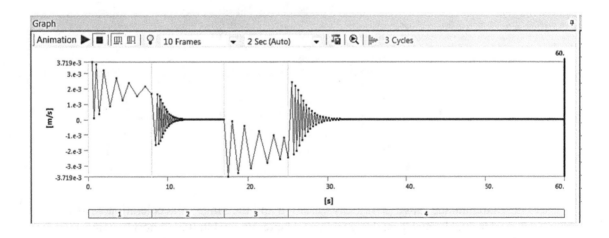

Tabular Data			
	Time [s]	☑ Minimum [m/s]	☑ Maximum
1	0.5	3.719e-003	3.719e-003
2	0.75	8.7084e-005	8.7084e-005
3	1.	3.5663e-003	3.5663e-003
4	1.375	3.5068e-004	3.5068e-004
5	1.9375	3.1687e-003	3.1687e-003
6	2.7813	8.6633e-004	8.6633e-004
7	3.5812	2.6703e-003	2.6703e-003
8	4.3812	1.2489e-003	1.2489e-003
9	5.1812	2.3672e-003	2.3672e-003
10	6.1812	1.5132e-003	1.5132e-003
11	7.1812	2.1376e-003	2.1376e-003
12	8.	1.6747e-003	1.6747e-003
13	8.5	-1.6745e-003	-1.6745e-003

The graph of Acceleration vs Time curve is shown below. The oscillation pattern clearly depicts that the transient and steady-state stages of the cam-follower system during the motion.

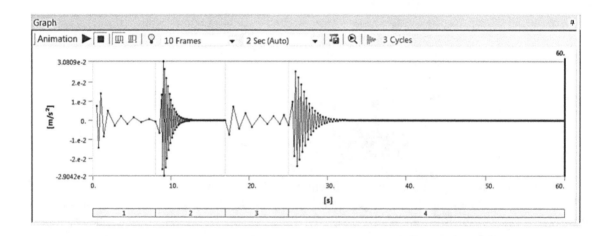

Tabular Data			
	Time [s]	☑ Minimum [m/s²]	☑ Maximum
1	0.5	7.438e-003	7.438e-003
2	0.75	-1.4528e-002	-1.4528e-002
3	1.	1.3917e-002	1.3917e-002
4	1.375	-8.5749e-003	-8.5749e-003
5	1.9375	5.0098e-003	5.0098e-003
6	2.7813	-2.7287e-003	-2.7287e-003
7	3.5812	2.2549e-003	2.2549e-003
8	4.3812	-1.7768e-003	-1.7768e-003
9	5.1812	1.3979e-003	1.3979e-003
10	6.1812	-8.5406e-004	-8.5406e-004
11	7.1812	6.2446e-004	6.2446e-004
12	8.	-5.6542e-004	-5.6542e-004
13	8.5	-6.6983e-003	-6.6983e-003

Users may tabulate the range of velocity variation and the range of acceleration variation for both the original cam design and the new cam design to gain a better understanding of how to control the dynamic motion of the follower during the operation. Also, if users are interested, the time scale can also be varied. The case study presented in this case study, a revolution of 60 rpm is assumed.

	Original Design	New Design
Velocity Range (mm/s)	(-3.97 - 3.97)	(-3.73 - 3.92)
Acceleration Range (mm/s^2)	(-32.9 - 34.8)	(-29.0 - 30.8)

At this moment, we conclude our discussion on using transient structural analysis to study the motion of an object for mechanism design. From the top menu, click **File > Close Mechanical**. In this way, we return to the main page of Workbench. From the top menu, click **File > Save Project**.

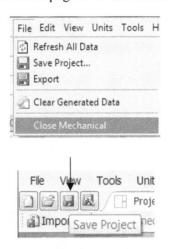

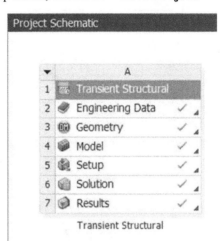

5.8 References

1. ANSYS Parametric Design Language Guide, Release 15.0, Nov. 2013, ANSTS, Inc.
2. ANSYS Workbench User's Guide, Release 15.0, Nov. 2013, ANSYS, Inc.
3. X. L. Chen and Y. J. Liu, Finite Element Modeling and Simulation with ANSYS Workbench, 1st edition, Barnes & Noble, January 2004.
4. J. W. Dally and R. J. Bonnenberger, Problems: Statics and Mechanics of Materials, College House Enterprises, LLC, 2010.
5. J. W. Dally and R. J. Bonnenberger, Mechanics II Mechanics of Materials, College House Enterprises, LLC, 2010.
6. X. S. Ding and G. L. Lin, ANSYS Workbench 14.5 Case Studies (in Chinese), Tsinghua University Publisher, Feb. 2014.
7. G. L. Lin, ANSYS Workbench 15.0 Case Studies (in Chinese), Tsinghua University Publisher, October 2014.
8. K. L. Lawrence. ANSYS Workbench Tutorial Release 13, SDC Publications, 2011.
9. K. L. Lawrence. ANSYS Workbench Tutorial Release 14, SDC Publications, 2012
10. Huei-Huang Lee, Finite Element Simulations with ANSYS Workbench 14, Theory, Applications, Case Studies, SDC Publications, 2012.
11. Huei-Huang Lee, Finite Element Simulations with ANSYS Workbench 16, Theory, Applications, Case Studies, SDC Publications, 2015.
12. Jack Zecher, ANSYS Workbench Software Tutorial with Multimedia CD Release 12, Barnes & Noble, 2009.
13. G. M. Zhang, Engineering Design and Creo Parametric 3.0, College House Enterprises, LLC, 2014.
14. G. M. Zhang, Engineering Analysis with Pro/Mechanica and ANSYS, College House Enterprises, LLC, 2011.

5.9 Exercises

1. The engineering drawing of a cantilever beam is shown below. The total length of the beam is 100 mm. The right side of the beam is fixed to the ground. The left end is set to free. The cross section is a square with each side equal to 10 mm.

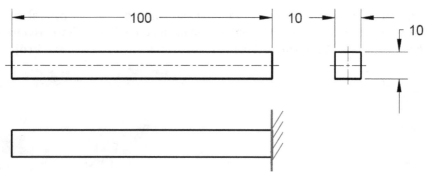

The material type is structural steel: mass density is 7850 kg/m^3, Young's modulus is 2.0 x 10^{11} N/m^2, and Poisson's ratio is 0.30. Based on the analytical formula in the vibration theory, the frequencies of the first 5 modes are listed below.

$$f_i = \frac{(2i-1)}{4L}\sqrt{\frac{E}{\rho}} = \frac{2i-1}{4*0.1}\sqrt{\frac{2x10^{11}}{7850}} = 12619(2i-1) \text{ Hz}$$

1	2	3	4	5
12619 Hz	37857 Hz	63094 Hz	88332 Hz	113570 Hz

The following steps are suggested for creating a 3D solid model in DesignModeler and run FEA.
Step 1: Click Modal and select Millimeter unit. Select XYPlane as the sketching plane to sketch a square with 10 x10 mm. Afterwards, extrude the sketch 100 mm. create a bar feature with a square section.

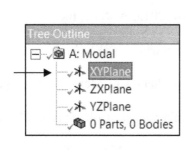

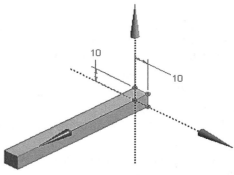

Step 2: In Mechanical, select the SI unit Metric (m, kg, N, s, V, A). Define a Frictionless Support. Select the 3 planes (3 rectangles): X = 0, Y = 0 and Z = 0.

Details of "Frictionless Support"	
Scope	
Scoping Method	Geometry Selection
Geometry	3 Faces
Definition	
Type	Frictionless Support
Suppressed	No

Click Analysis Settings and specify 5 as Max Modes to Find. ZXPlane as the sketching plane to create the cut feature.

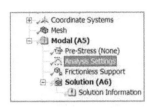

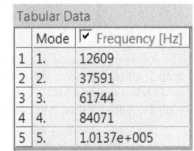

Highlight Solution (A6) and pick Solve. The 5 modes are shown in Graph, and the numerical values of the 5 mode frequencies are listed in Tabular Data.

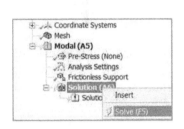

Tabular Data

	Mode	Frequency [Hz]
1	1.	12609
2	2.	37591
3	3.	61744
4	4.	84071
5	5.	1.0137e+005

Step 3: While holding down the Ctrl key to pick the 5 numerical values and right-click to pick Create Mode Shape Results. The deformation of each of the five modes is listed. The mode shapes of Mode 2 and Mode 4 are shown below.

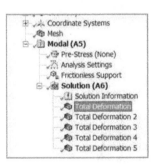

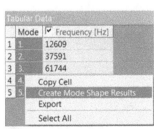

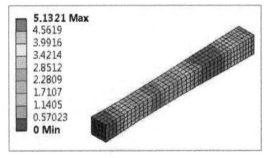

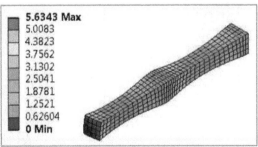

2. The engineering drawing of a cantilever beam is shown below. The total length of the beam is 150 mm. The right and left sides of the beam are fixed to the ground. The cross section is a square with each side equal to 10 mm. The material type is structural steel: mass density is 7850 kg/m^3, Young's modulus is 2.0×10^{11} N/m^2, and Poisson's ratio is 0.30. On the top surface, a force of 200 N is acting on.

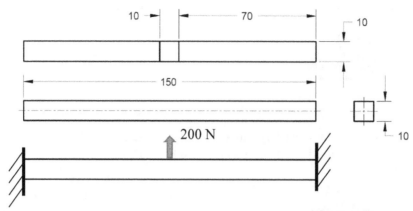

The following steps are suggested for creating a 3D solid model in DesignModeler and run FEA.
Step 1: Click Harmonic Response and select Millimeter unit. Select XYPlane as the sketching plane
to sketch a square with 10 x10 mm. Afterwards, extrude the sketch 150 mm to create a bar feature
with a square section. Afterwards create the Imprint Faces, as shown.

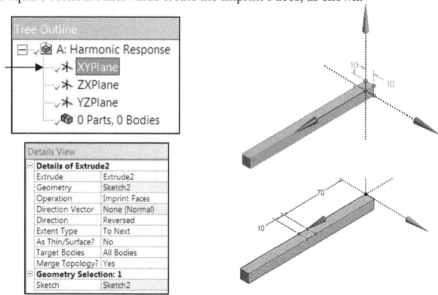

Step 2: In Mechanical, select the SI unit Metric (m, kg, N, s, V, A). Insert a force of 200 N in the
Y direction. Insert a fixed support constraint at each of the two ends.

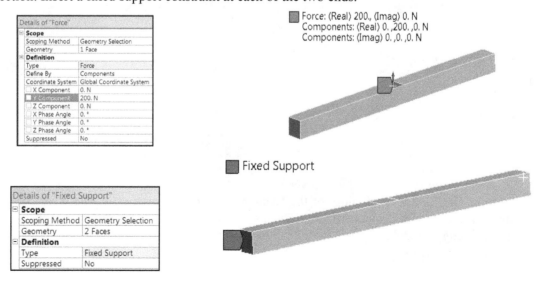

Click Analysis Settings and specify 8000 as Range Maximum and 80 as Solution Intervals.

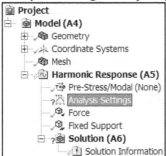

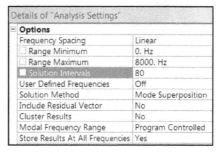

Highlight Solution (A6) and click Insert > Frequency Response > Stress. Pick the vertex, as shown and click Apply. Select the normal stress along the Z Axis.

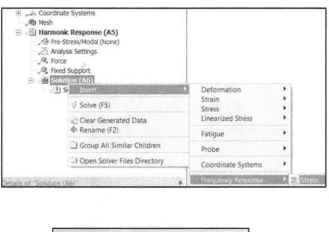

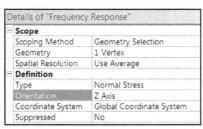

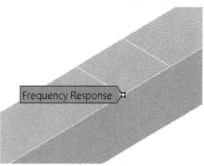

Highlight Solution (A6) and click Insert > Frequency Response > Deformation. Pick the vertex, as shown and click Apply. Select the directional deformation along the Y Axis.

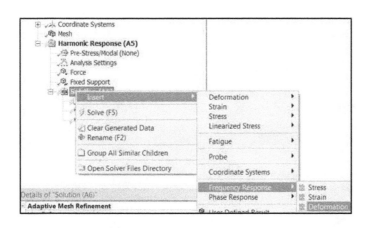

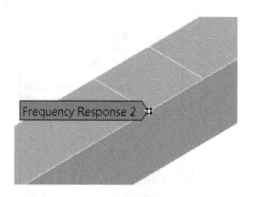

Click **Solve**. The Frequency Response for the normal stress in the Z Axis direction is shown below.

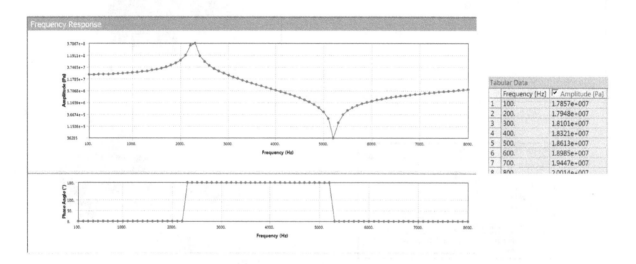

Click **Solve**. The Frequency Response 2 for the deformation in the Y Axis direction is shown below.

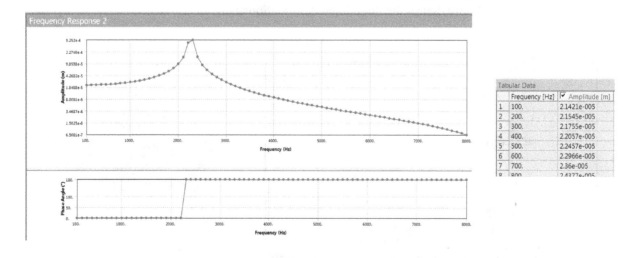

3.   A diving board used in the swimming pool is shown below. The material type is aluminum alloy. The three dimensions of the diving board are 800 x 120 x 20 mm. The diving board is cantilevered out from the wall. On the top surface, there is a surface region, 120 x 60 mm, marked as standing area, assuming a static load of force is acting on it. The magnitude of the force is 500 N, representing the weight of a diving person. Create the geometric model of the diving board with the surface region, and perform FEA. You need to perform two (2) analyses. The first analysis is "static structural". The second analysis is "modal". Mesh size requirements: element size = 10 mm and click Generate Mesh.

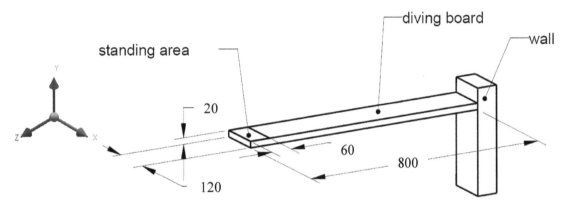

4.   A cylindrical string used in a guitar instrument is shown below. The material type is magnesium alloy. You are asked to create the geometry and perform FEA. You need to perform static structural analysis to determine the max total deformation and the max value of von Mises stress. Perform modal analysis to determine the natural frequencies of the first 8 modes. Assume that the string length is 600 mm. The cross-section is a circular shape and its diameter is 1 mm. The right end is fixed to the ground. The left end is subjected to a displacement constraint with X =Y = 0, and Z is set free. The magnitude of the tension force acting on the left end is 100 N. Mesh size requirement: Element Size = 10 mm.

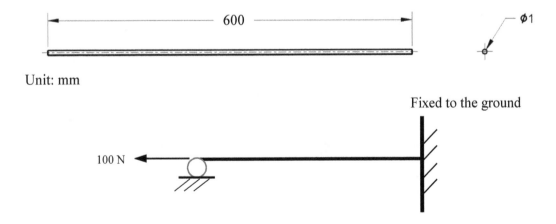

CHAPTER 6

Concept Modeling: Line and Surface Bodies

6.1   Introduction

In this chapter, the focus is concept modeling. Concept modeling focuses on creating points, creating lines through connecting points, and creating a sketch consisting of a set of lines. Afterwards, we convert the created lines to line bodies for analyzing mechanical structures, such as trusses and beams. Concept modeling provides users with a set of pre-defined cross sections to facilitate the modeling of rectangular beams, circular beams and I-beams. Concept modeling also allows users to create surface bodies through line bodies and sketches. This approach allows users to work with a 2D model for plane stress or plane strain analysis. Section 6.2 presents a 5-bar truss structure where we introduce the procedure to define line bodies through points and lines. We also introduce the procedure to determine the axial force in each of the five bars and determine the reaction forces at joints and rollers. Section 6.3 presents a truss structure consisting of 13 bars. We demonstrate the effectiveness of using line bodies and a rectangular section to define each bar to form a truss structure, and the easiness of determining the axial force for each of the 13 bars. We also discuss the effects of the defined section area on the magnitudes of the axial forces. Section 6.4 and Section 6.5 present two case studies in beams. The beam under study in Section 6.4 is a 6061-T6 square tube with one end fixed to the ground. The procedure to construct the shear-moment diagram is detailed in step by step. Section 6.5 is a case study dealing with the support structure of a ceiling fan. The support structure consists of two beams with different material types. We also incorporate the weight of the two beams into the FEA simulation. Section 6.6 presents a case study where a 3D hexagon wrench is modeled through a 2D surface model. We perform a plane stress analysis with the thickness value specified. We present a steady-thermal analysis of a chimney in Section 6.7. We construct both a 2D surface model and a 3D model. We compare the results obtained from both models to validate the use of a 2D surface model to real-life applications.

6.2   A Five Bar Truss Structure Subjected to a Vertical Load

Users may have already noticed that, on the top menu of DesignModeler, there is a module called Concepts, as shown below. This module provides a variety of operations related to defining points, and defining line and surface bodies.

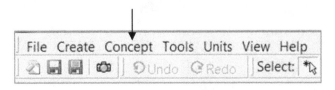

In this case study, we work with a planar truss structure. This truss structure consists of 5 straight bars. The cross section of the 5 bars is a square with the side dimension equal to 10 mm. Assume that all 5 bars are two-force members. The load acting on this truss structure is a force. This force acts on the top of the truss structure. Its magnitude is 1000 N downward. The node on the low-left is constrained with a joint. The node on the low-right is constrained with a roller.

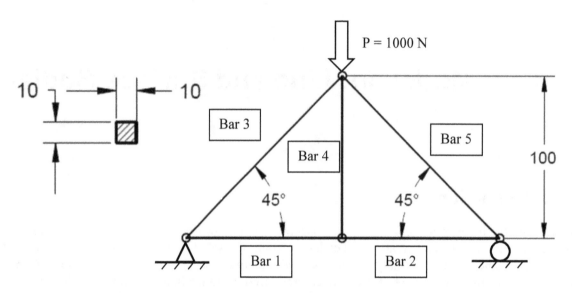

Based on the static equilibrium condition, the reaction force acting on the left joint is 500 N and the reaction force acting on the right support is also 500 N because of the nature of geometrical symmetry. The axial force acting on each of the 5 bars and the stress developed in the axial direction for each of the 5 bars are listed below.

	Axial Force (N)	Direct Stress (MPa)
Bar 1	500	5
Bar 2	500	5
Bar 3	-707.1	-7.07
Bar 4	0	0
Bar 5	-707.1	-7.07

There are 3 objectives of this case study. The first objective is to learn the procedure of creating this planar truss structure. The second objective is to obtain the axial force data and the reaction force data through FEA. The third objective is to validate the results obtained from FEA after comparing them with the numerical values obtained from the calculations based on the static equilibrium condition.

Step 1: Open Workbench and Save the File Name.
From the start menu, click **ANSYS 19** > **Workbench 19**. Users may directly click Workbench 19 when the symbol of Workbench 19 is on display.

 Workbench 19.2

From the top menu, click the icon of **Save Project**. Specify **planar truss structure 1** as the file name, and click **Save**.

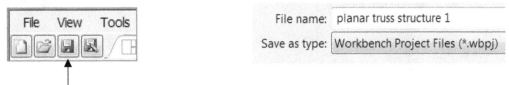

Step 2: From **Toolbox**, highlight **Static Structural** and drag it to the main screen.

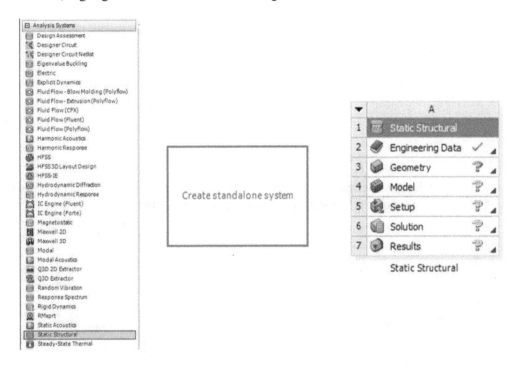

Step 3: In Design Modeler, create a 2D sketch for the truss structure.

Right-click the **Geometry** cell and pick **New DesignModeler Geometry** to start **Design Modeler**. Users may double click **Geometry** to directly enter **Design Modeler** if **New SpaceClaim Geometry** is not present. Click the **Units** drop down on the top menu, and select **Millimeter**.

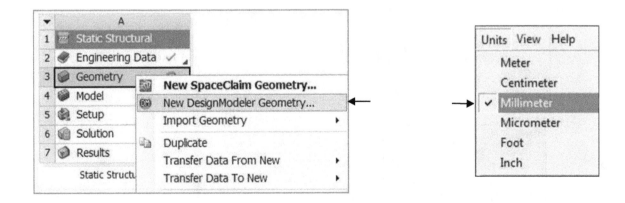

In **Tree Outline**, select **XYPlane** as the sketching plane. Click Look At to orient the sketching plane for defining 5 straight lines.

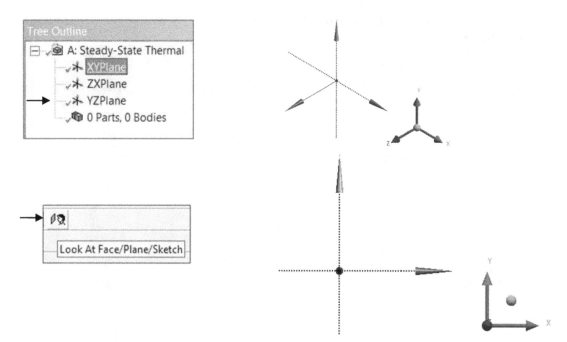

Click Sketching, and click Draw. In the Draw window, click Line. Sketch a vertical line from the origin. Make sure P is shown when starting the sketch, as shown below.

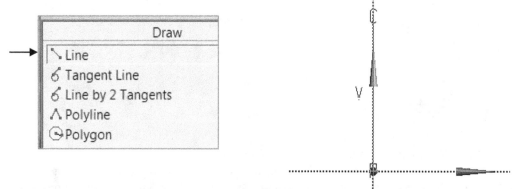

Click **Dimensions** and from the **Dimensions** panel, select **General**. Pick the sketched line and make a left click to position the size dimension. Specify 100 as its value.

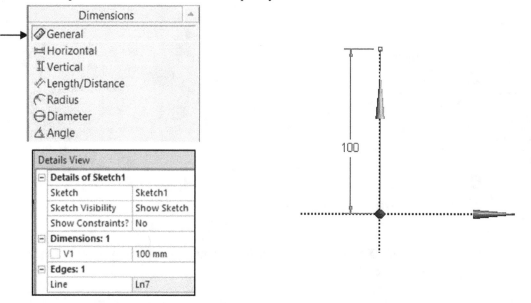

In the Draw window, click **Line**. Sketch 2 horizontal lines from the origin. Make sure P is shown when starting the sketch, as shown below.

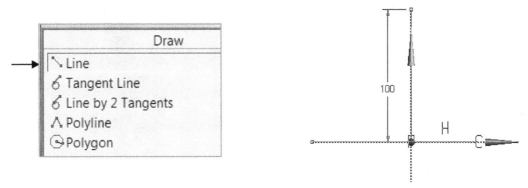

In the Constraints window, click **Equal Length**. Left click the vertical line and left click on of the two horizontal lines to set them equal. Repeat this procedure to set the other horizontal line equal to the length of the vertical line.

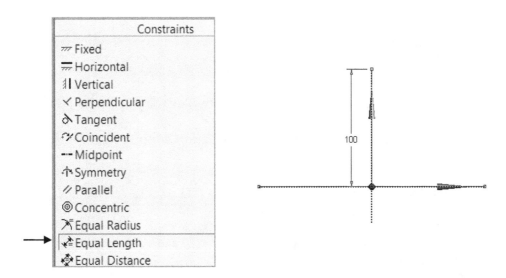

In the Draw window, click **Line**. Connect the 2 vertex points to create a line, as shown. Make sure that P is on display when connecting. Repeat this procedure to connect the other 2 vertex points to create the line on the other side, as shown below.

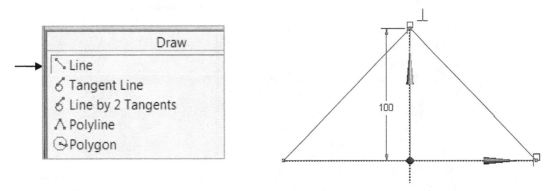

Now let us define the lines to be used for the truss structure. From the top menu, click **Concept** > **Lines from Sketches**. In Tree Outline, highlight Sketch1 and click **Apply**. Click **Generate**. **Line1** is created. An item called **Line Body** is also listed.

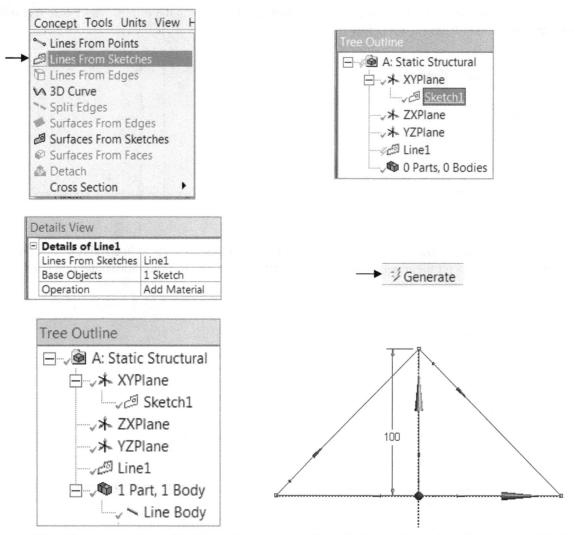

Click **Concept > Cross Section > Rectangular**. Specify 10 and 10 as the dimensions of 2 sides. Rec1 is created and listed in Tree Outline.

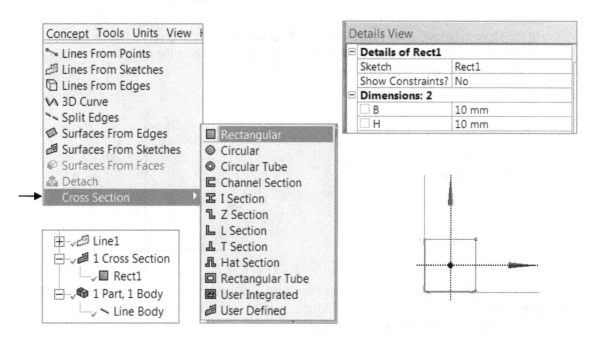

To assign the defined section Rect1 to Line1, highlight the **Line Body** listed under **1 Part**, **1 Body**. In Details View, select Rect1 in Cross Section.

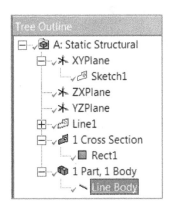

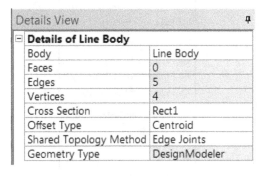

To visualize the created 3D cross section solids for the 2 3D Curves, from the top menu, click **View** > **Cross Section Solids**. The created solid model is on display.

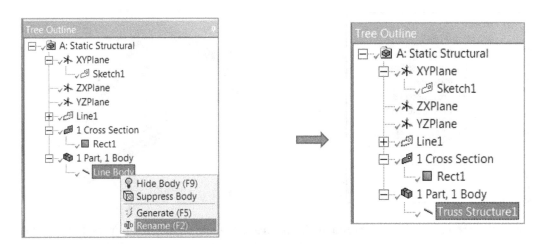

In Tree Outline, **1 Part**, **1 Body** is listed. Under **1 Part**, **1 Body**, highlight Line Body, right-click to pick Rename. Type Truss Structure1.

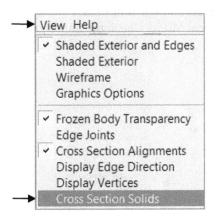
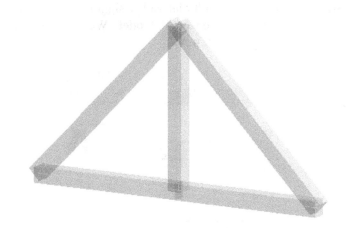

Upon completing the creation of this 3D solid model, we need to close Design Modeler. Click **File** > Close **DesignModeler**. We go back to Project Schematic. Click **Save.**

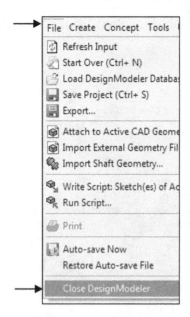

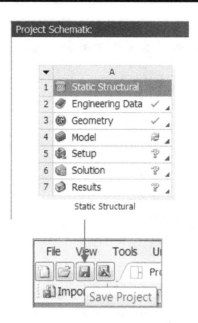

Step 4: Accept the Default Material Assignment of Structural Steel and Perform FEA.

Double-click the icon of **Model**. We enter the Mechanical Mode. Check the unit system. Select Metric (mm, t, N, s, mV, mA).

From the project model tree, highlight **Mesh,** and right-click to pick **Generate Mesh**.

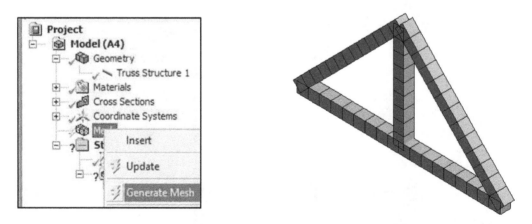

From the project tree, highlight **Static Structural (A5)** and right-click to pick **Insert**. Afterwards, select **Displacement**. From the screen, pick the low-left vertex, as shown. Set zero for X, Y, and Z Components. Users may need to select Vertex from the filter listed on the top menu before making the selection.

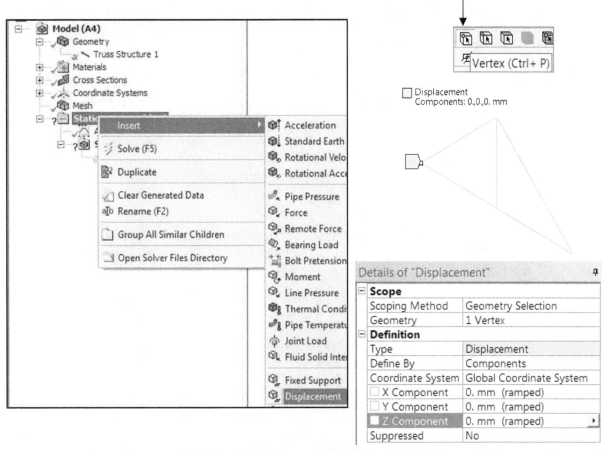

From the project tree, highlight **Static Structural (A5)** and right-click to pick **Insert**. Afterwards, select **Displacement**. From the screen, pick the low-right vertex, as shown. Keep Free for X Component and Set zero for Y and Z Components.

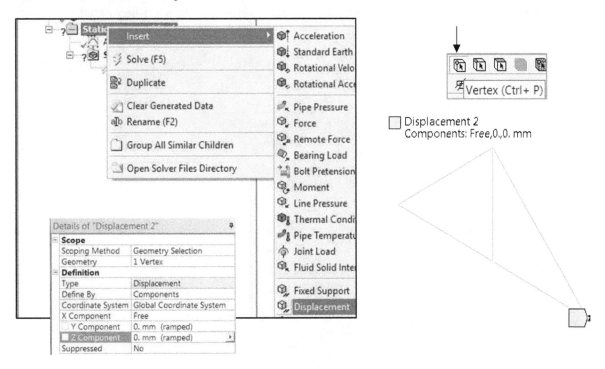

From the project tree, highlight **Static Structural (A5)** and right-click to pick **Insert**. Afterwards, select **Fixed Rotation**. From the screen, pick the low-left vertex, as shown. Keep **Fixed** for X and Y Components and Set **Free** for Z Components.

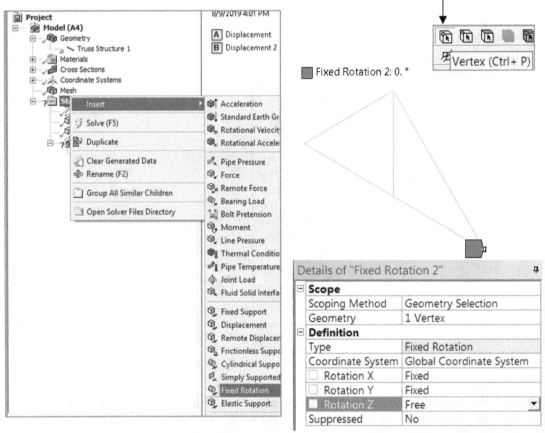

From the project tree, highlight **Static Structural (A5)** and right-click to pick **Insert**. Afterwards, select **Fixed Rotation**. From the screen, pick the low-right vertex, as shown. Keep **Fixed** for X and Y Components and Set **Free** for Z Components.

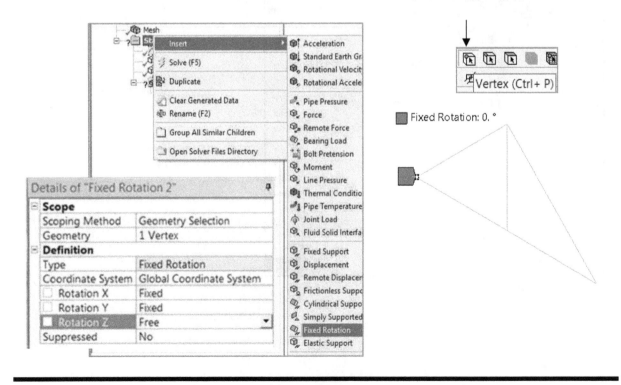

From the project tree, highlight **Static Structural (A5)** and right-click to pick **Insert**. Afterwards, select **Force**. From the screen, pick the vertex at the top, as shown. Specify -1000 N as the magnitude of the force load and downward.

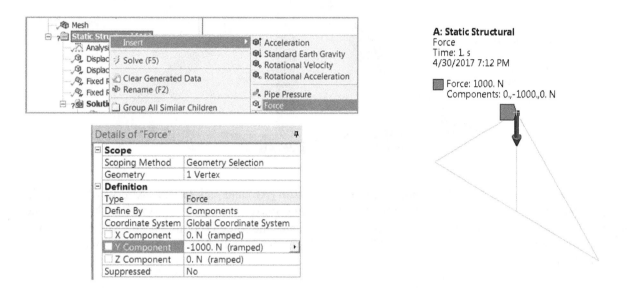

From the project tree, highlight **Solution (A6)** and right-click to pick **Solve**. Workbench system is running FEA.

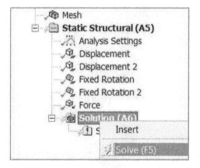

To plot the displacement distribution, highlight **Solution (A6)** and pick **Insert**. Afterwards, highlight **Deformation** and pick **Total**. Afterwards, right click to pick **Evaluate All Results**. The maximum value of deformation at the top is 0.0098 mm. It is important to note that the deformation distribution is not symmetric about the vertical axis or the Y axis. The deformation at the low left vertex or corner is zero. However, the deformation at the low right vertex or corner is 0 mm.

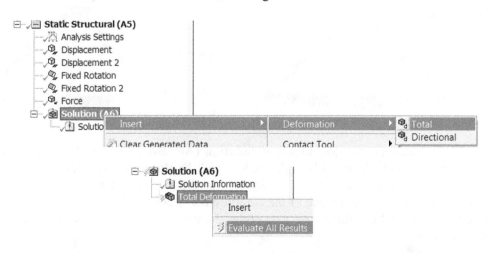

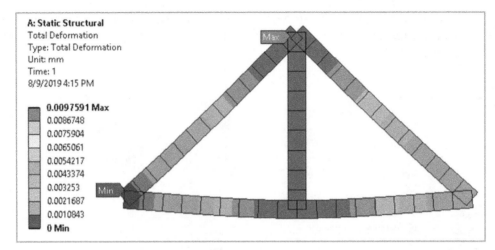

To plot the axial force distribution, highlight **Solution (A6)** and pick **Insert**. Afterwards, highlight **Beam Results > Axial Force.** Afterwards, click Probe and make a left click on Bar 5 or the bar on the right side. -697.6 N is on display. The negative sign indicates the force is compressive. Make a left click on Bar 3 or the bar on the left side. -697.6 N is on display. Make a left click on the other three bars. All axial forces are on display.

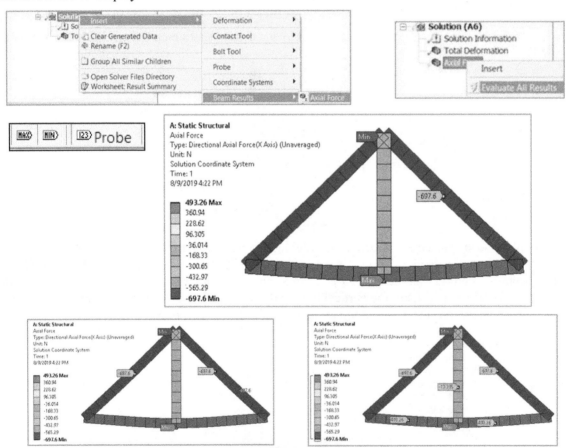

To plot the stress distribution, highlight **Solution (A6)** and pick **Insert**. Afterwards, highlight **Beam Tools > Beam Tool.** Afterwards, right click to pick **Evaluate All Results**. Click **Direct Stress.** The maximum value of the tensile stress is 4.98 MPa at the low middle location or the origin of the coordinate system. The maximum value of the compressive stress is 6.976 MPa at the two low corners. Users may click Probe and use the mouse to go over the truss structure to view the stress value at the different locations.

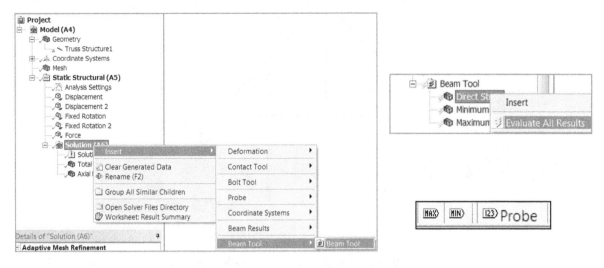

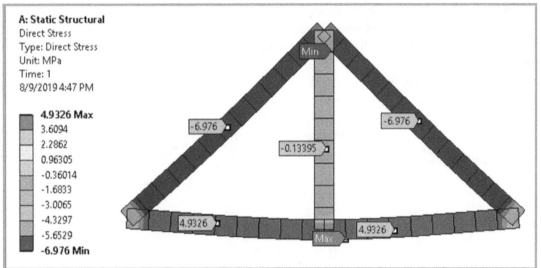

To plot the reaction forces at the joint, highlight **Solution (A6)** and pick **Insert > Probe > Force Reaction.** In Details of "Force Reaction". Click **Displacement** under Boundary Condition. An item called Force Reaction is listed under Beam Tool. Highlight **Force Reaction,** and right click to pick **Evaluate All Results**. Click the listed **Force Reaction.** The vertical reaction force is on display with the magnitude equal to 500 N.

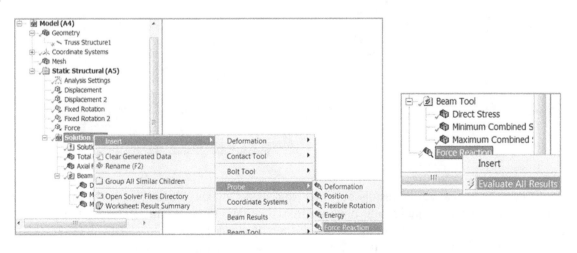

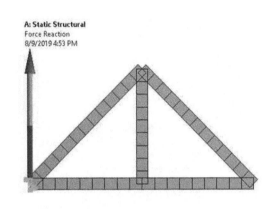

To plot the reaction forces at the right side, highlight **Solution (A6)** and pick **Insert > Probe > Force Reaction.** In Details of "Force Reaction". Click Displacement 2 under Boundary Condition. An item called **Force Reaction 2** is listed under Beam Tool. Highlight **Force Reaction 2,** and right click to pick **Evaluate All Results**. Click the listed **Force Reaction.** The vertical reaction force is on display with the magnitude equal to 500 N.

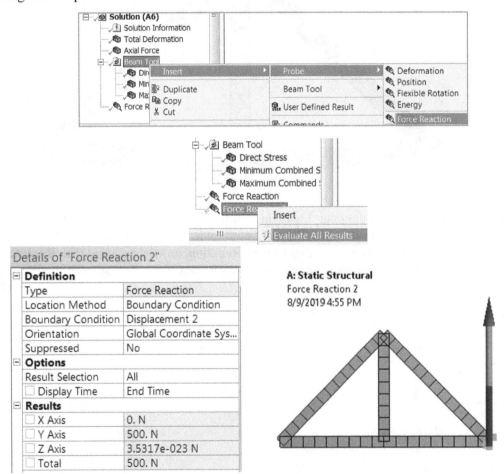

When we summarize the results obtained from FEA run, the reaction forces at the joint and the support well match the calculated results obtained from the Static Equilibrium condition. In the following, we tabulate the axial force for each of the 5 bars. Generally speaking, the numerical values obtained from the FEA run match the calculated results obtained from the Static Equilibrium condition. The differences are at the level of 2%.

	FEA Axial Force (N)	FEA Direct Stress (MPa)
Bar 1	493.26	4.933
Bar 2	493.26	4.933
Bar 3	-697.6	-6.976
Bar 4	-13.395	0.134
Bar 5	-697.6	-6.976

At this moment, we conclude our discussion on the first truss structure case. From the top menu, click **File** > **Close Mechanical**. In this way, we return to the main page of Workbench. Click the icon of **Save Project**.

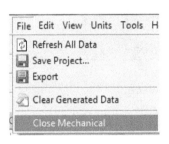

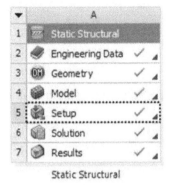

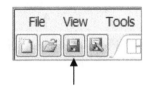

6.3 A Planar Truss Structure with 13 Bars

A planar truss structure is shown below. There are 13 bars connecting together. The key dimensions of this truss structure are shown in the following drawing. The section area of each bar is a square. Each side is 0.4 meter. This truss structure is constrained by a joint at the left side, and a roller at the right side. Assume that the material type is Structural Streel. There are two force loads acting on this truss structure. At node C, the force has a magnitude equal to 5000 N downwards. At node D, the force has a magnitude equal to 3000 N. The first objective of this study is to determine 3 sets of values: the reaction forces at the joint and the roller, the maximum value of deformation, and the axial force acting on each of the 13 bars. The second objective of this study is to investigate the effect of sectional area on the magnitude of the axial force acting on each of the 13 bars.

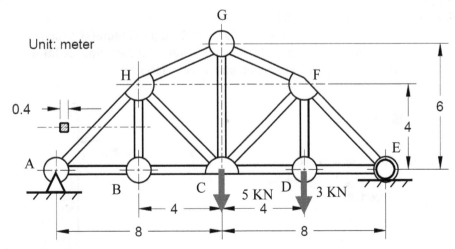

Step 1: Open Workbench and Save the File Name.
From the start menu, click **ANSYS 19** > **Workbench 19**. Users may directly click Workbench 19 when the symbol of Workbench 19 is on display.

From the top menu, click the icon of **Save Project**. Specify **planar truss structure 2** as the file name, and click **Save**.

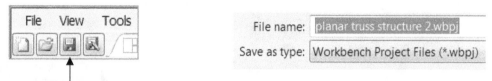

Step 2: From **Toolbox**, highlight **Static Structural** and drag it to the main screen.

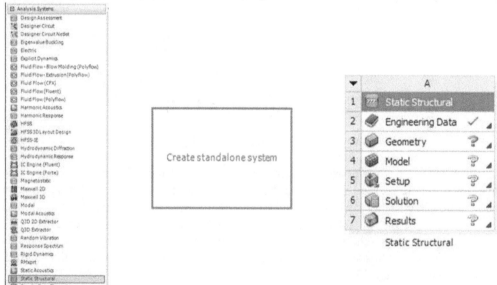

Step 3: In Design Modeler, create a 2D sketch for the truss structure.

Right-click the **Geometry** cell and pick **New DesignModeler Geometry** to start **Design Modeler**. Users may double click **Geometry** to directly enter **Design Modeler** if **New SpaceClaim Geometry** is not present. Click the **Units** drop down on the top menu, and select **Meter**.

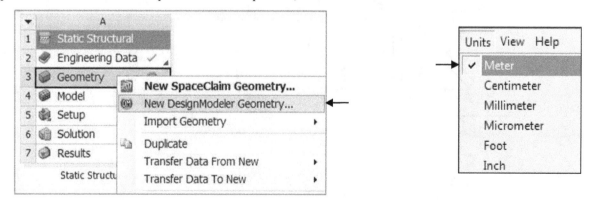

In **Tree Outline**, select **XYPlane** as the sketching plane. Click **Look At** to orient the sketching plane.

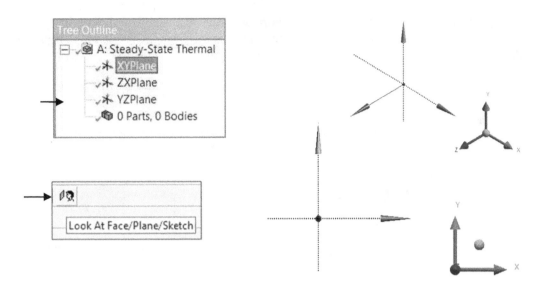

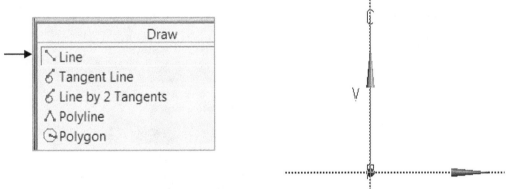

Click **Sketching**, and click **Draw**. In the Draw window, click **Line**. Sketch a vertical line from the origin. Make sure P is shown when starting the sketch, as shown below.

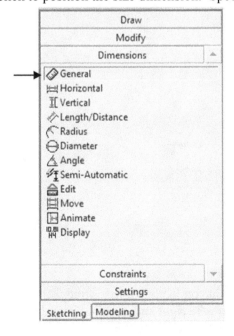

Click **Dimensions** and from the **Dimensions** panel, select **General**. Pick the sketched and make a left click to position the size dimension. Specify 6 as its value.

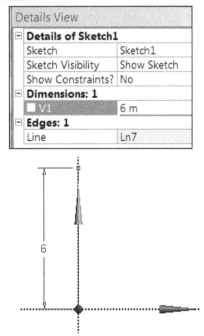

In the Draw window, click **Line**. Sketch a horizontal line starting from the origin to the left side. Make sure that P is shown. Sketch the second horizontal line starting from the origin to the right side. Make sure P is shown when starting the sketch, as shown below.

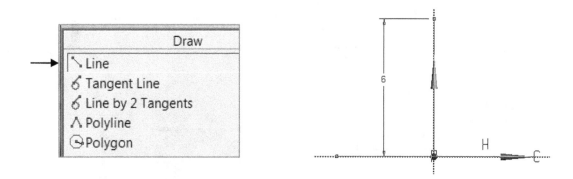

Click **Dimensions** and from the **Dimensions** panel, select **General**. Pick one of the 2 sketched lines and specify 4 as its value. Repeat this process to specify 4 as the length for the other sketched line.

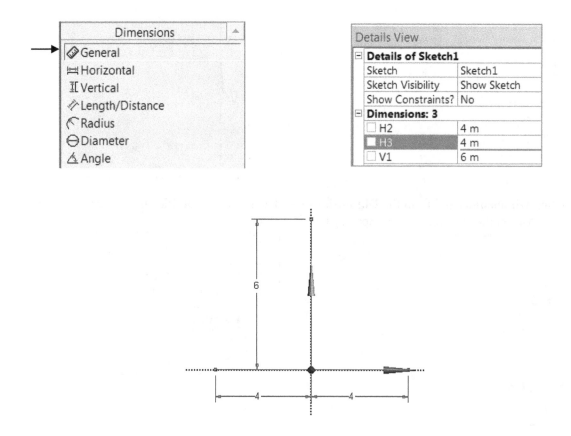

In the Draw window, click **Line**. Sketch a horizontal line on the left side starting from the vertex of the sketched line on the left side. Sketch a horizontal line on the right side starting from the vertex of the sketched line on the right side. Make sure that P is shown when sketching this line.

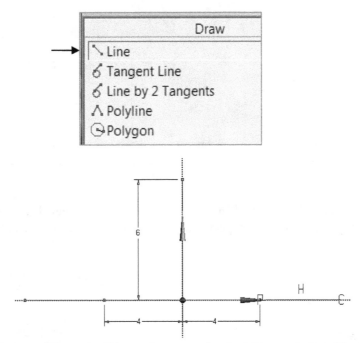

Click **Dimensions** and from the **Dimensions** panel, select **General**. Specify 4 as the length for the 2 sketched lines.

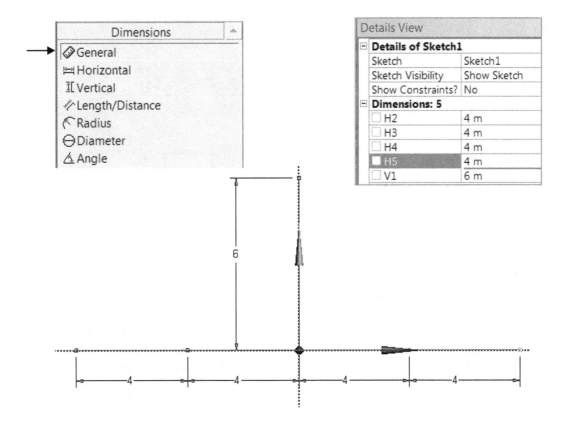

In the Draw window, click **Line**. Sketch 2 vertical lines. The first vertical line starts at the connecting points of the 2 lines on the left side. The second vertical line starts at the connecting point of the 2 lines on the right side, as shown below. Make sure that P is shown when picking a node and V is shown when sketching a vertical line.

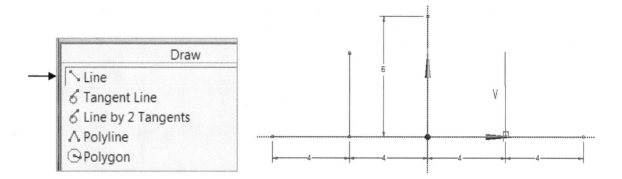

Click **Dimensions** and from the **Dimensions** panel, select **General**. Pick one of the 2 sketched vertical lines and specify 4 as its value. Repeat this process to specify 4 as the length for the other vertical lines.

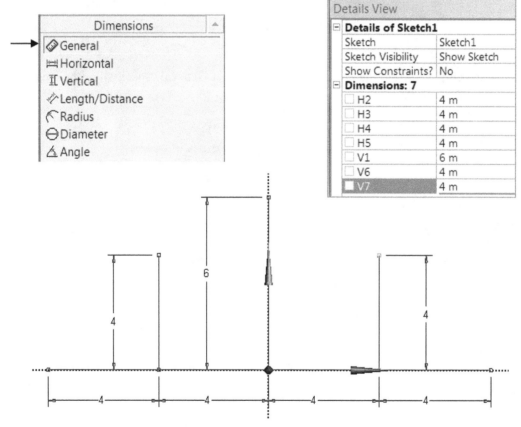

In the Draw window, click **Line**. Connect 2 vertex points to create a line, as shown. Make sure that P is on display when connecting. Repeat this procedure to connect the other pairs of 2 vertex points to create the 6 lines, as shown below. Make sure that P is shown when picking a node.

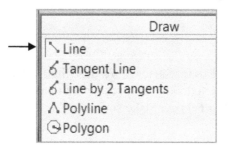

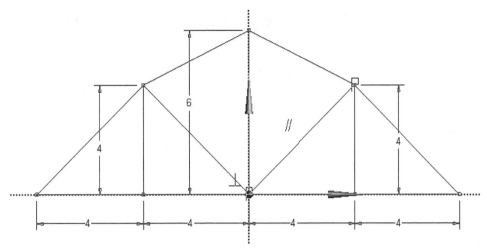

When completing this sketch, let us define the lines to be used for the truss structure. From the top menu, click **Concept > Lines from Sketches**. In Tree Outline, highlight Sketch1 and click **Apply**. Click **Generate**. **Line1** is created. An item called Line Body is also listed.

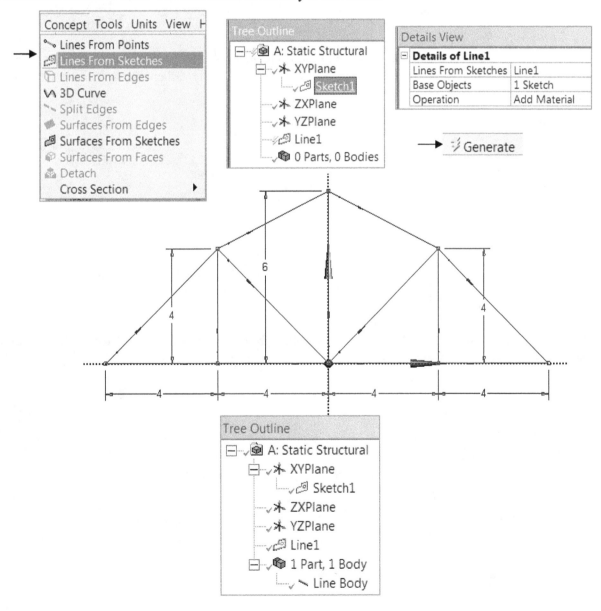

Click **Concept** > **Cross Section** > **Rectangular**. Specify 0.4 and 0.4 as the dimensions of 2 sides. Rec1 is created and listed in Tree Outline.

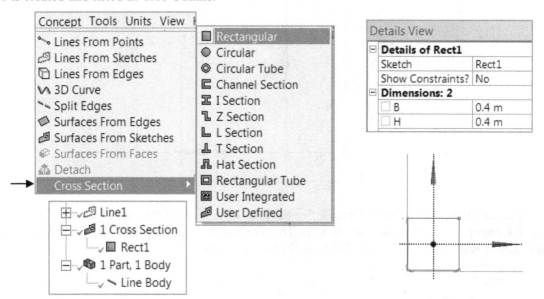

To assign the defined section Rect1 to Line Body, highlight the **Line Body** listed under 1 Part, 1 Body. In Details View, select Rect1 in Cross Section. In Details View, there are 8 vertex points and 13 edges listed.

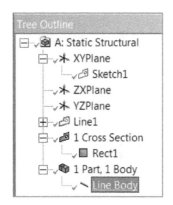

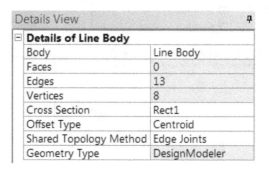

To visualize the created 3D cross section solids, from the top menu, click **View** > **Cross Section Solids**. The created solid model is on display.

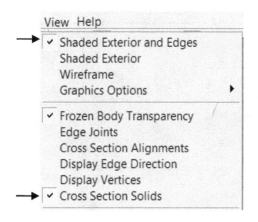

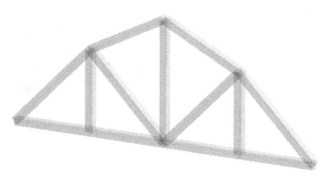

In Tree Outline, 1 Part, 1 Body is listed. Right click to pick Rename. Change Line Body to Truss Structure2.

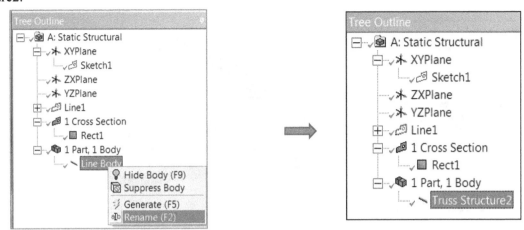

Upon completing the creation of this truss structural model, we need to close Design Modeler. Click **File** > Close **DesignModeler**. We go back to Project Schematic. Click **Save.**

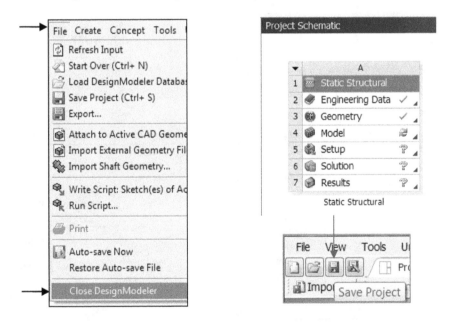

Step 4: Accept the Default Material Assignment of Structural Steel and Perform FEA.

Double-click the icon of **Model**. We enter the Mechanical Mode. Check the unit system. Select SI unit or Metric (m, kg, N, s, V, A).

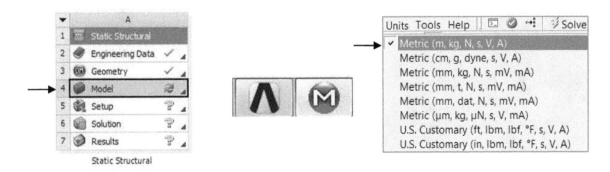

From the project model tree, highlight **Mesh,** and right-click to pick **Generate Mesh**.

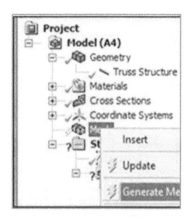
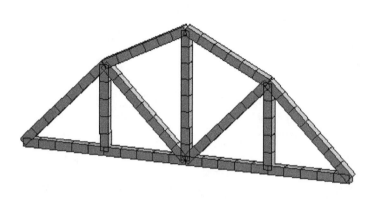

From the project tree, highlight **Static Structural (A5)** and right-click to pick **Insert**. Afterwards, select **Displacement**. From the screen, pick the low-left vertex, as shown. Set zero for X, Y, and Z Components. Click the Vertex filter before picking the point.

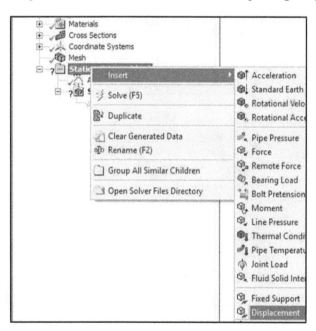

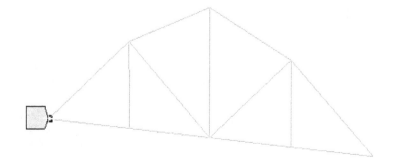

Displacement
Components: 0.,0.,0. m

From the project tree, highlight **Static Structural (A5)** and right-click to pick **Insert**. Afterwards, select **Displacement**. From the screen, pick the low-right vertex, as shown. Keep Free for X Component and Set zero for Y and Z Components.

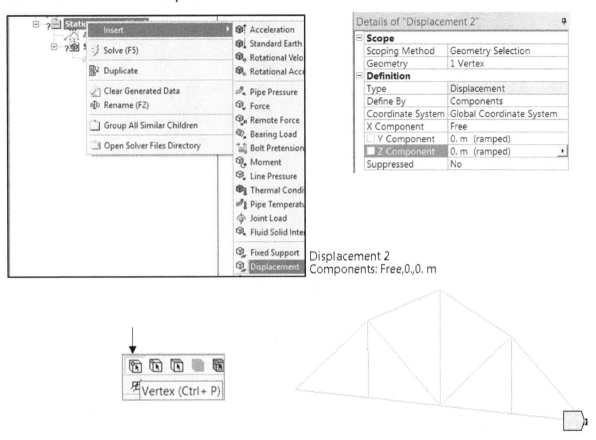

From the project tree, highlight **Static Structural (A5)** and right-click to pick **Insert**. Afterwards, select **Fixed Rotation**. From the screen, pick the low-left vertex, as shown. Keep **Fixed** for X and Y Components and Set **Free** for Z Components.

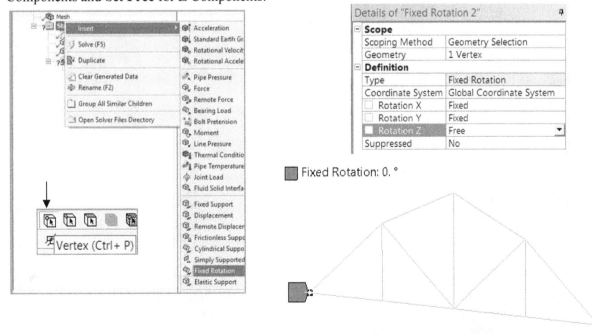

From the project tree, highlight **Static Structural (A5)** and right-click to pick **Insert**. Afterwards, select **Fixed Rotation**. From the screen, pick the low-right vertex, as shown. Keep **Fixed** for X and Y Components and Set **Free** for Z Components.

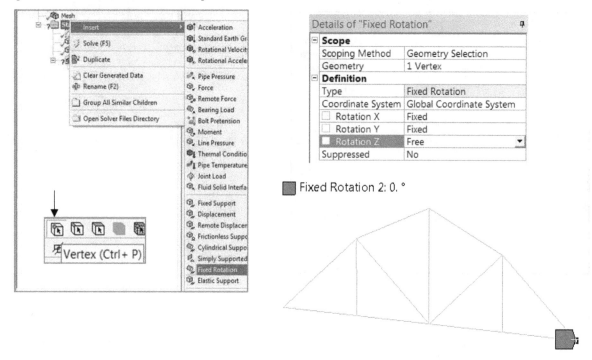

From the project tree, highlight **Static Structural (A5)** and right-click to pick **Insert**. Afterwards, select **Force**. From the screen, pick the vertex at the middle of the bottom, or Node C, as shown. Specify -5000 N as the magnitude of the force load and downward.

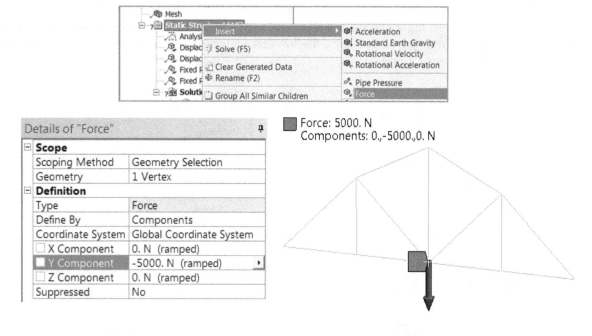

From the project tree, highlight **Static Structural (A5)** and right-click to pick **Insert**. Afterwards, select **Force**. From the screen, pick the vertex next to Node C in the positive X direction, as shown. Specify -3000 N as the magnitude of the force load and downward.

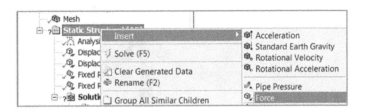

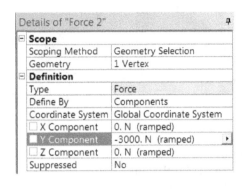

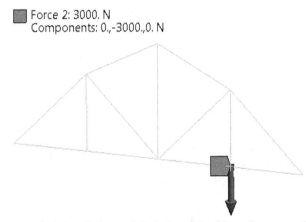

From the project tree, highlight **Solution (A6)** and right-click to pick **Solve**. Workbench system is running FEA.

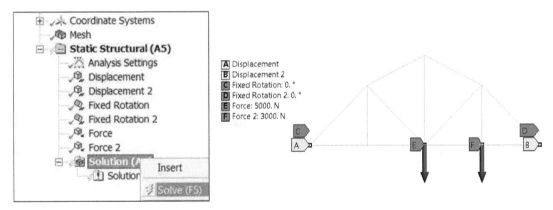

To plot the displacement distribution, highlight **Solution (A6)** and pick **Insert**. Afterwards, highlight **Deformation** and pick **Total**. Afterwards, right click to pick **Evaluate All Results**. The maximum value of deformation at the top is 0.0043 mm.

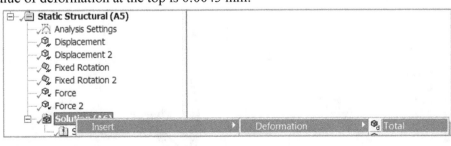

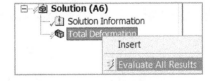

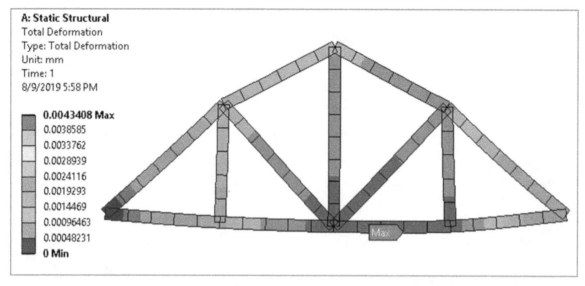

To plot the axial force distribution, highlight **Solution (A6)** and pick **Insert**. Afterwards, highlight **Beam Results > Axial Force**. Afterwards, right click to pick **Evaluate All Results**. The axial force values range from -6629 N (compressive) to 4688.2 N (tensile). Use Probe to show them all.

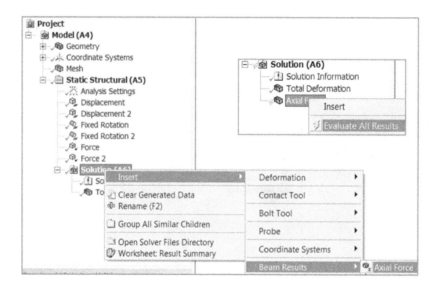

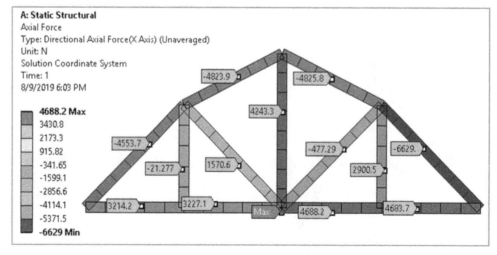

The following graphs show the axial force for each of the 13 bars.

bar AB: 3214 N (tensile)

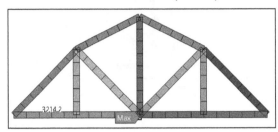

bar BC: 3227 N (tensile)

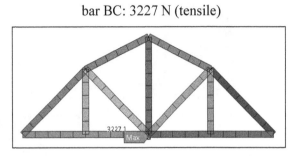

bar CD: 4688 N (tensile)

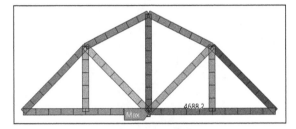

bar DE: 4684 N (tensile)

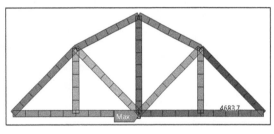

bar EF: -6629 N (compressive)

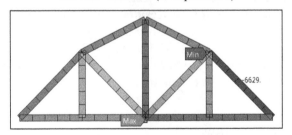

bar FG: -4826 N (compressive)

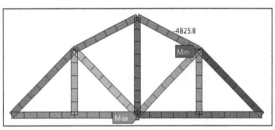

bar GH: -4824 N (compressive)

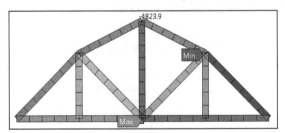

bar HA: -4554 N (compressive)

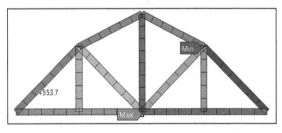

bar HB: -21 N (compressive)

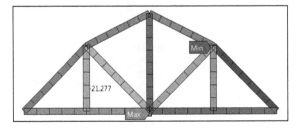

bar HC: 1571 N (tensile)

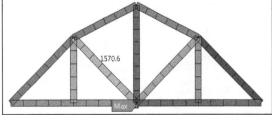

bar GC: 4243 N (tensile)   bar FC: 477 N (tensile)

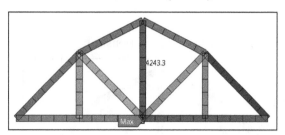

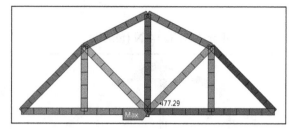

bar FD: 2901 N (tensile)

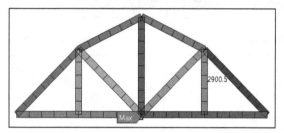

To plot the stress distribution, highlight **Solution (A6)** and pick **Insert**. Afterwards, highlight **Beam Tools > Beam Tool.** Afterwards, right click to pick **Evaluate All Results**. Click **Direct Stress.** The maximum value of the tensile stress is 29401 Pa at the low middle location or the origin of the coordinate system. The maximum value of the compressive stress is 41431 Pa near Node F. Users may click Probe and use the mouse to go over the truss structure to view the stress value at the different locations. Make left clicks on Bar CD and Bar EF to vilify these values with the calculations.

$$Direct\_Stress\_CD = \frac{Axial\_Force\_CD}{Sectional\_Area} = \frac{4688}{(0.4)(0.4)} = 29300 \ \text{(Pascal)}$$

$$Direct\_Stress\_EF = \frac{Axial\_Force\_EF}{Sectional\_Area} = \frac{-6629}{(0.4)(0.4)} = -41431 \ \text{(Pascal)}$$

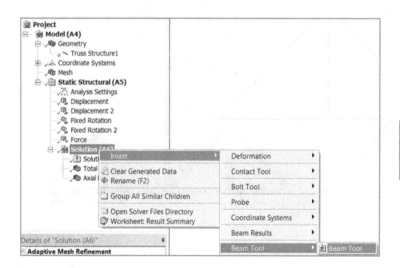

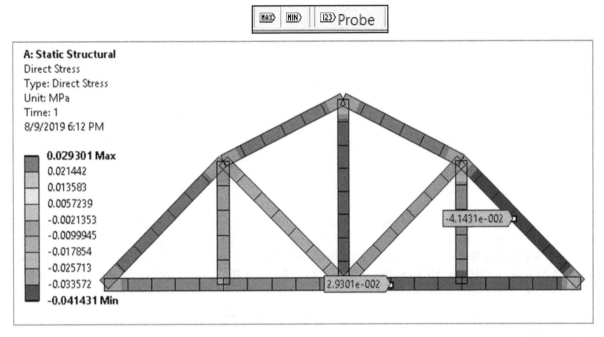

To plot the reaction forces at the joint, highlight **Solution (A6)** and pick **Insert > Probe > Force Reaction.** In Details of "Force Reaction". Click Displacement under Boundary Condition. An item called Force Reaction is listed under Beam Tool. Highlight **Force Reaction,** and right click to pick **Evaluate All Results**. In Details of "Force Reaction", expand the box called Boundary Condition and click Displacement. Afterwards, highlight Force Reaction and right-click to pick Evaluate All Results. The reaction force at the joint on the left side is on display with the magnitude equal to 3250 N.

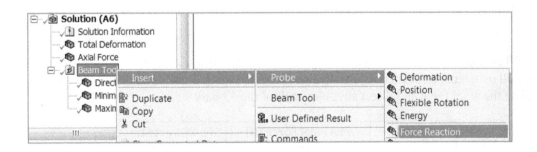

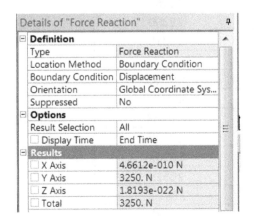

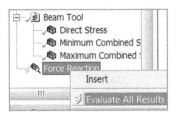

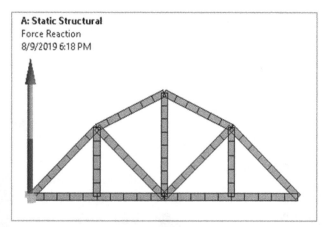

To plot the reaction forces at the roller support on the right side, highlight **Solution (A6)** and pick **Insert > Probe > Force Reaction.** In Details of "Force Reaction". Click Displacement 2 under Boundary Condition. An item called Force Reaction 2 is listed under Beam Tool. Highlight **Force Reaction 2,** and right click to pick **Evaluate All Results.** The reaction force at the roller support on the right side is on display with the magnitude equal to 4750 N.

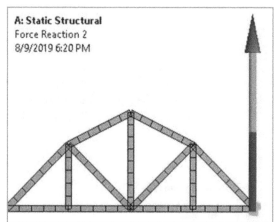

When we summarize the results obtained from FEA run, the reaction forces at the joint and the support well match the calculated results obtained from the Static Equilibrium condition. In the following, we tabulate the axial force and direct stress for each of the 13 bars.

	FEA Axial Force (N)	FEA Direct Stress (Pa)
Bar AB	3214	20088
Bar BC	3227	20169
Bar CD	4688	29300
Bar DE	4684	29275
Bar EF	-6629	-41431
Bar FG	-4826	-30163
Bar GH	-4824	-30150
Bar HA	-4554	-28463
Bar HB	-21.00	-131.25
Bar HC	-1571	-9818.8
Bar GC	4243	26519
Bar FC	477.0	2981.3
Bar FD	2901	18131

At this moment, we conclude our discussion on the second truss structure case. From the top menu, click **File > Close Mechanical**. In this way, we return to the main page of Workbench. Click the icon of **Save Project**.

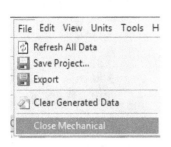

It is important to note that in the science of mechanics, we assume that the axial force for each bar in a truss structure is not related to its sectional area. When we use an FEA approach, users have to pay attention to the effect of the sectional area used in the process of modeling the truss structure. In the following, we make a change to the sectional area from the current 0.16 m^2 to 0.01 m^2, namely, the side dimension of a square is changed from 0.4 m to 0.1 m. We keep all other conditions unchanged, and run FEA to see what have not been changed and what have been changed.

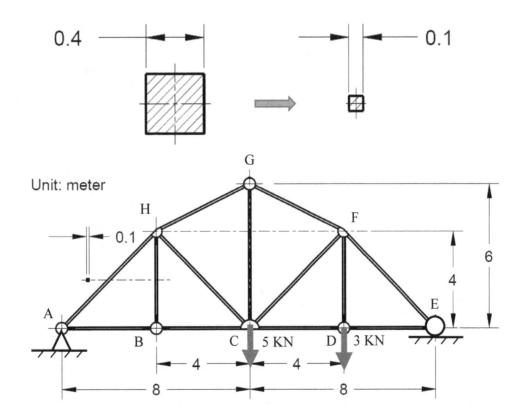

In the window of **Project Schematic**, highlight **Static Structural A1**, and right-click to pick **Duplicate**. Static Structural B is on display. Let us change the name Copy of Static Structural to Area Reduction. Afterwards, double click Geometry B3 to enter DesignModeler.

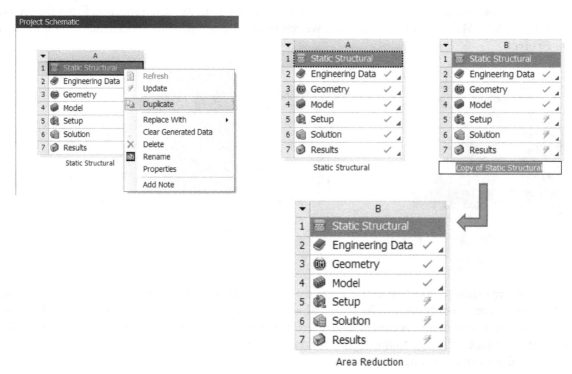

In Tree Outline, highlight **Rect1** and change 0.4 m to 0.1 m. Click **Generate**. A new trust structure model is created, as shown.

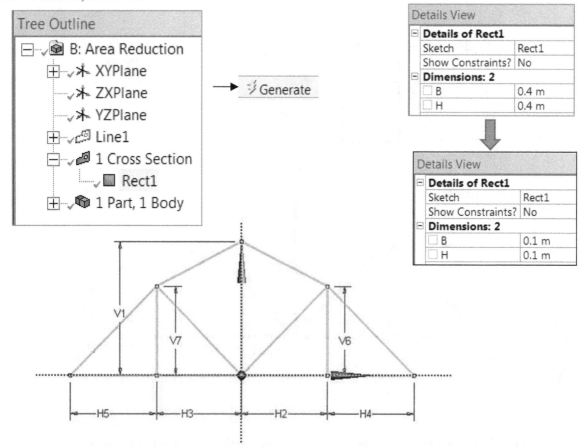

To close Design Modeler, click **File** > **Close DesignModeler**. We go back to Project Schematic. Right click on the screen to pick **Update Project**. Afterwards, double click Solution B6 to enter Mechanical.

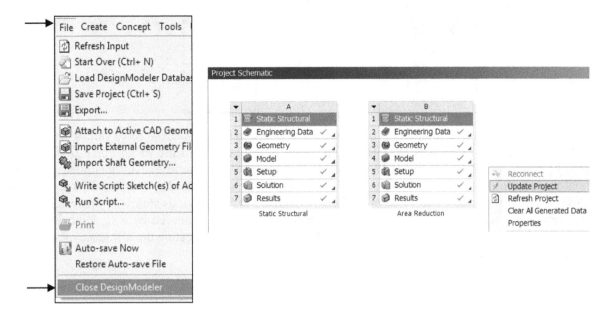

It is important to note that there is no need to run FEA again. The process of Update Project has already updated all the information. From the Project tree, highlight Force Reaction under Solution (B6). The magnitude of the reaction force at the left side is 3250 N, remaining unchanged.

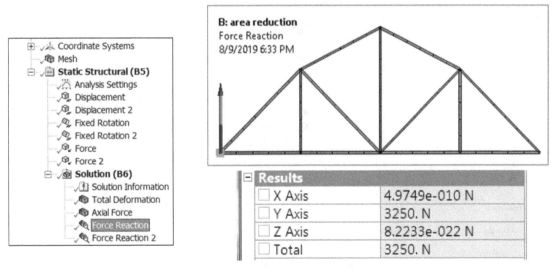

From the Project tree, highlight Force Reaction 2 under Solution (B6). The magnitude of the reaction force at the right side is 4750 N, remaining unchanged.

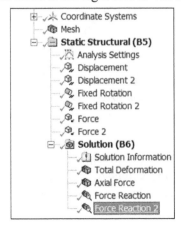

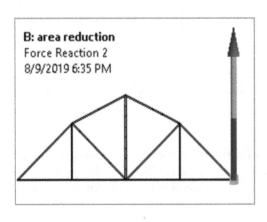

Results	
☐ X Axis	0. N
☐ Y Axis	4750. N
☐ Z Axis	-1.8738e-021 N
☐ Total	4750. N

From the Project tree, highlight Axial Force under Solution (B6). The axial force values range from -6712 N (compressive) to 4746 N (tensile). This range is different from the range obtained from the previous FEA run, which was from -6629 N (compressive) to 4688.2 N (tensile). As illustrated below is the axial force acting on bar CF where the axial force magnitude is -581.83 N. This axial force value is also different from the axial force value of -477.29 N obtained in the previous study. If interested, users may reference to a textbook titled Engineering Mechanics STATICS, authored by R.C. Hibbeler. The theoretical analysis comes out the axial force along Bar CF is -589 N, indicating the influence of sectional area on the axial force calculation.

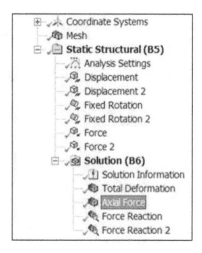

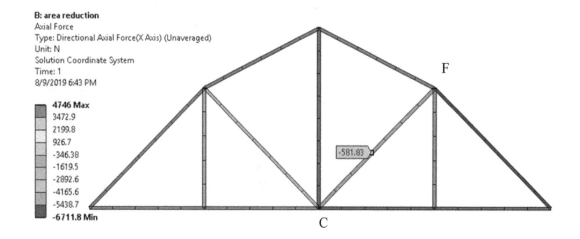

The following table lists the axial force and direct stress for each of the 13 bars. These listed values are different from those values listed in the similar table. Certainly, the values of FEA direct stress are much larger than those listed in the table for the previous run as the area values are reduced dramatically.

	FEA Axial Force (N)	FEA Direct Stress (Pa)
Bar AB	3248	324800
Bar BC	3249	324900
Bar CD	4746	474600
Bar DE	4746	474600
Bar EF	-6712	-671200
Bar FG	-4844	-484400
Bar GH	-4844	-484400
Bar HA	-4594	-459400
Bar HB	-1.44	-144.00
Bar HC	1535	153500.0
Bar GC	4328	432800
Bar FC	582.0	58200.0
Bar FD	2995	299500

Although the values are different, the tensile and compressive stress pattern remains unchanged. As a summary to this study, we learned the axial force acting on each bar in a truss structure varies as the section area changes its value when using ANSYS Workbench. A small value of sectional area leads to a large value of axial force, consequently, a larger value of direct stress. The axial force acting on each bar calculated based on the science of mechanics serves as the limiting value of the axis force acting on each bar.

At this moment, from the top menu, click **File > Close Mechanical**. In this way, we return to the main page of Workbench. Click the icon of **Save Project**.

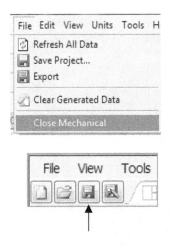

Static Structural   Area Reduction

6.4   A Square Tube Beam Subjected to Two Force Loads

In this case study, we work with a beam structure. The beam is a 6061-T6 square tube object. As shown below, the side dimension of the square is 3 inch. The thickness value is 0.125 inch. The material type is aluminum alloy. As shown in the right figure, there are two force loads. The magnitudes of these 2 force loads are 80 lbf downwards and 50 lbf downwards, respectively. The beam length is 24 inch. The left side of the beam is fixed to the wall, and the right side of the beam is set free. Our objectives are to determine the reaction force and moment at the fixed end and the maximum deformation at the free end. We also construct the shear and moment diagrams for the beam structure.

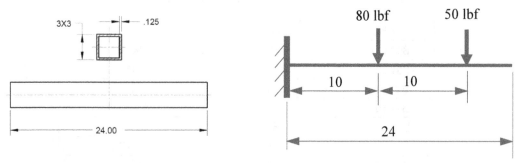

Based on the static equilibrium condition, the reaction force at the fixed end is 130 lbf and the moment at the fixed end is 1800 lbf-inch with counterclockwise direction.

Step 1: Open Workbench and Save the File Name.

From the start menu, click **ANSYS 19 > Workbench 19.** Users may directly click Workbench 19 when the symbol of Workbench 19 is on display.

From the top menu, click the icon of **Save Project**. Specify **AL square tube** as the file name, and click **Save**.

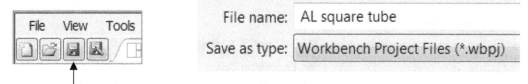

Step 2: From **Toolbox**, highlight **Static Structural** and drag it to the main screen.

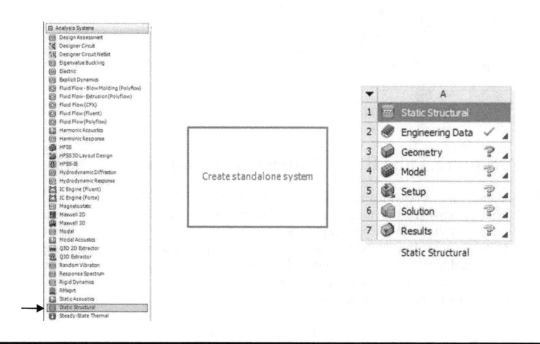

Step 3: In Design Modeler, create a 2D sketch for the square tube.

Right-click the **Geometry** cell and pick **New DesignModeler Geometry** to start **Design Modeler**. Users may double click **Geometry** to directly enter **Design Modeler** if **New SpaceClaim Geometry** is not present. Click the **Units** drop down on the top menu, and select U.S. Customary Unit System: Inch.

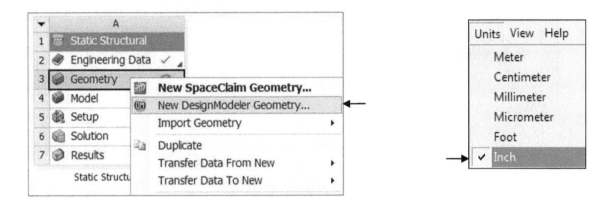

In **Tree Outline**, select **XYPlane** as the sketching plane. Click Look at to orient the sketching plane for defining 3 straight lines.

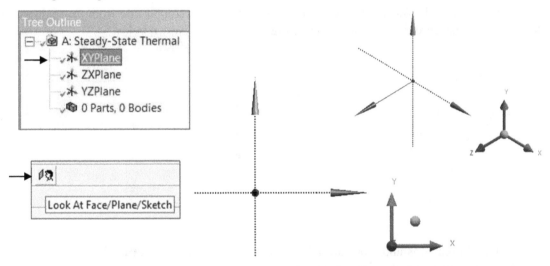

Click Sketching, and click Draw. In the Draw window, click Line. Sketch a horizontal line from the origin. Make sure P is shown when starting the sketch, as shown below.

Click **Dimensions** and from the **Dimensions** panel, select **General**. Pick the sketched line and make a left click to position the size dimension. Specify 10 as its value.

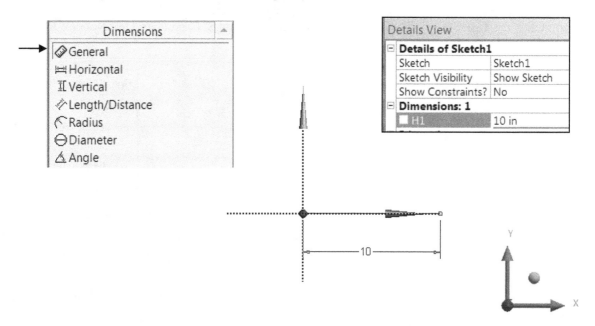

In the Draw window, click **Line**. Sketch the second horizontal line from the right vertex of the sketched line. Make sure P is shown when starting the sketch, as shown below.

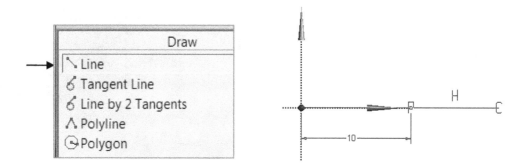

Click **Dimensions** and from the **Dimensions** panel, select **General**. Pick the sketched line and make a left click to position the size dimension. Specify 10 as its value.

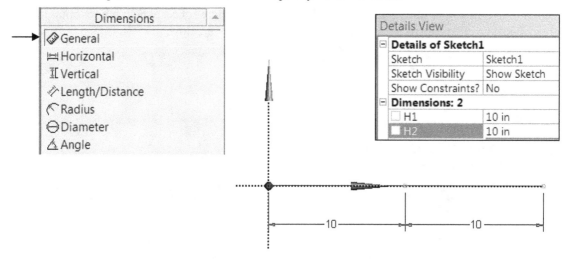

In the Draw window, click **Line**. Sketch the 3rd line starting from the right vertex of the second sketched line. Make sure that P is on display when sketching.

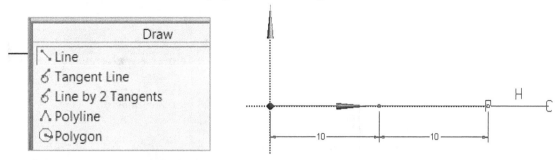

Click **Dimensions** and from the **Dimensions** panel, select **General**. Pick the sketched line and make a left click to position the size dimension. Specify 4 as its value.

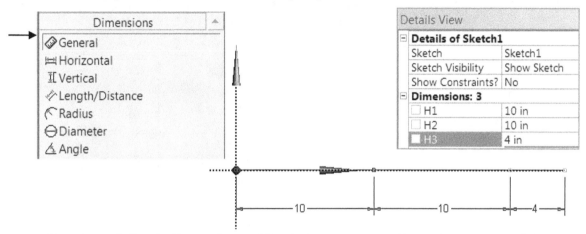

Now let us define the lines to be used for the beam structure. From the top menu, click **Concept > Lines from Sketches**. In Tree Outline, highlight Sketch1 and click **Apply**. Click **Generate**. **Line1** is created. An item called **Line Body** is also listed.

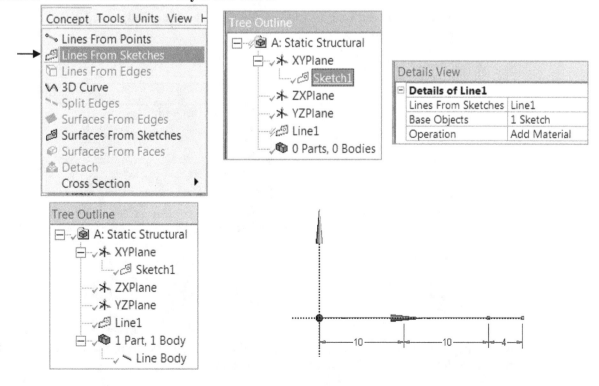

Click **Concept** > **Cross Section** > **Rectangular Tube**. Specify 3 and 3 as the dimensions of 2 sides. Specify 0.125 as the thickness value. Rec1 is created and listed in Tree Outline.

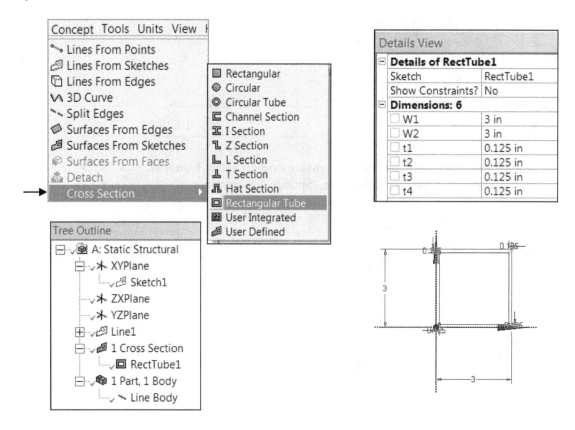

To assign the defined section **RectTube1** to **Line1**, highlight the **Line Body** listed under **1 Part**, **1 Body**. In Details View, select **RectTube1** in Cross Section.

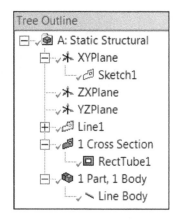

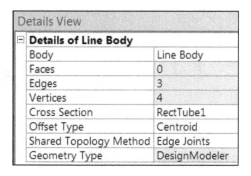

To visualize the created 3D cross section solids for the 2 3D Curves, from the top menu, click **View** > **Cross Section Solids**. The created solid model is on display.

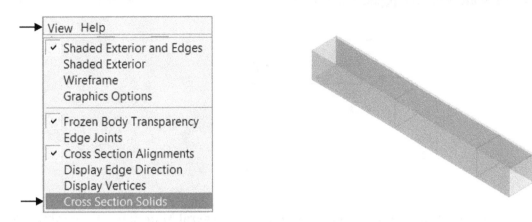

In Tree Outline, **1 Part**, **1 Body** is listed. Under **1 Part**, **1 Body**, highlight **Line Body**, right-click to pick Rename. Type Rectangular Tube.

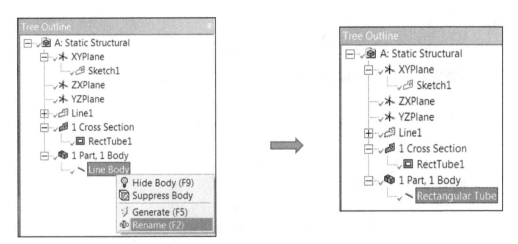

Upon completing the creation of this 3D solid model, we need to close Design Modeler. Click **File** > Close **DesignModeler**. We go back to Project Schematic. Click **Save.**

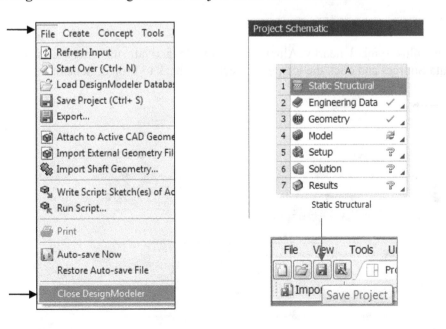

Step 4: Activate the Aluminum Alloy in the Engineering Data Sources

In the Static Structural panel, double click Engineering Data. We enter the Engineering Data Mode.

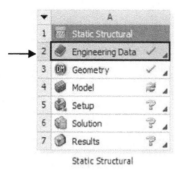

Static Structural

Click the icon of **Engineering Data Sources**. Click **General Materials**. We are able to locate Aluminum Alloy.

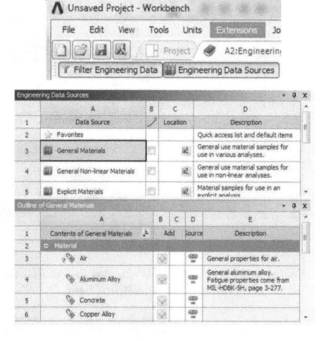

Click the plus symbol nearby Aluminum Alloy to activate it. Afterwards, click the icon of Engineering Data Sources and click the icon of Project to go back to Project Schematic.

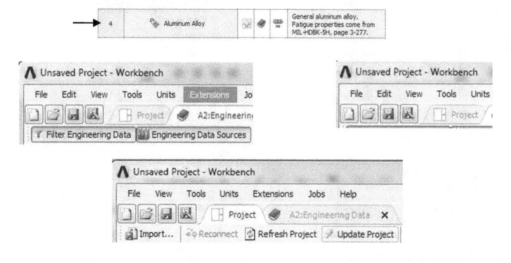

Step 6: Assign Aluminum Alloy to the rectangular tube Model

Double-click the icon of **Model**. We enter the Mechanical Mode. Check the unit system. Select U.S. Customary Unit System (in, lbm, lbf, °F, s, V, A).

Now let us do the material assignment. From the Project tree, expand Geometry. Highlight Rectangular Tube. In Details of "Rectangular Tube", click material assignment, pick Aluminum Alloy to replace Structural Steel, which is the default material assignment.

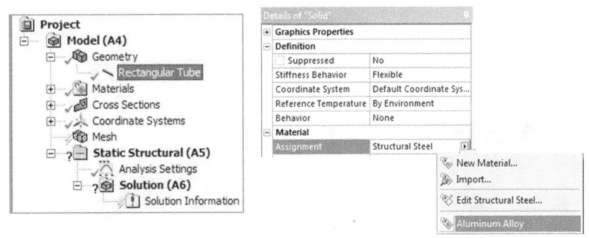

Step 7: Generate Mesh and Perform FEA.

From the project model tree, highlight **Mesh,** and right-click to pick **Generate Mesh**.

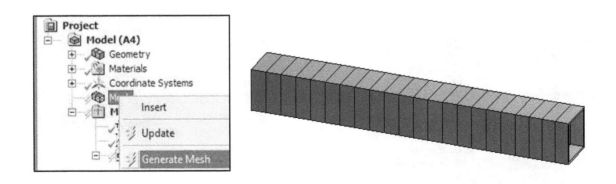

From the project tree, highlight **Static Structural (A5)** and right-click to pick **Insert**. Afterwards, select **Fixed Support**. From the screen, pick the low-left vertex, as shown.

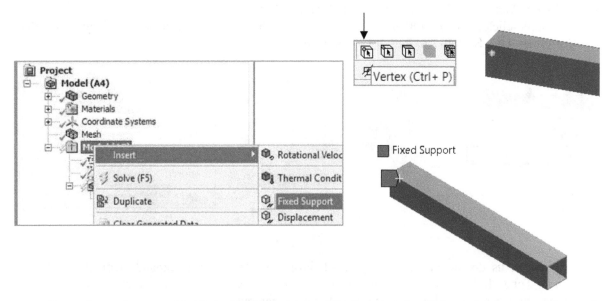

From the project tree, highlight **Static Structural (A5)** and right-click to pick **Insert**. Afterwards, select **Force**. From the screen, pick the right vertex of the first sketched line, as shown. Select Y Component and specify -80 lbf as the magnitude of the force load and downward.

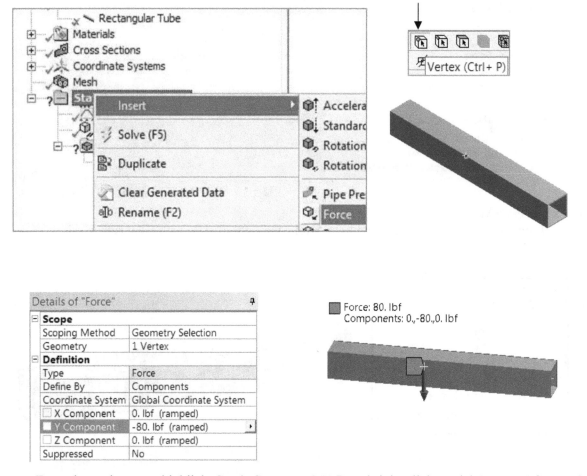

From the project tree, highlight **Static Structural (A5)** and right-click to pick **Insert**. Afterwards, select **Force**. From the screen, pick the right vertex of the second sketched line, as shown. Select Y Component and specify -50 lbf as the magnitude of the force load and downward.

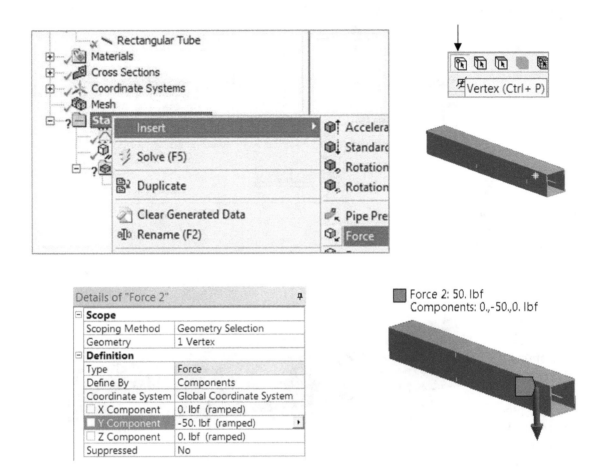

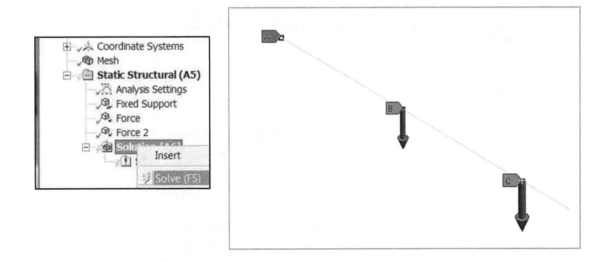

From the project tree, highlight **Solution (A6)** and right-click to pick **Solve**. Workbench system is running FEA.

To plot the displacement distribution, highlight **Solution (A6)** and pick **Insert**. Afterwards, highlight **Deformation** and pick **Total**. Afterwards, right click to pick **Evaluate All Results**. The maximum value of deformation at the free end is 0.013 inch.

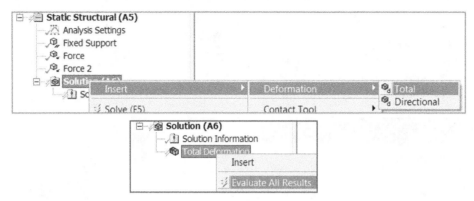

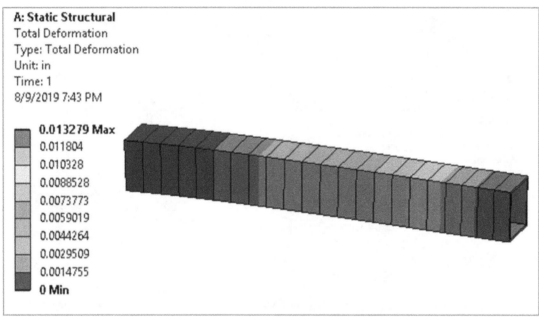

To plot the reaction force at the fixed end, highlight **Solution (A6)** and pick **Insert**. Afterwards, highlight **Probe > Force Reaction.** In Details of "Force Reaction", select Fixed Support in Boundary Condition. Afterwards, right click to pick **Evaluate All Results**. The total force in Y direction is 130 lbf, as shown in Details of "Force Reaction".

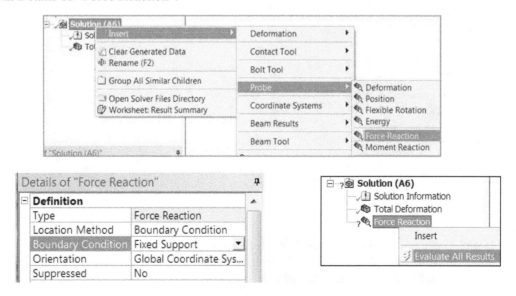

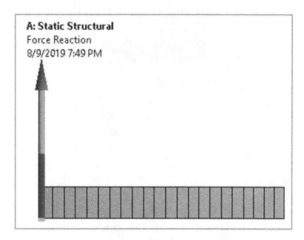

Results	
Maximum Value Over Time	
☐ X Axis	0. lbf
☐ Y Axis	130. lbf
☐ Z Axis	-8.5354e-020 lbf
☐ Total	130. lbf

To plot the moment reaction at the fixed end, highlight **Solution (A6)** and pick **Insert**. Afterwards, highlight **Probe > Moment Reaction.** In Details of "Moment Reaction", select Fixed Support in Boundary Condition. Afterwards, right click to pick **Evaluate All Results**. The total moment in Z direction is 1800 lbf-in, as shown in Details of "Moment Reaction".

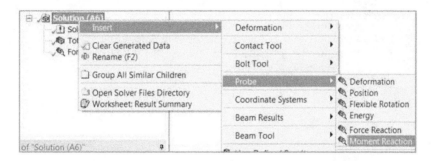

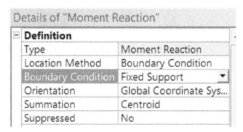

Details of "Moment Reaction"

Definition	
Type	Moment Reaction
Location Method	Boundary Condition
Boundary Condition	Fixed Support
Orientation	Global Coordinate Sys...
Summation	Centroid
Suppressed	No

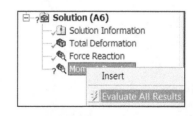

Results	
Maximum Value Over Time	
☐ X Axis	-8.3369e-008 lbf·in
☐ Y Axis	3.7269e-014 lbf·in
☐ Z Axis	1800. lbf·in
☐ Total	1800. lbf·in

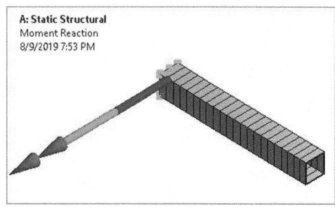

To plot the shear-moment diagram, we need to define a path beforehand. How to define a path? In the Project Tree, highlight Model (A4). Right-click to pick **Insert > Construction Geometry.** In the Project Tree, Construction Geometry is listed.

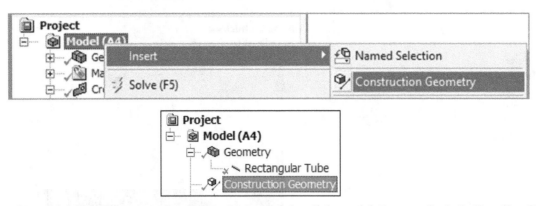

Now highlight **Construction Geometry** and right-click to pick Insert > Path. In Details of "Path", select Edge in Path Type. From the top menu, select Edge Section. While holding down the **Ctrl** key, pick the 3 sketched lines. Afterwards, click Apply. An item called Path is created and listed in the Project tree.

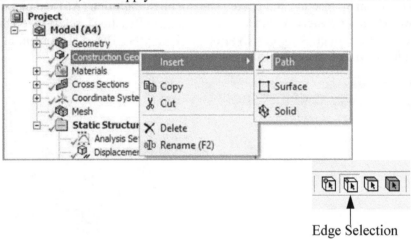

Edge Selection

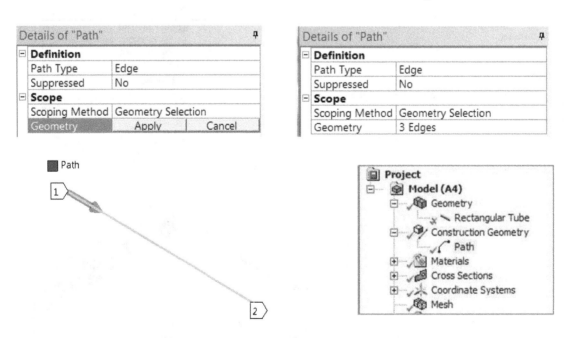

Now let us construct the shear-moment diagram. Highlight **Solution (A6)** and pick **Insert**. Afterwards, highlight **Beam Results > Shear-Moment Diagram.** In Details of "Total Shear-Moment Diagram", in the box called Path, select Path. Afterwards, right click to pick **Evaluate All Results**. The Total Shear-Moment Diagram is on display.

Click **Direct Stress.** The maximum value of the tensile stress is 4.98 MPa at the low middle location or the origin of the coordinate system. The maximum value of the compressive stress is 6.976 MPa at the two low corners. Users may click Probe and use the mouse to go over the truss structure to view the stress value at the different locations.

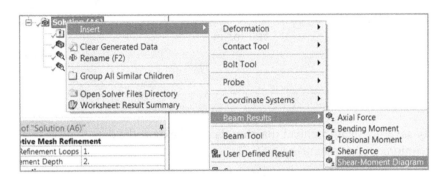

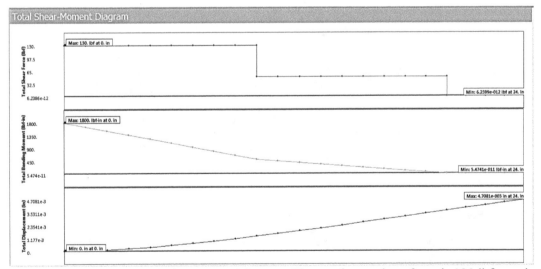

Let us examine the Total Shear Force Diagram. The maximum shear force is 130 lbf, starting at the fixed end, and extends 10 inches to the right. Afterwards, the shear force drops to 50 lbf, extending 10 inches to the right. Afterwards, the shear force drops to 0 lbf, extending to the free end.

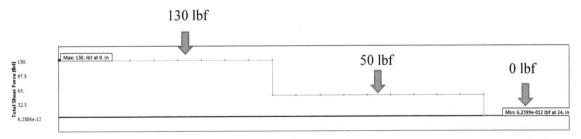

Let us examine the Total Moment Diagram. The maximum moment is 1800 lbf-in, starting at the fixed end. The magnitude of this bending moment linearly drops its magnitude as it extends 10 inches to the right with a slope equal to -130 lbf-in/inch. At the 10 inch point, the moment magnitude is 500 lbf-in.

Its magnitude continuously drops with a rate of -50 lbf-in/inch as it extends another 10 inches to the right. At the 20 inch point, the moment magnitude is zero and keeps at zero as it extends to the free end.

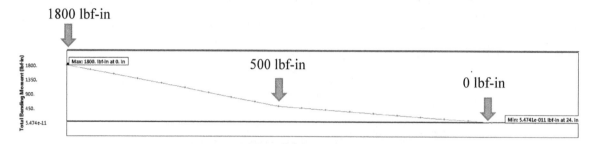

Users may save the shear-moment diagram data and export the date to an EXCEL file. Users may construct the shear-moment diagram using EXCEL.

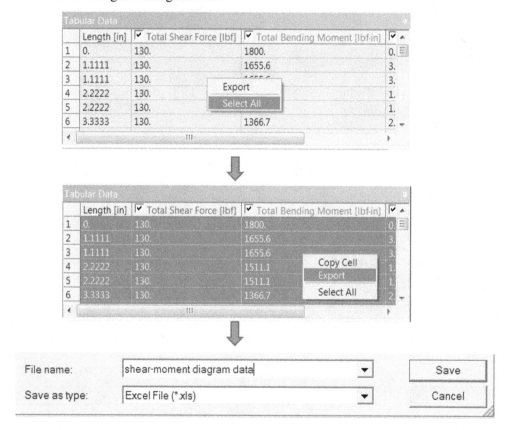

At this moment, we conclude our discussion on the rectangular tube case study. From the top menu, click **File > Close Mechanical**. In this way, we return to the main page of Workbench. Click the icon of **Save Project**.

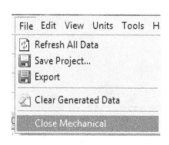

Static Structural

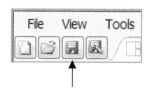

6.5   A Simple Support Rectangular Beam Subjected to Loads

A simple supported beam with a pin connection is shown below. The beam material is structural steel. The left end is a pin connection and the right end is simple-supported. A uniformly distributed load is acting at the left portion and the total magnitude is 200 N and the direction is downward. A point load is acting at PNT3. Its magnitude is 100 N and downwards. The cross section of the beam is a rectangle and the two dimensions are 25 mm and 50 mm, respectively. Determine the reaction forces at the two ends. Determine the total deformation along the beam and construct the shear-moment diagram.

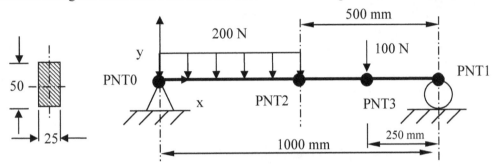

Step 1: Open Workbench and Save the File Name.

From the start menu, click **ANSYS 19 > Workbench 19.** Users may directly click Workbench 19 when the symbol of Workbench 19 is on display.

From the top menu, click the icon of **Save Project**. Specify **AL square tube** as the file name, and click **Save**.

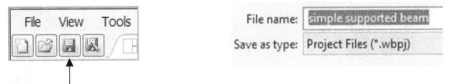

Step 2:From **Toolbox**, highlight **Static Structural** and drag it to the main screen.

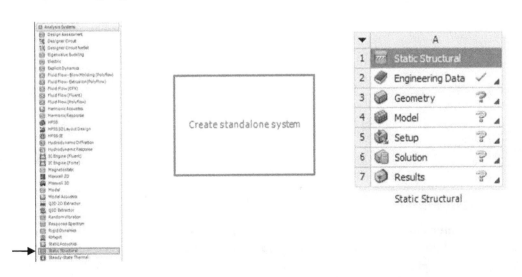

Step 3: In Design Modeler, create a 2D sketch for the square tube.

Right-click the **Geometry** cell and pick **New DesignModeler Geometry** to start **Design Modeler**. Users may double click **Geometry** to directly enter **Design Modeler** if **New SpaceClaim Geometry** is not present. Click the **Units** drop down on the top menu, and select U.S. Customary Unit System: Millimeter.

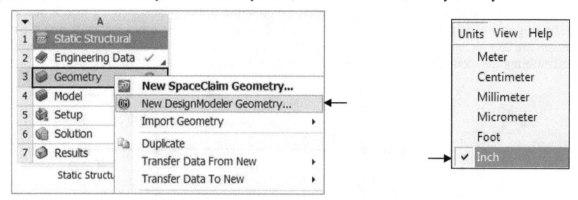

In **Tree Outline**, select **XYPlane** as the sketching plane. Click Look at to orient the sketching plane for defining 3 straight lines.

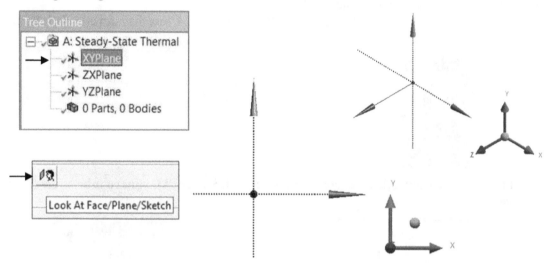

Click Sketching, and click Draw. In the Draw window, click Line. Sketch a horizontal line from the origin. Make sure P is shown when starting the sketch, as shown below.

Click **Dimensions** and from the **Dimensions** panel, select **General**. Pick the sketched line and make a left click to position the size dimension. Specify 500 as its value.

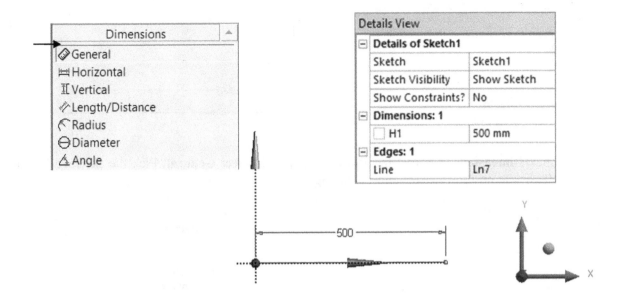

In the Draw window, click **Line**. Sketch the second horizontal line from the right vertex of the sketched line. Make sure P is shown when starting the sketch, as shown below.

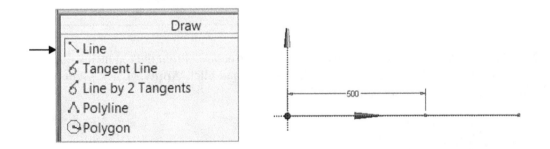

Click **Dimensions** and from the **Dimensions** panel, select **General**. Pick the sketched line and make a left click to position the size dimension. Specify 250 as its value.

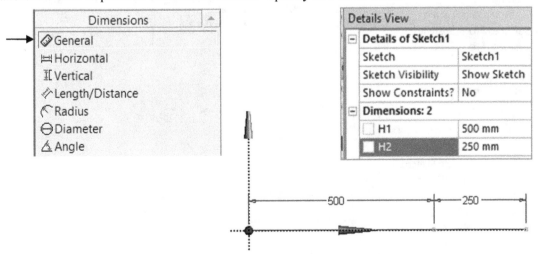

In the Draw window, click **Line**. Sketch the 3rd line starting from the right vertex of the second sketched line. Make sure that P is on display when sketching.

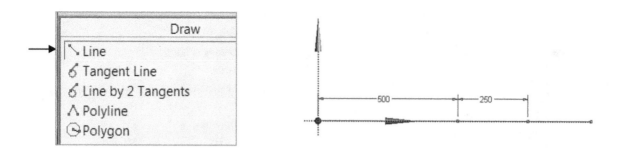

Click **Dimensions** and from the **Dimensions** panel, select **General**. Pick the sketched line and make a left click to position the size dimension. Specify 250 as its value.

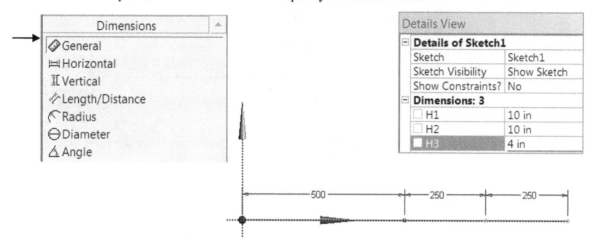

Now let us define the lines to be used for the beam structure. From the top menu, click **Concept** > **Lines from Sketches**. In Tree Outline, highlight Sketch1 and click **Apply**. Click **Generate**. **Line1** is created. An item called **Line Body** is also listed.

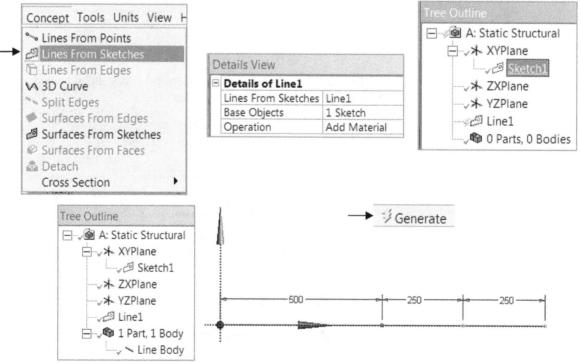

Click **Concept** > **Cross Section** > **Rectangular**. Specify 50 and 25 as the dimensions of 2 sides. Rec1 is created and listed in Tree Outline.

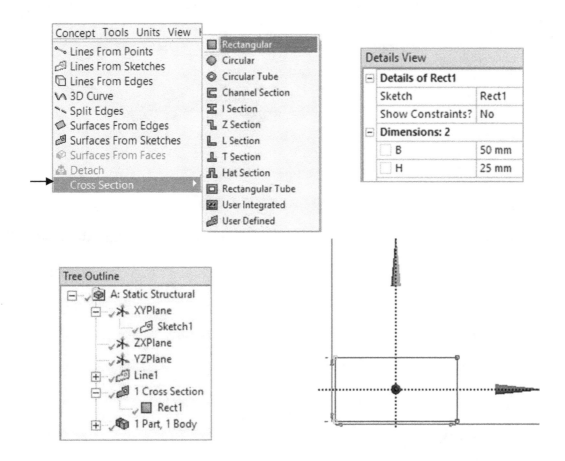

To assign the defined section **Rect1** to **Line1**, highlight the **Line Body** listed under **1 Part, 1 Body**. In Details View, select **Rect1** in Cross Section.

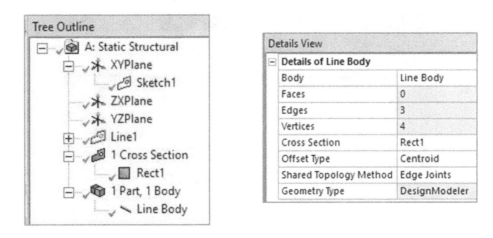

To visualize the created 3D cross section solids for the 2 3D Curves, from the top menu, click **View > Cross Section Solids**. The created solid model is on display.

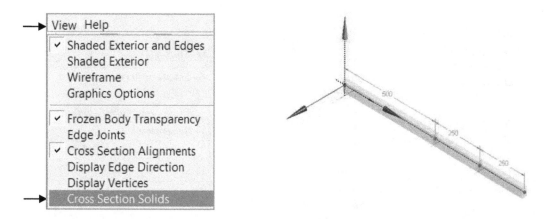

In Tree Outline, **1 Part**, **1 Body** is listed. Under **1 Part**, **1 Body**, highlight **Line Body**, right-click to pick **Rename**. Type Rectangular.

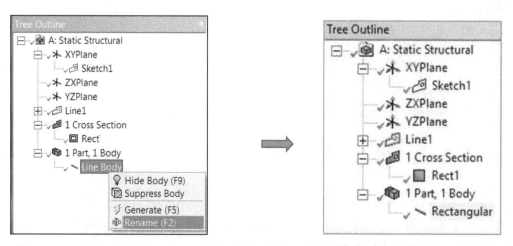

Upon completing the creation of this 3D solid model, we need to close Design Modeler. Click **File** > Close **DesignModeler**. We go back to Project Schematic. Click **Save.**

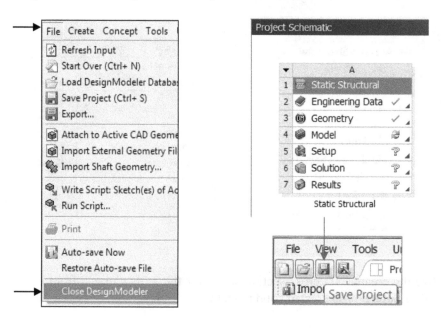

Step 4: Assign Structural Steel by Default, Generate Mesh and Perform FEA.

From the project model tree, highlight **Mesh,** and right-click to pick **Generate Mesh**.

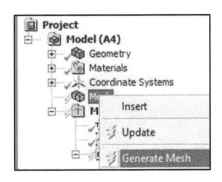

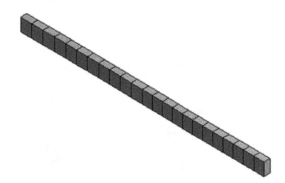

From the project tree, highlight **Static Structural (A5)** and right-click to pick **Insert**. Afterwards, select **Displacement**. From the screen, pick the low-left vertex, as shown, as shown. Set zero for X, Y, and Z Components. Click the Vertex filter before picking the point

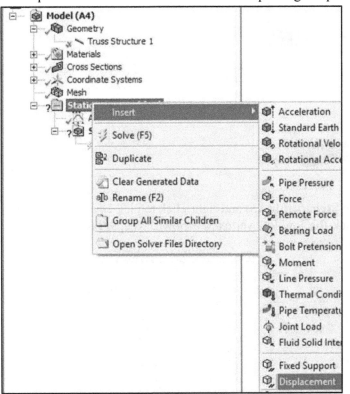

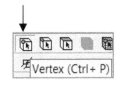

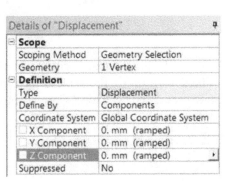

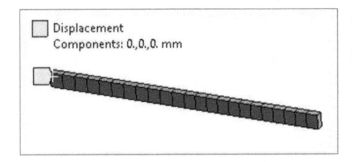

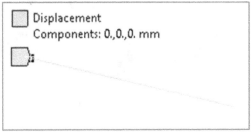

From the project tree, highlight **Static Structural (A5)** and right-click to pick **Insert**. Afterwards, select **Displacement**. From the screen, pick the low-right vertex, as shown, as shown. Set free for X, and zero for Y, and Z Components. Click the Vertex filter before picking the point

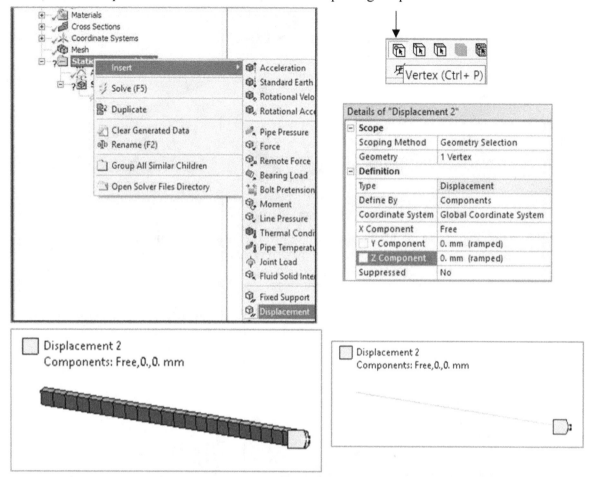

From the project tree, highlight **Static Structural (A5)** and right-click to pick **Insert**. Afterwards, select **Fixed Rotation**. From the screen, pick the low-left vertex, as shown. Keep **Fixed** for X and Y Components and Set **Free** for Z Components.

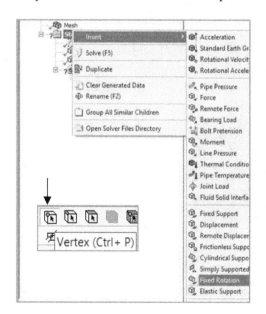

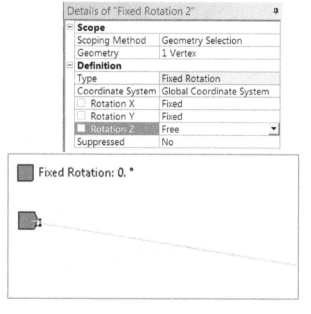

From the project tree, highlight **Static Structural (A5)** and right-click to pick **Insert**. Afterwards, select **Fixed Rotation**. From the screen, pick the low-right vertex, as shown. Keep **Fixed** for X and Y Components and Set **Free** for Z Components.

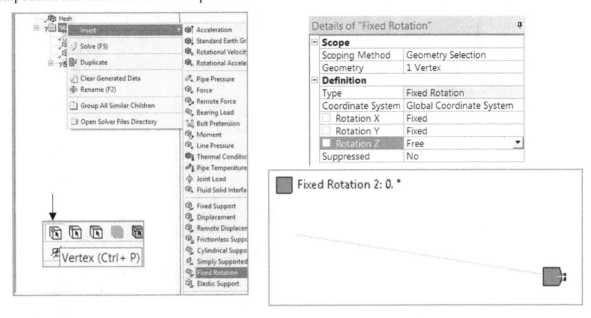

From the project tree, highlight **Static Structural (A5)** and right-click to pick **Insert**. Afterwards, select **Force**. From the screen, pick the edge on the left side or the first sketched line, as shown. Select Y Component and specify -200 N as the magnitude of the force load and downward.

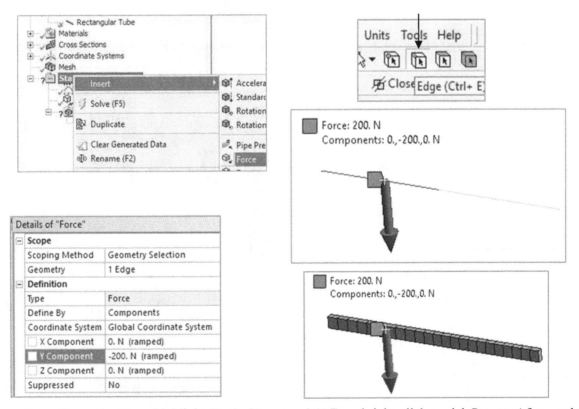

From the project tree, highlight **Static Structural (A5)** and right-click to pick **Insert**. Afterwards, select **Force**. From the screen, pick the vertex between the second and third sketched lines, as shown. Select Y Component and specify -100 N as the magnitude of the force load and downward.

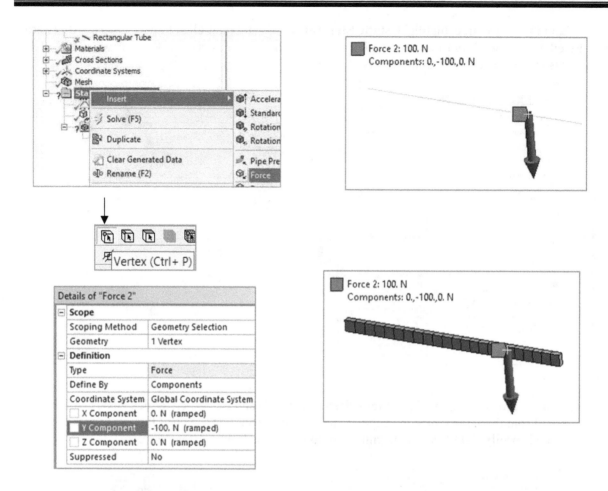

From the project tree, highlight **Solution (A6)** and right-click to pick **Solve**. Workbench system is running FEA.

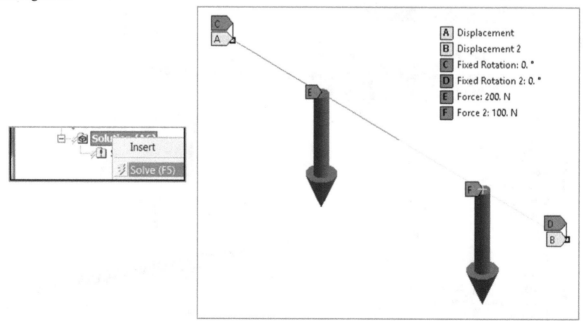

To plot the displacement distribution, highlight **Solution (A6)** and pick **Insert**. Afterwards, highlight **Deformation** and pick **Total**. Afterwards, right click to pick **Evaluate All Results**. The maximum value of deformation at the free end is 0.078 mm.

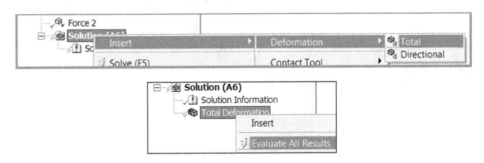

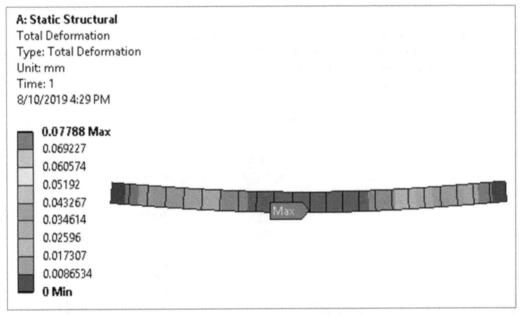

To plot the reaction force at the fixed end, highlight **Solution (A6)** and pick **Insert**. Afterwards, highlight **Probe > Force Reaction**. In Details of "Force Reaction", select Displacement in Boundary Condition. Afterwards, right click to pick **Evaluate All Results**. The total force in Y direction is 175 N, as shown in Details of "Force Reaction".

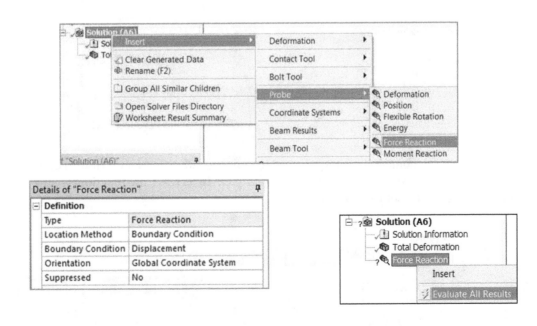

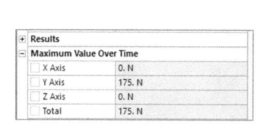

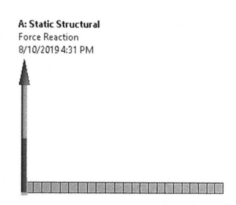

To plot the moment reaction at the fixed end, highlight **Solution (A6)** and pick **Insert**. Afterwards, highlight **Probe > Force Reaction.** In Details of "Force Reaction", select Displacement 2 in Boundary Condition. Afterwards, right click to pick **Evaluate All Results**. The total force in Y direction is 125 N, as shown in Details of "Force Reaction".

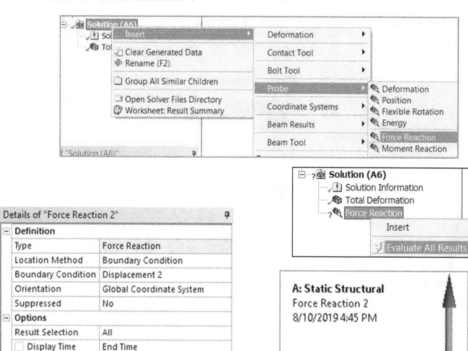

To plot the shear-moment diagram, we need to define a path beforehand. How to define a path? In the Project Tree, highlight Model (A4). Right-click to pick **Insert > Construction Geometry.** In the Project Tree, Construction Geometry is listed.

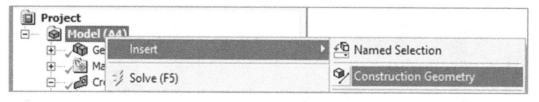

Now highlight **Construction Geometry** and right-click to pick Insert > Path. In Details of "Path", select Edge in Path Type. From the top menu, select Edge Section. While holding down the **Ctrl** key, pick the 3 sketched lines. Afterwards, click Apply. An item called Path is created and listed in the Project tree.

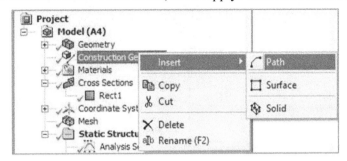

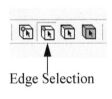

Edge Selection

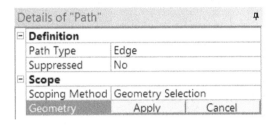

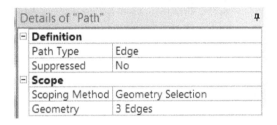

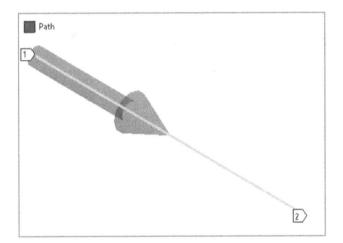

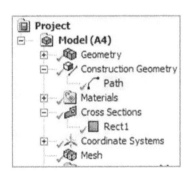

Now let us construct the shear-moment diagram. Highlight **Solution (A6)** and pick **Insert**. Afterwards, highlight **Beam Results > Shear-Moment Diagram.** In Details of "Total Shear-Moment Diagram", in the box called Path, select Path. Afterwards, right click to pick **Evaluate All Results**. The Total Shear-Moment Diagram is on display.

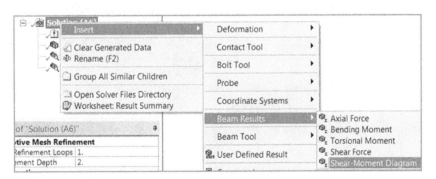

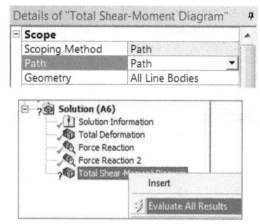

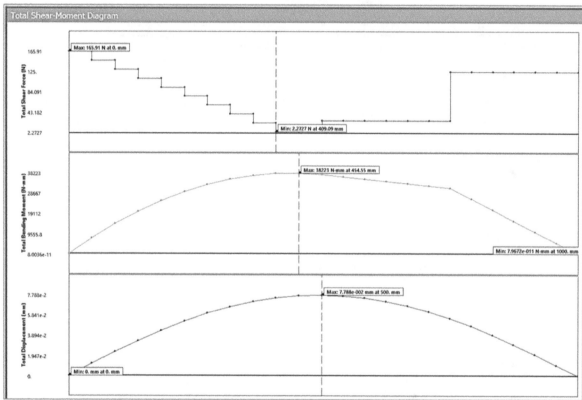

Let us examine the Total Shear Force Diagram. The maximum shear force is 175 N starting at the left end, its value decreases towards the middle of the beam. The shear force is 25 N for 250 mm. The shear force value jumps to 125 N at 750 mm, and maintains 125 N to the right end.

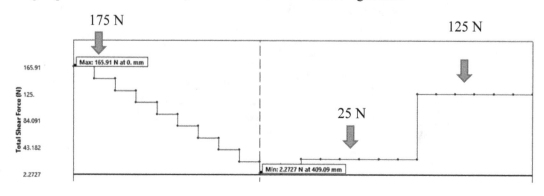

Users may save the shear-moment diagram data and export the date to an EXCEL file. Users may construct the shear-moment diagram using **EXCEL**.

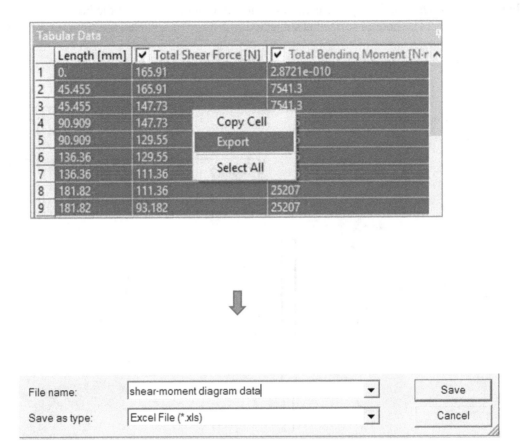

At this moment, we conclude our discussion on the rectangular beam case study. From the top menu, click **File > Close Mechanical**. In this way, we return to the main page of Workbench. Click the icon of **Save Project**.

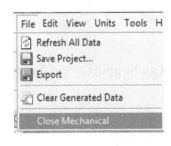

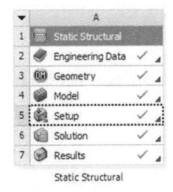

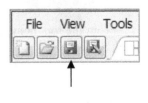

6.6 Using 3D Curve to Model Ceiling Fan Support

In this case study, we use the ceiling fan support component as an example to demonstrate the procedures of using 3D curves to create the ceiling fan support. Assume that the support (Structural Steel) consists of two bars: a squared bar on the top and a circular bar (Aluminum Alloy) on the bottom. The top surface is fixed to the ceiling. The bottom surface is subjected to a force of 500 Newton, imitating the weight of the ceiling fan.

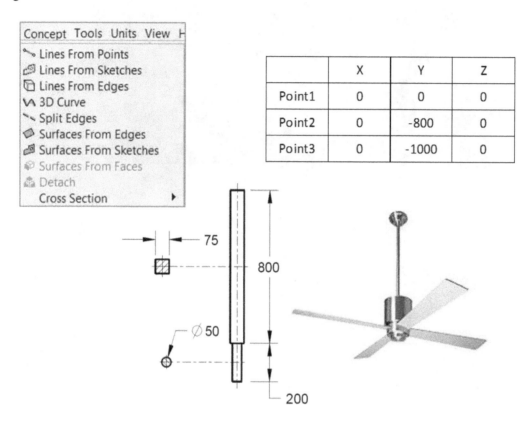

	X	Y	Z
Point1	0	0	0
Point2	0	-800	0
Point3	0	-1000	0

Step 1: Open Workbench and Save the File Name.

From the start menu, click **ANSYS 18.0 > Workbench 18.0**. Users may directly click Workbench 18.0 when the symbol of Workbench 18.0 is on display.

From the top menu, click the icon of **Save Project**. Specify **Ceiling Fan Support** as the file name, and click **Save**.

Step 2:From **Toolbox**, highlight **Static Structural** and drag it to the main screen.

Step 3: In Design Modeler, create three datum points and 2 lines.

Right-click the **Geometry** cell and pick **New DesignModeler Geometry** to start **Design Modeler**. Users may double click **Geometry** to directly enter **Design Modeler** if **New SpaceClaim Geometry** is not present. Click the **Units** drop down on the top menu, and select **Millimeter**.

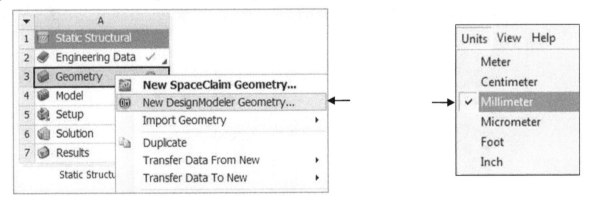

In **Tree Outline**, select **XYPlane** as the sketching plane.

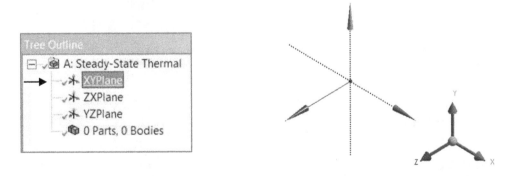

Click **Look At** to orient the sketching plane for defining 3 data points.

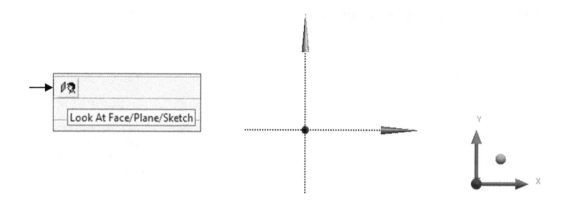

Let us create Point1. From the top menu, click the icon of **Point**. In Details of View, select Manual Input and specify (0, 0, 0) as the coordinates of Point1. Click **Generate.**

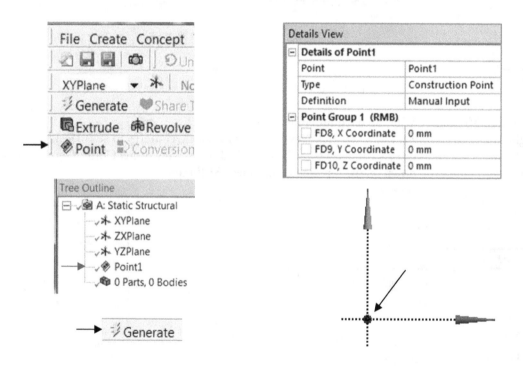

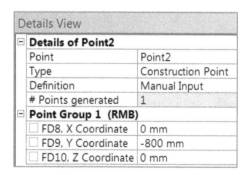

Repeat the above procedure 2 more times to define 2 more datum points: Point2 (0, -800, 0), Point3 (0, -1000, 0), as shown below.

Details of Point2	
Point	Point2
Type	Construction Point
Definition	Manual Input
# Points generated	1
Point Group 1 (RMB)	
☐ FD8, X Coordinate	0 mm
☐ FD9, Y Coordinate	-800 mm
☐ FD10, Z Coordinate	0 mm

Details of Point3	
Point	Point3
Type	Construction Point
Definition	Manual Input
# Points generated	1
Point Group 1 (RMB)	
☐ FD8, X Coordinate	0 mm
☐ FD9, Y Coordinate	-1000 mm
☐ FD10, Z Coordinate	0 mm

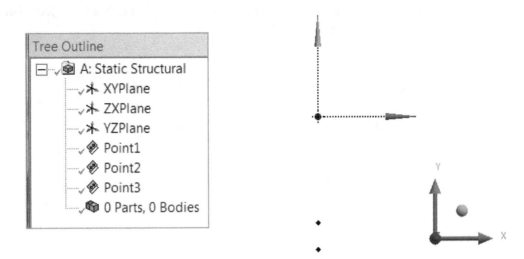

Click Concept > Lines from Points. While holding down the **Ctrl** key, pick Point1 and Point2. Click **Apply**. Click **Generate**. Line1 is created. An item called Line Body is also listed.

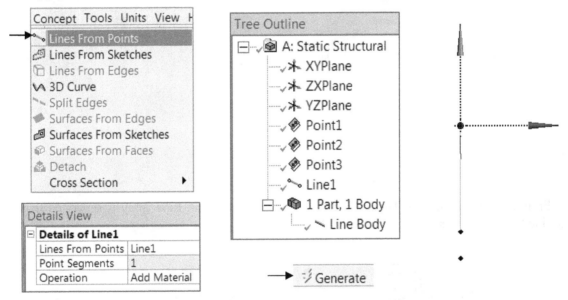

Click **Concept** > **Cross Section** > **Rectangular**. Specify 75 and 75 as the dimensions of 2 sides. Rec1 is created and listed in Tree Outline.

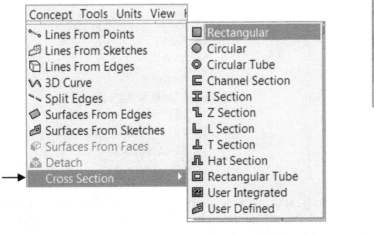

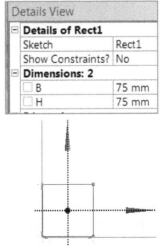

To assign the defined section Rect1 to the Curve1, highlight the first Line Body listed under 2 Parts, 2 Bodies. In Details View, select Rect1 in Cross Section.

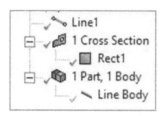

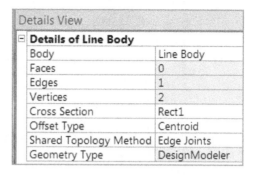

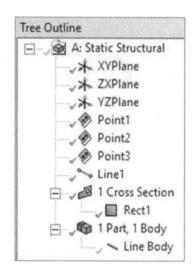

From the top menu, click Tools > Freeze. In this way, the first Line Body is separated from the second Line Body to be created.

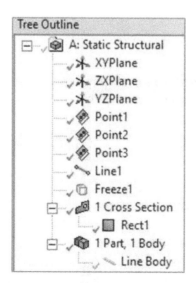

Now let us define the second Line Body, starting with creating a line. Click **Concept > Lines from Points**. While holding down the **Ctrl** key, pick Point2 and Point3. Click **Apply**. Click **Generate**. Line2 is created. The second **Line Body** is listed in Tree Outline. Also listed is **2 Parts, 2 Bodies**.

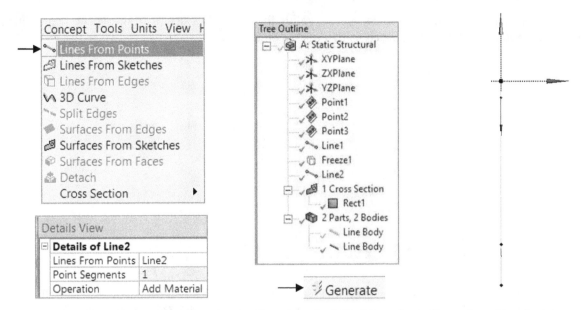

Click **Concept** > **Cross Section** > **Circular**. Specify 25 as the radius value. Circular1 is created and listed in Tree Outline.

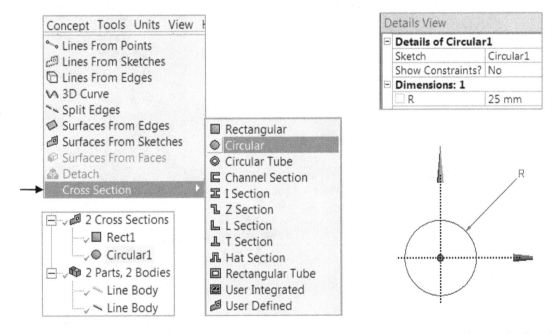

To assign the defined section Circular1 to the Curve2, highlight the second Line Body listed under 2 Parts, 2 Bodies. In Details View, select Circular1 in Cross Section.

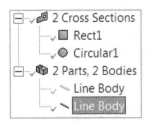
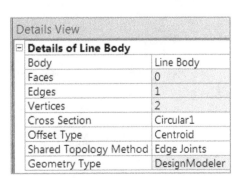

To visualize the created 3D cross section solids for the 2 3D Curves, from the top menu, click View > Cross Section Solids. The created solid model is on display.

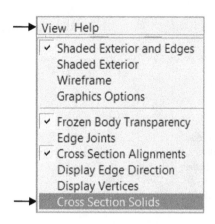

Let us examine the information listed in Tree Outline. We have two Line Bodies. Therefore, we have 2 parts. Now let us highlight the first Line Body and right-click to pick **Rename** to change Line Body to Top Bar, and change the second Line Body to Bottom Bar.

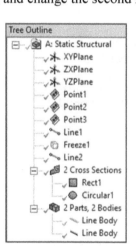

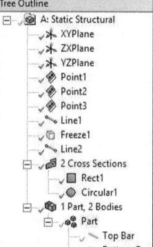

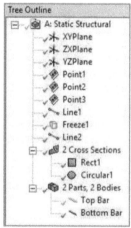

In Tree Outline, 2 Parts, 2 Bodies, holding down the **Ctrl** key, left click Top Bar and Bottom Bar. Afterwards, right-click to pick Form New Part. A new part is formed, which consists of both Top Bar and Bottom Bar.

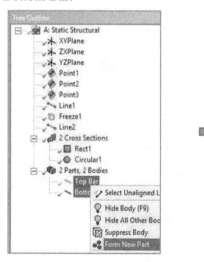

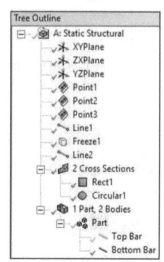

Upon completing the creation of this 3D solid model, we need to close Design Modeler. Click **File** > Close **DesignModeler**. We go back to Project Schematic. Click Save.

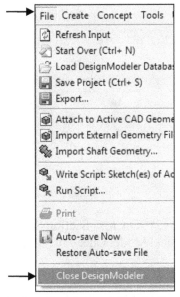
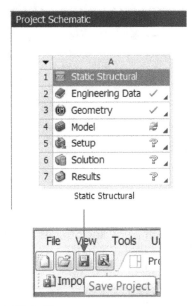

Step 4: Activate the Aluminum Alloy in the Engineering Data Sources
In the Static Structural panel, double click Engineering Data. We enter the Engineering Data Mode.

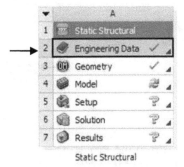

Click the icon of **Engineering Data Sources**. Click **General Materials**. We are able to locate Aluminum Alloy.

Click the plus symbol nearby Aluminum Alloy to activate it. Afterwards, click the icon of **Engineering Data Sources** and click the icon of **Project** to go back to Project Schematic.

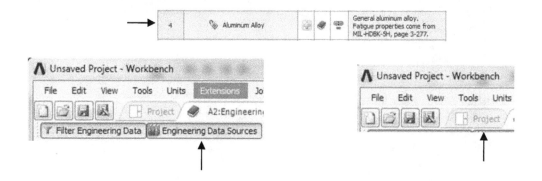

Step 5: Assign Aluminum Alloy and Structural Steel to the two (2) 3D Cross Section Solids.

Double-click the icon of **Model**. We enter the Mechanical Mode. Check the unit system. Select Metric (mm, t, N, s, mV, mA).

Now let us do the material assignment. From the Project tree, expand Geometry. Highlight **Top Bar**. In Details of "Top Bar", Structural Steel is listed in Assignment by the system default. Therefore, keep this default material assignment.

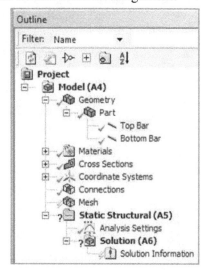

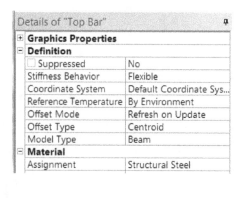

From the Project tree, highlight Bottom Bar. In Details of "Bottom Bar", click material assignment, pick **Aluminum Alloy** to replace Structural Steel, which is the default material assignment.

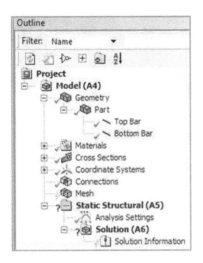

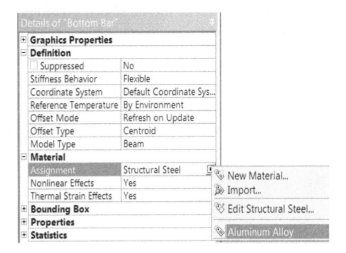

Step 6: Mesh generation and Perform FEA.

From the project model tree, highlight **Mesh,** and right-click to pick **Generate Mesh**.

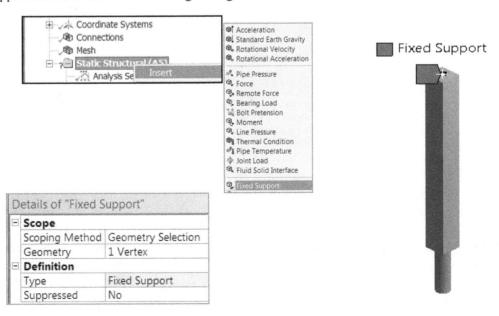

From the project tree, highlight **Static Structural (A5)** and right-click to pick **Insert**. Afterwards, select **Fixed Support**. From the screen, pick Point1, which is on the top of Top Bar, as shown. As a result, this support model is fixed to the ceiling through Point1.

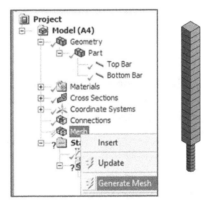

From the project tree, highlight **Static Structural (A5)** and right-click to pick **Insert**. Afterwards, select **Force**. From the screen, pick Point3 from the bottom of Bottom Bar, as shown. Specify 500 N as the magnitude of the force load.

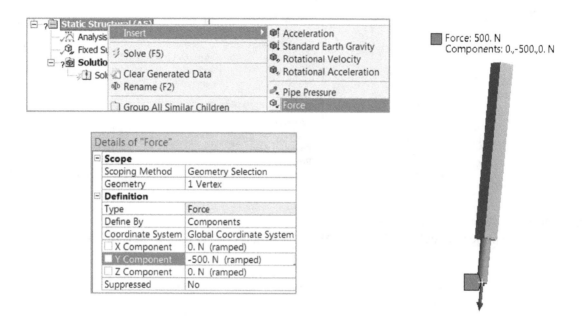

From the project tree, highlight **Static Structural (A5)** and right-click to pick **Insert**. Afterwards, select **Standard Earth Gravity**. In Details of "Standard Earth Gravity", set Direction to −Y. The numerical value of -9806.6 mm/s² is automatically shown.

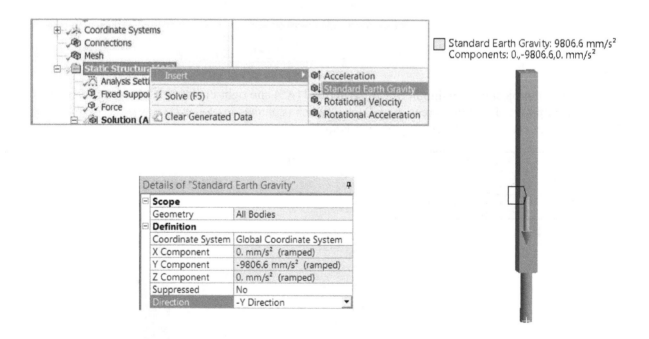

From the project tree, highlight **Solution (A6)** and right-click to pick **Solve**. Workbench system is running FEA.

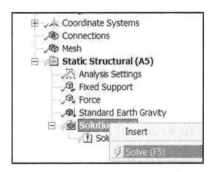

To plot the displacement distribution, highlight **Solution (A6)** and pick **Insert**. Afterwards, highlight **Deformation** and pick **Total.** Afterwards, right click to pick Evaluate All Results. The maximum value of deformation at the free end is 0.0012 mm.

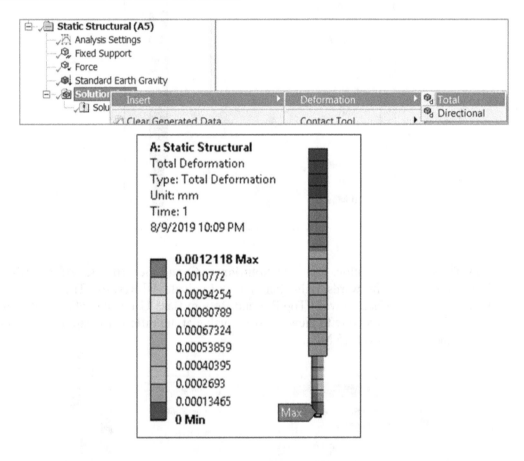

To plot the axial force distribution, highlight **Solution (A6)** and pick **Insert**. Afterwards, highlight **Beam Results > Axial Force.** Afterwards, right click to pick Evaluate All Results. The maximum value of the axial force at the fixed end is 857.08 N (500 N + weights of the top and bottom bars). The minimum value of the axial force at the free end is 500 N (weight of the fan).

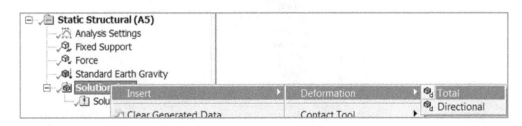

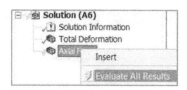

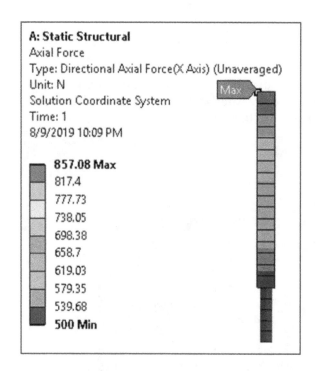

To plot the stress distribution, highlight **Solution (A6)** and pick **Insert**. Afterwards, highlight **Beam Tools > Beam Tool.** Afterwards, right click to pick Evaluate All Results. The maximum value of stress is 0.26 MPa at the interface between Top Bar and Bottom Bar. Users may click Probe and use the mouse to go over the support structure to view the stress value at the different locations. As an example, the stress value at the fixed end is 0.15 MPa.

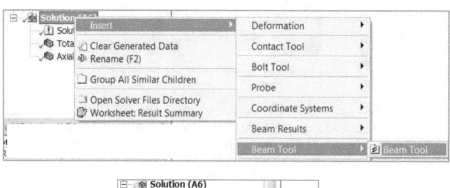

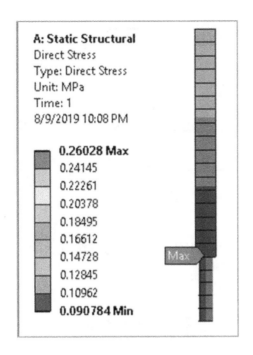

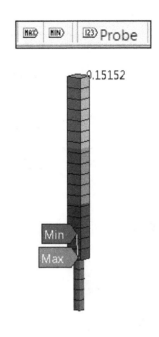

At this moment, we conclude our discussion on the ceiling fan support structure. From the top menu, click **File** > **Close Mechanical**. In this way, we return to the main page of Workbench. Click the icon of Save Project.

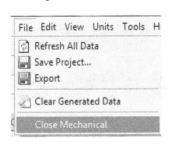

Static Structural

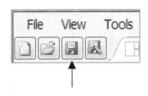

6.7   Plane Stress with Thickness Model for Hexagon Wrench

A hexagon wrench is shown below. This wrench is made from sheet metal with thickness value equal to 3 mm. In this case study, we create a model using a surface body. When we perform FEA, we specify the thickness value. Such a model is called a plane stress with thickness model as we assume that the stress along the vertical direction is near zero. In this study, we first create a surface body model in the 2D space. We assume the material type is structural steel and the wrench model is subjected to a pressure load with magnitude equal to 0.05 MPa. We will perform an FEA run.

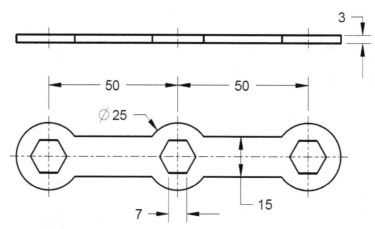

Step 1: Open Workbench and Save the File Name.

From the start menu, click **ANSYS 19 > Workbench 189**. Users may directly click Workbench 19 when the symbol of Workbench 19 is on display.

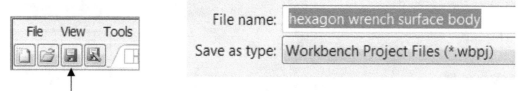

From the top menu, click the icon of **Save Project**. Specify **Hexagon wrench surface body** as the file name, and click **Save**.

Step 2: From **Toolbox**, highlight **Static Structural** and drag it to the main screen.

Step 3: In Design Modeler, create three datum points and 2 lines.

Right-click the **Geometry** cell and pick **New DesignModeler Geometry** to start **Design Modeler**. Users may double click **Geometry** to directly enter **Design Modeler** if **New SpaceClaim Geometry** is not present. Click the **Units** drop down on the top menu, and select **Millimeter**.

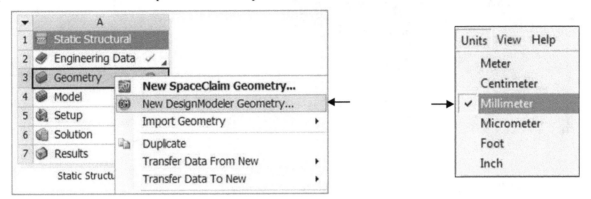

In **Tree Outline**, select **XYPlane** as the sketching plane. Click **Look At** to orient the sketching plane.

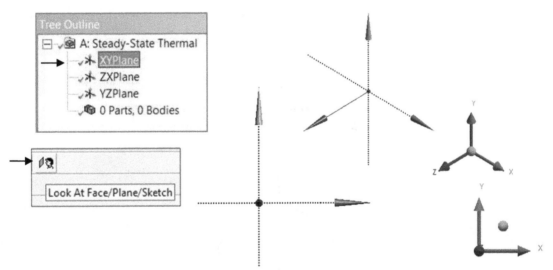

Click the icon of **Sketching**. Click **Draw**. Select **Circle**. Sketch a circle, as shown.

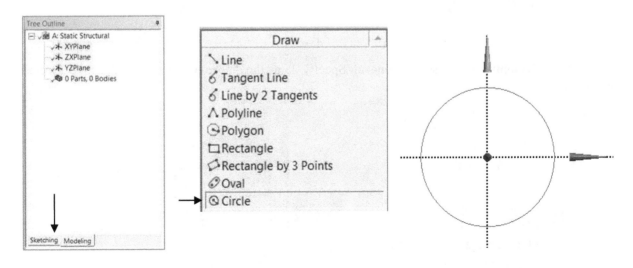

Click **Dimensions**, select **General**. Specify 25 as the diameter value, as shown.

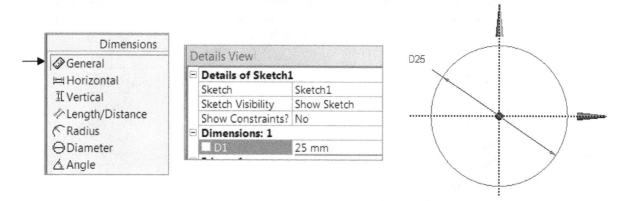

Click **Draw**. Select **Circle**. Sketch 2 circles, as shown.

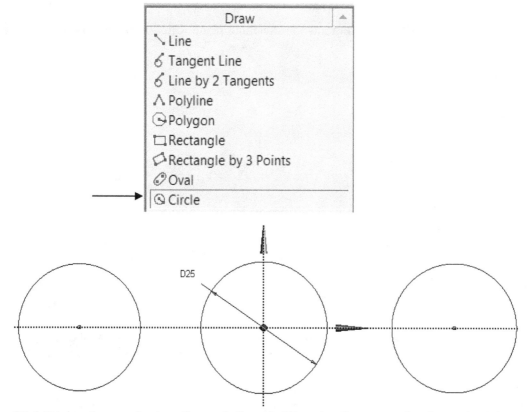

Click **Dimensions** and select **General**. Specify 25 as the diameter value for each of the 2 circles, as shown.

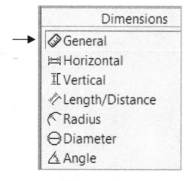

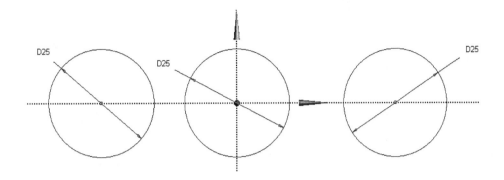

Click **Dimensions** and select **General**. Specify 50 as the distance value between the centers of the 2 neighboring circles, as shown.

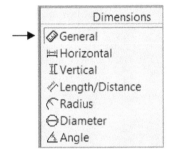

Details View	
□ **Details of Sketch1**	
Sketch	Sketch1
Sketch Visibility	Show Sketch
Show Constraints?	No
□ **Dimensions: 5**	
□ D1	25 mm
□ D2	25 mm
□ D3	25 mm
□ H4	50 mm
■ H5	50 mm

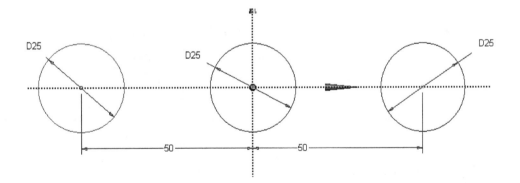

Click **Draw**. Select **Line**. Sketch 2 horizontal lines, as shown.

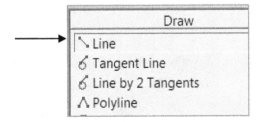

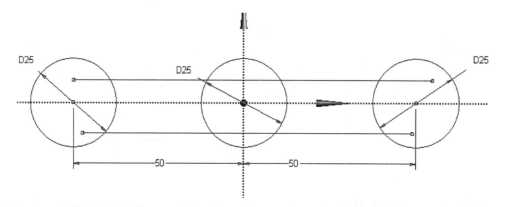

Click **Constraints**. Select **Symmetry**. Make 3 left clicks. The first left is the horizontal axis. The second and third left clicks are the 2 sketched horizontal lines.

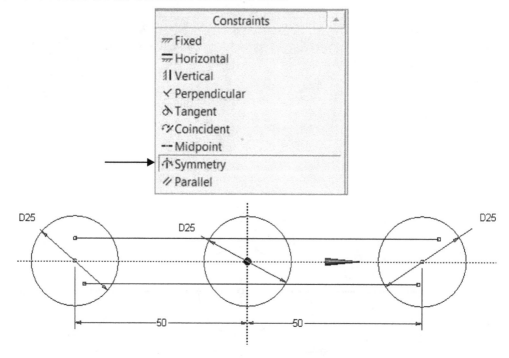

Click **Dimensions** and select **Vertical**. Specify 15 as the distance value between the 2 sketched horizontal lines, as shown.

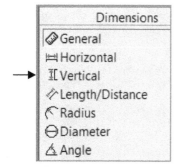

Details View	
⊟ **Details of Sketch1**	
Sketch	Sketch1
Sketch Visibility	Show Sketch
Show Constraints?	No
⊟ **Dimensions: 6**	
☐ D1	25 mm
☐ D2	25 mm
☐ D3	25 mm
☐ H4	50 mm
☐ H5	50 mm
☐ V7	15 mm

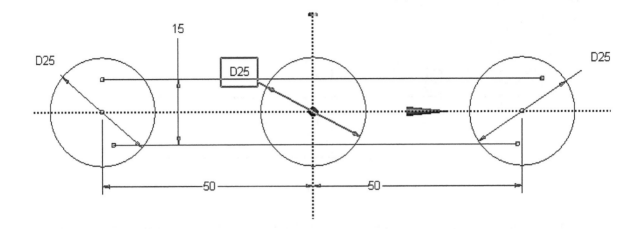

Click **Modify**. In the **Modify** window, click **Trim**. Make left-clicks on those line segments and arcs so that they are deleted from the sketch, as shown.

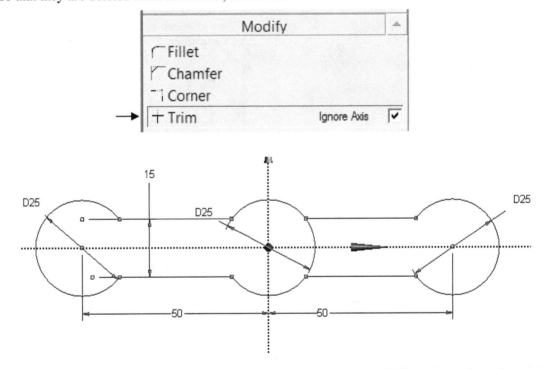

Click **Draw**. Select **Polygon** and specify 6 as hexagon. At the center of this sketch, sketch the first hexagon, as shown.

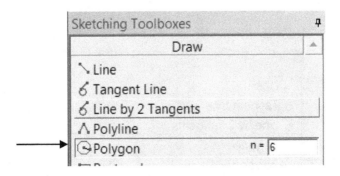

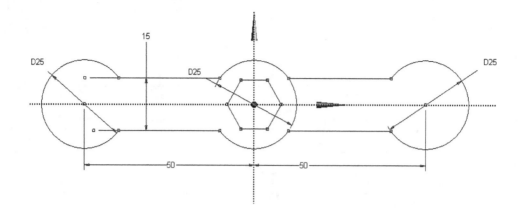

Click **Dimensions** and select **General**. Specify 7 as the size of the sketched hexagon, as shown.

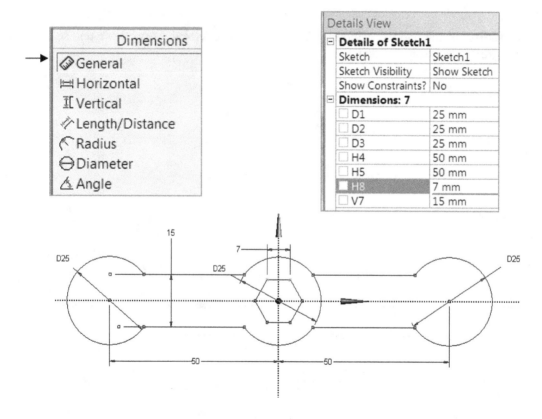

Click **Draw**. Select **Polygon** and specify 6 as hexagon. Sketch the 2 more hexagons, one is on the left side and the other is on the right side, as shown.

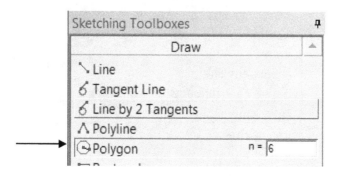

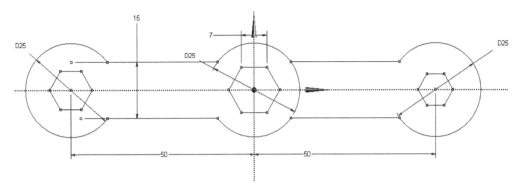

Click **Dimensions** and select **General**. Specify 7 as the size of the 2 sketched hexagons, as shown.

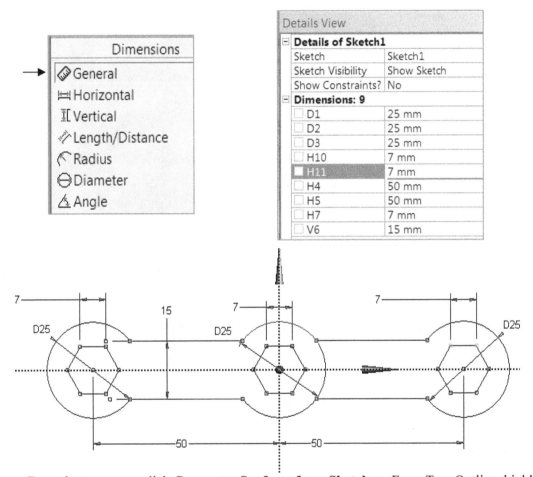

From the top menu, click **Concept > Surfaces from Sketches**. From Tree Outline, highlight Sketch1, click **Apply**. Specify 3 as the thickness value. Click **Generate** and the surface body is created. In Tree Outline, 1 Part, 1 Body is shown and **Surface Body** is listed.

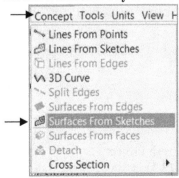

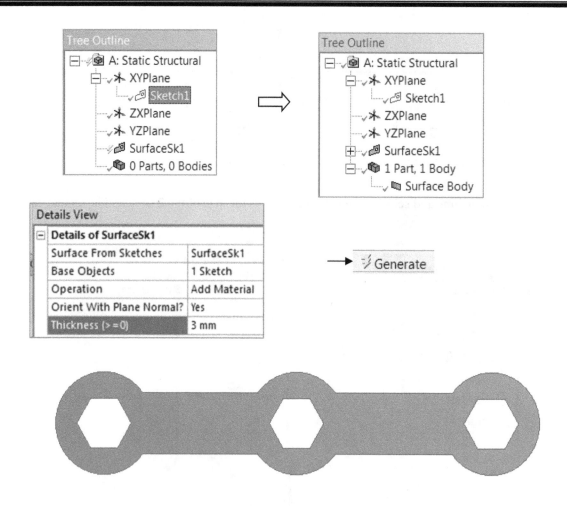

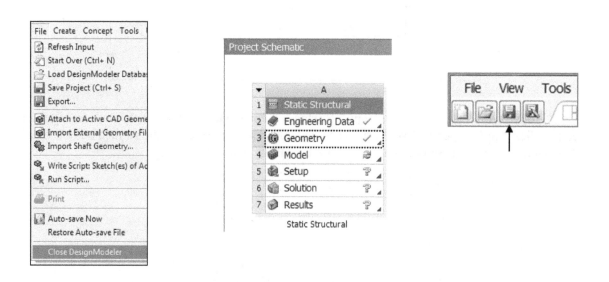

Upon completing the process of creating the surface body, click **File** > Close **Design Modeler**, and go back to **Project Schematic**. Click Save Project.

Step 4: Specify 2D Properties and Refresh Project.
Right-click the Geometry cell (**A3**) in Static Structural. Right-click to pick **Properties**.

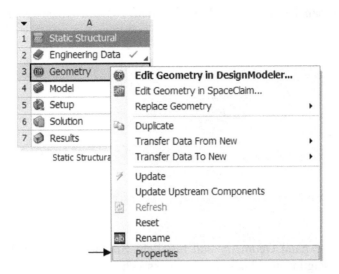

In the Properties window, change 3D to 2D in Analysis Type. Click the close symbol to go back to the Project Schematic page.

		A	B
		Property	Value
1			
2	□ General		
3		Component ID	Geometry
4		Directory Name	SYS
5	□ Notes		
6		Notes	
7	□ Used Licenses		
8		Last Update Used Licenses	
9	□ Geometry Source		
10		Geometry File Name	C:\Users\zhang\Documents\Workbench summer 2016\Workbench Notes\hexagon wrench surface body 1_files\dp0\SYS\DM\SYS.agdb
11	□ Advanced Geometry Options		
12		Analysis Type	2D
13		Compare Parts On Update	No

Properties of Schematic A3: Geometry

On the screen, right-click to pick **Refresh Project**.

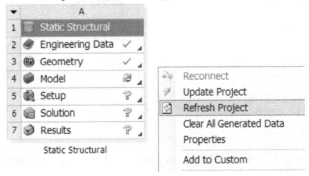

Step 5: Accept the Default Material Assignment of Structural Steel and Perform FEA.

In the Static Structural panel, double click **Model**. Check the unit system: Metric (mm, t, N, s, mV, mA).

From the Project tree, expand Geometry and highlight **Surface Body**. In the Details window, Structural Steel is shown in Assignment. Check the thickness value of 3. Structural Steel is the material assignment.

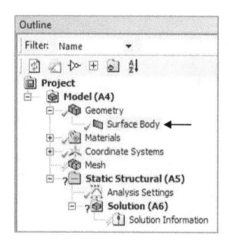
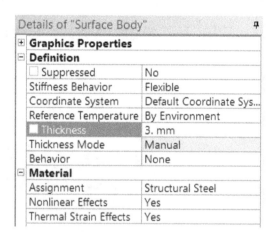

Now highlight Geometry, in Details of "Geometry", Plane Stress is defined by the system default, as shown in **2D Behavior**.

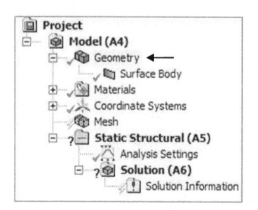
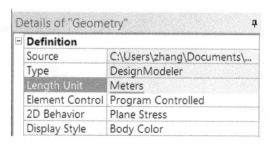

From the Project tree, Highlight **Mesh**. In Details of "Mesh", specify 2 as Element Size. Right-click to pick **Generate Mesh**.

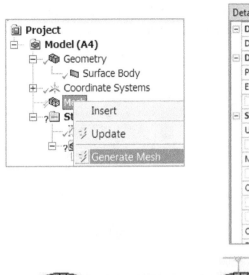

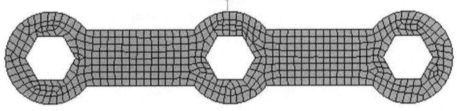

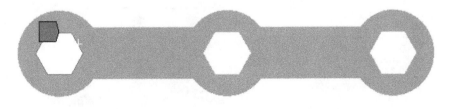

To define the boundary conditions, first highlight **Static Structural (A5),** and right-click to pick **Insert > Fixed Support**. From the top menu, pick Edge Selection. While holding down the Ctrl key, pick the 6 edges from the left hexagon, as shown.

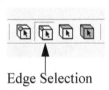

Edge Selection

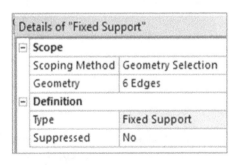

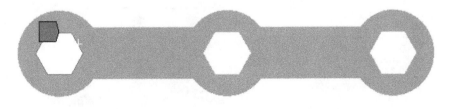

To define the pressure load, highlight **Static Structural (A5),** and right-click to pick **Insert > Pressure**. From the top menu, pick Edge Selection. Pick the edge, as shown. Specify 1.0 MPa as the pressure magnitude.

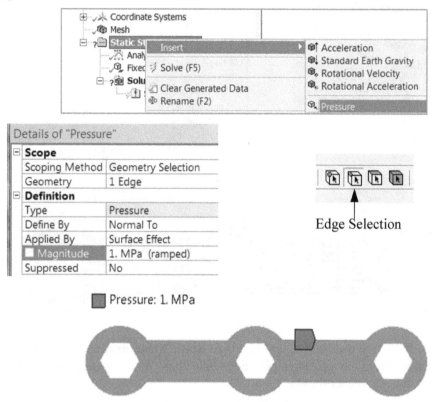

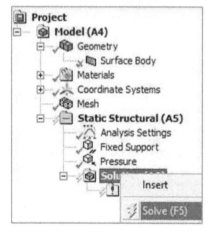

Highlight **Solution (A6)** and right-click to pick **Solve**.

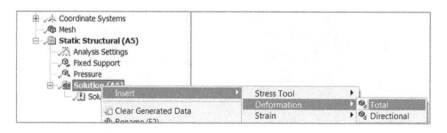

To plot the displacement distribution, highlight **Solution** (A6) and pick **Insert**. Afterwards, highlight **Deformation** and pick **Total.** Afterwards, right click to pick **Evaluate All Results**. The maximum value of deformation is 0.11 mm.

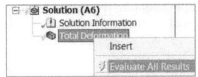

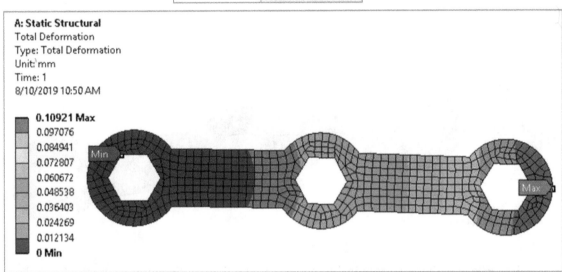

To plot the von Mises distribution, highlight **Solution (A6)** and pick **Insert**. Afterwards, highlight **Stress > Equivalent Stress**. Afterwards, right click to pick **Evaluate All Results**. The maximum value of von Mises stress is 76.5 MPa.

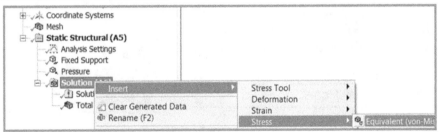

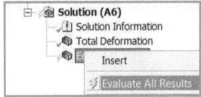

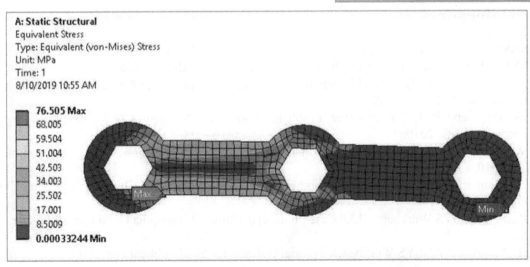

Users may press down the middle button of the mouse. A 3D view of the wrench model is shown with the von Mises distribution on display.

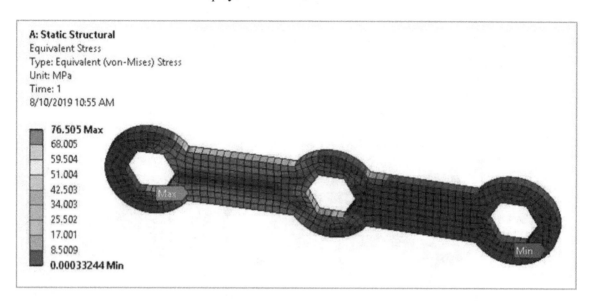

Click **File** and **Close Mechanical** to go back to the **Project Schematic** screen. Click the icon of **Save Project** to complete this study.

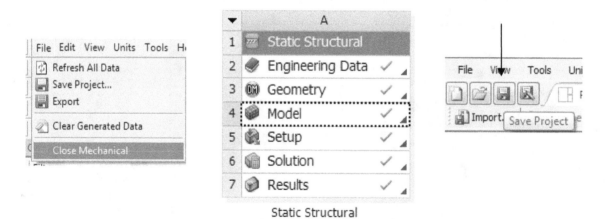

Static Structural

6.8   References

1. ANSYS Parametric Design Language Guide, Release 15.0, Nov. 2013, ANSTS, Inc.
2. ANSYS Workbench User's Guide, Release 15.0, Nov. 2013, ANSYS, Inc.
3. X. L. Chen and Y. J. Liu, Finite Element Modeling and Simulation with ANSYS Workbench, 1st edition, Barnes & Noble, January 2004.
4. J. W. Dally and R. J. Bonnenberger, Problems: Statics and Mechanics of Materials, College House Enterprises, LLC, 2010.
5. J. W. Dally and R. J. Bonnenberger, Mechanics II Mechanics of Materials, College House Enterprises, LLC, 2010.
6. X. S. Ding and G. L. Lin, ANSYS Workbench 14.5 Case Studies (in Chinese), Tsinghua University Publisher, Feb. 2014.
7. G. L. Lin, ANSYS Workbench 15.0 Case Studies (in Chinese), Tsinghua University Publisher, October 2014.
8. K. L. Lawrence. ANSYS Workbench Tutorial Release 13, SDC Publications, 2011.

9. K. L. Lawrence. ANSYS Workbench Tutorial Release 14, SDC Publications, 2012
10. Huei-Huang Lee, Finite Element Simulations with ANSYS Workbench 14, Theory, Applications, Case Studies, SDC Publications, 2012.
11. Huei-Huang Lee, Finite Element Simulations with ANSYS Workbench 16, Theory, Applications, Case Studies, SDC Publications, 2015.
12. Jack Zecher, ANSYS Workbench Software Tutorial with Multimedia CD Release 12, Barnes & Noble, 2009.
13. G. M. Zhang, Engineering Design and Creo Parametric 3.0, College House Enterprises, LLC, 2014.
14. G. M. Zhang, Engineering Analysis with Pro/Mechanica and ANSYS, College House Enterprises, LLC, 2011.

6.9 Exercises

1.   A simple supported beam with a pin connection is shown below. The beam material is structural steel. The left end is a pin connection and the right end is simple-supported. The load is acting at the middle section and the magnitude is 100 N and the direction is downward. The cross section of the beam is a rectangle and the two dimensions are 25 mm and 50 mm, respectively.

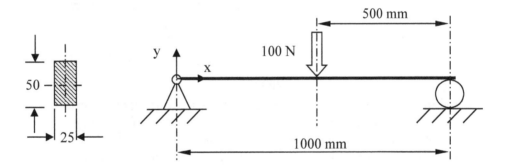

Determine the reaction forces at the two ends with two plots. Determine the total deformation along this beam with a plot. Construct the total shear-moment diagram.

2.   A simple supported beam with a pin connection is shown below. The beam material is structural steel. The left end is a pin connection and the right end is simple-supported. A uniformly distributed load is acting on the part of the beam with the magnitude of 10 N/mm. A torque is acting at PNT3. Its magnitude is 2 N-meter and counter clockwise direction. The cross section of the beam is a rectangle and the two dimensions are 20 mm and 40 mm, respectively.

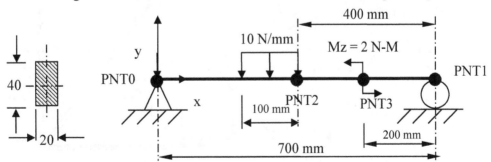

Determine the reaction forces at the two ends with two plots. Determine the total deformation along this beam with a plot. Construct the total shear-moment diagram.

3.   The following drawing shows a planar truss structure. Note that the dimension unit is Meter. This truss structure consists of 5 straight bars (AB, BC, BD, DC and AC). The cross section of the 5 bars is a square with the side dimension equal to 0.1 Meter. There is a force load of 400 N acting

on this truss structure. As shown. The node marked as A is constrained with a joint. The node marked as B is constrained with a roller. Determine the axial force developed in the five bars. Determine the max value of deformation total and the directional stress developed in each of the five bars.

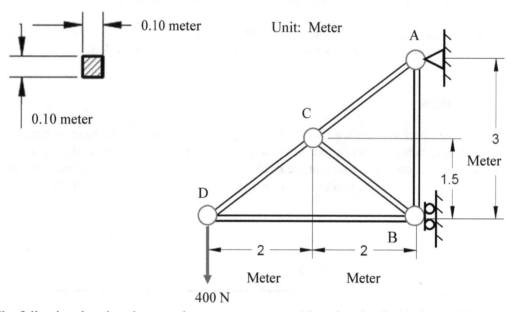

4.   The following drawing shows a planar truss structure. Note that the dimension unit is Meter. This truss structure consists of 5 straight bars. The cross section of the 5 bars is a square with the side dimension equal to 0.1 Meter. There are two loads acting on this truss structure: 400 N and 600 N. The node marked as A is constrained with a joint. The node marked as B is constrained with a roller. Determine the axial force developed in the five bars. Determine the max value of deformation total and the directional stress developed in each of the five bars.

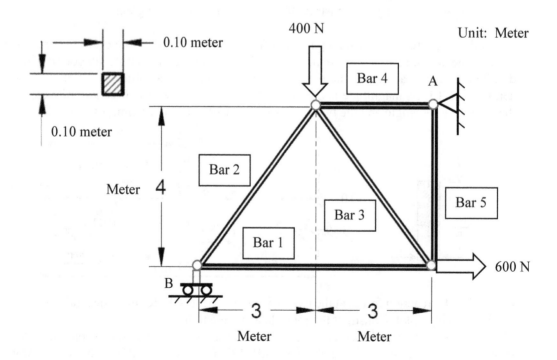

5.   A chimney structure is shown below. The unit of the dimensions shown is inch. The inner layer is constructed from concrete material. The outer layer is constructed from bricks. Assume that the temperature on the surfaces of the inner walls or the concrete layer is 140° F with the coefficient of convection equal to 0.037 Btu/(s.in^2.°F). The outside surfaces are exposed to the surrounding air, the temperature of which, is at 10° F with the coefficient of convection equal to 0.012 Btu/(s.in^2 °F). Determine the temperature distribution under a steady-state condition, and plot the heat fluxes through these 2 layers.

Type	Thermal Conductivity
Brick	0.47 W/m/°C
Concrete	0.72 W/m/°C

Unit: inch

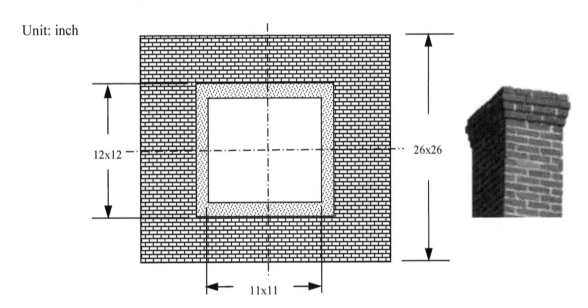

Create a 3D model for this assembly first (make sure that the user selects XYPlane to create the first feature). Afterwards, modify the 3D model to a 2D model. The operation, such as Extrude may need to be deleted so that Concept > Surfaces from Sketches can be performed. Perform a Steady-State Thermal Analysis to determine the temperature distribution.

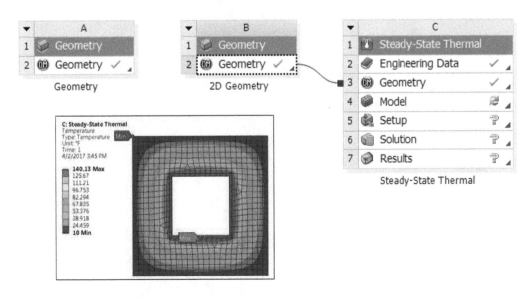

CHAPTER 7

FEA Applications

7.1 Introduction

In this chapter, our focus will be on FEA applications. Section 7.2 presents a case where the deflection under a load condition is large. The example is a clamp device. Workbench has a large deflection module for users to use. Section 7.3 is a case study where two ends of the clamp device need to touch each other. Workbench offers a specific type of contact called no separation configuration to deal with. Section 7.4 presents a case study where a support and a shaft are assembled with a clearance between them. The two contact surfaces slides under the load condition. Under such a circumstance, the no separation contact configuration may be necessary as compared to the bonded contact configuration. Section 7.5 is a case study with a building structure. The building structure consists of two parts: a building frame and a set of floors. The beams and columns of the building frame are modeled as a line body. This line body is assembled with the floors modeled as the surface bodies. Section 7.6 presents the subject of ANSYS CAD Integration. Our discussion focuses on importing a CAD file to DesignModeler.

7.2 Clamp Device with Large Deflection Subjected to Loading

A clamp device is shown below. There are two surface regions with the dimensions equal to 150 x 60 mm. One surface region is the area for defining the fixed to the ground constraint. The other surface region is the area for defining the pressure load. The material type is aluminum alloy. This clamp device is flexible when the pressure load is acting on the surface region on the top surface. The objective of this study is to determine the value of the pressure load so that the deformation at the opening reaches 60 mm. As shown in the drawing the size of the opening without the pressure load is 100 mm.

Use the Design Modeler of Workbench to create a 3D solid model and perform FEA to determine the magnitude of the pressure load.

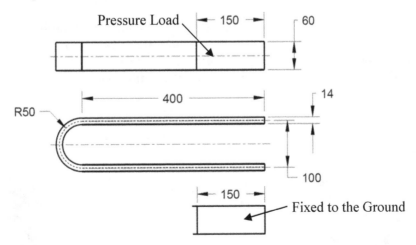

Step 1: Open Workbench.

From the start menu, click **ANSYS 19 > Workbench 19**. Users may directly click Workbench 19 when the symbol of Workbench 19 is on display.

Workbench 19.2

From the top menu, click the icon of **Save Project**. Specify **clamp device large displacement** as the file name, and click **Save**.

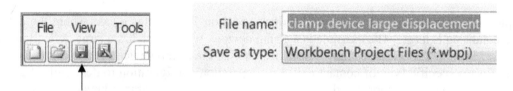

Step 2: From **Toolbox**, highlight **Static Structural** and drag it to the main screen.

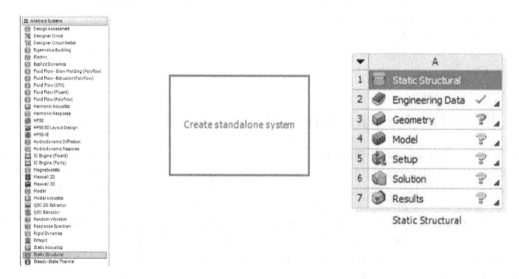

Step 3: In Design Modeler, Create 2 Sketches for the Profile and Path, Respectively.

Right-click the **Geometry** cell and pick **New DesignModeler Geometry** to start **Design Modeler**. Users may double click **Geometry** to directly enter **Design Modeler** if **New SpaceClaim Geometry** is not present. Click the **Units** drop down on the top menu, and select **Millimeter**.

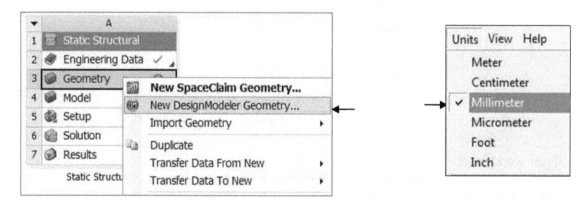

In **Tree Outline**, select **XYPlane** as the sketching plane. Click **Look At** to orient the selected sketching plane.

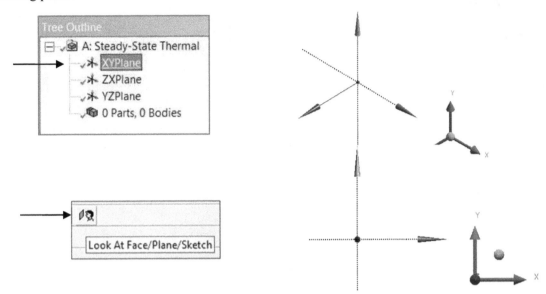

Click **Sketching**. Click **Draw**. Select **Rectangle** and sketch a rectangle, as shown.

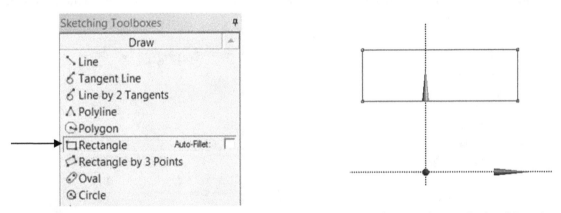

Click **Constraints**. Click **Symmetry**. Make a left click on the Y axis (vertical axis) and make a left click on each of the two vertical lines for symmetry.

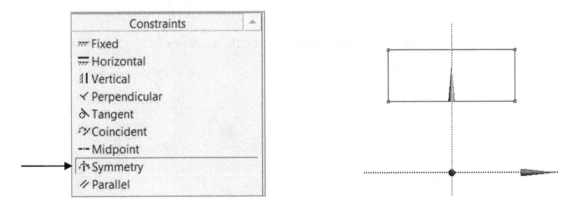

Click **Dimensions.** Select **General**. Specify 60 and 14 as the 2 size dimensions of the rectangle. Specify 43 as the distance between the horizontal axis (X axis) and the horizontal line, as shown. At this moment, we have completed the first sketch (profile).

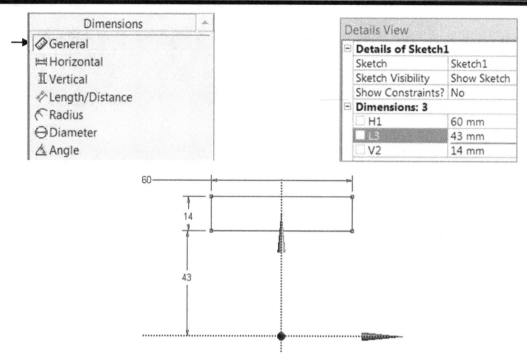

Upon completing this sketch (profile), click **Modeling** for creating the second sketch (path). In the **Tree Outline**, select **YZPlane**. Click the icon of **New Sketch** as the sketching plane. Click **Look At** to orient the selected sketching plane.

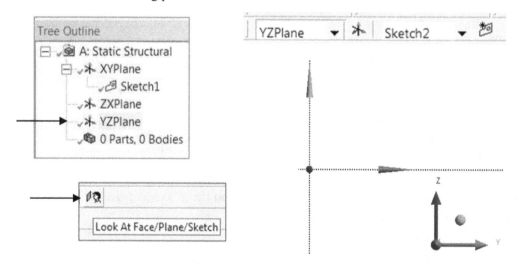

Click **Sketching** > **Draw**. Select **Circle** and sketch a circle, as shown.

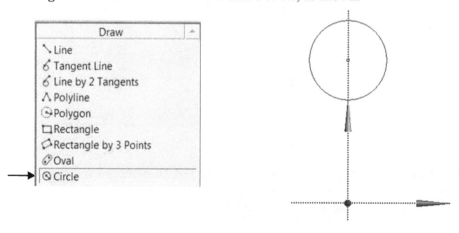

Click **Line** and sketch 2 vertical lines, as shown.

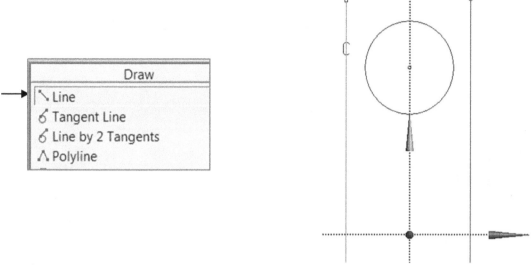

Click **Constraint > Tangent.** Click a vertical line and the sketched circle. Click the other vertical line and the sketched circle, as shown.

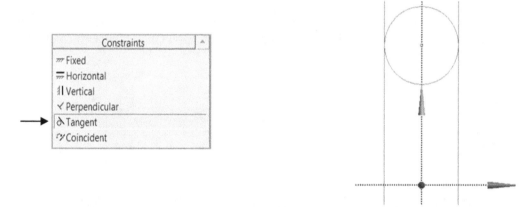

Click **Modify > Trim.** Delete the not-needed line segments and arc, as shown.

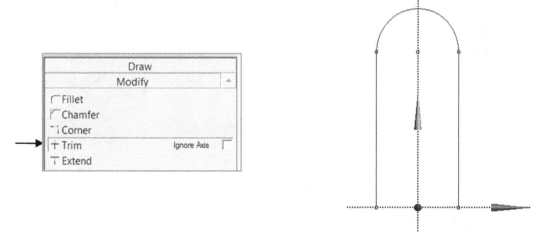

Click **Dimensions. S**elect **General**. Specify 50 as the radius value and 400 as the distance value, as shown. At this moment, we have completed the second sketch (path).

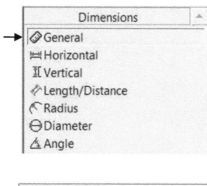

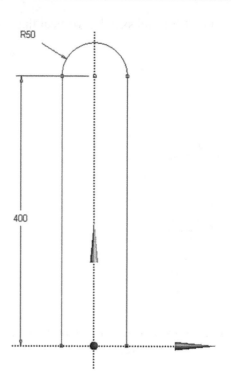

Upon completing the 2 sketches (profile and path), click **Modeling. Click Sweep**. Click Sketch1 (profile) first and click **Apply**. Afterwards, click Sketch2 (path) and **Apply.** Click **Generate** to obtain the 3D solid model of the clamp device, as shown.

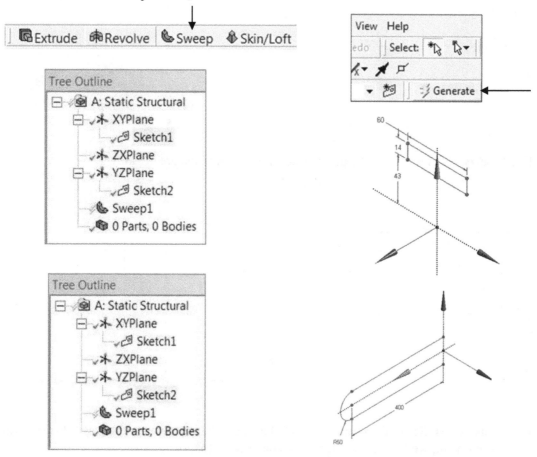

Details of Sweep1	
Sweep	Sweep1
Profile	Sketch1
Path	Sketch2
Operation	Add Material
Alignment	Path Tangent

Step 4: Create 2 rectangular surface regions through the use of imprinting 2 rectangular sketches to accommodate the pressure load and define the fixed support constraint.

First pick **ZXPlane** from Tree Outline. Click the icon of **New Plane**. In Details View, select From Face in the Type field. In the Base Plane field, left click to show **Apply** and pick the top surface of the clamp model. Click **Apply > Generate**. Plane4 is created.

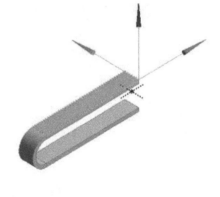

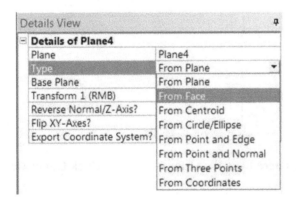

Details of Plane4	
Plane	Plane4
Type	From Plane
Base Plane	From Plane
Transform 1 (RMB)	From Face
Reverse Normal/Z-Axis?	From Centroid
Flip XY-Axes?	From Circle/Ellipse
Export Coordinate System?	From Point and Edge
	From Point and Normal
	From Three Points
	From Coordinates

Details of Plane4	
Plane	Plane4
Type	From Face
Subtype	Outline Plane
Base Face	Selected
Use Arc Centers for Origin?	Yes

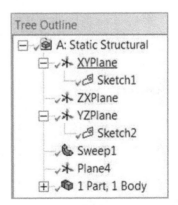

Now click the icon of **New Sketch > Look At**. Click **Sketching > Draw** and select Rectangle to sketch a corner rectangle, as shown.

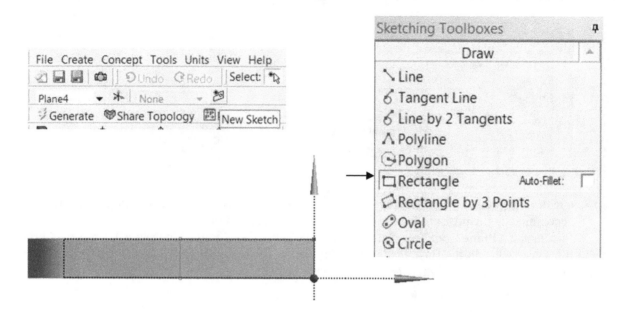

Click **Dimensions.** From the **Dimensions** panel, select **General**. Specify 150 as the length value.

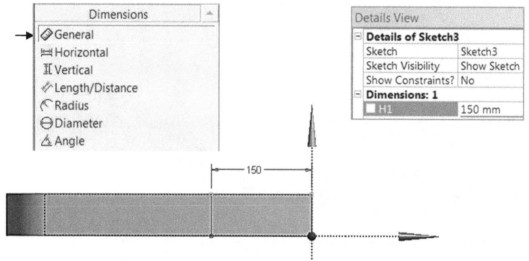

Click **Modeling** > **Extrude**. Select **Imprint Faces** from the **Operation** field. Click **Generate.**

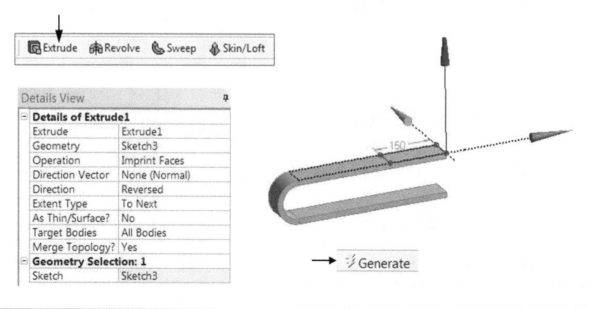

Repeat this process to create a second surface region. Click the icon of **New Plane**. In Details View, select From Face in the Type field. In the Base Plane field, left click to show **Apply** and pick the bottom surface of the clamp model. Click **Apply > Generate**. Plane5 is created.

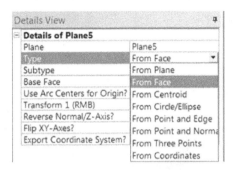

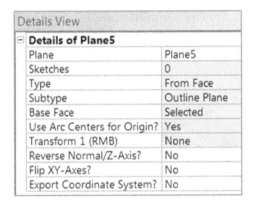

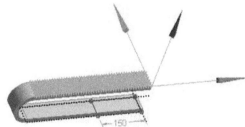

Now click the icon of **New Sketch > Look At**. Click **Sketching > Draw**, and select **Rectangle** to sketch a corner rectangle, as shown.

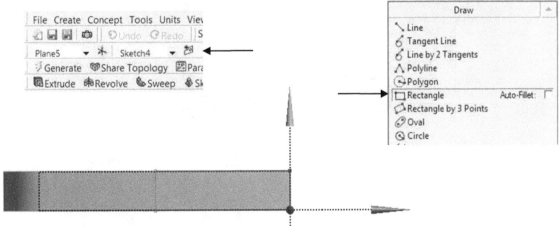

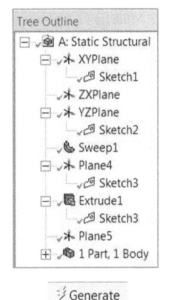

Click **Dimensions.** From the **Dimensions** panel, select **General**. Specify 150 as the length value.

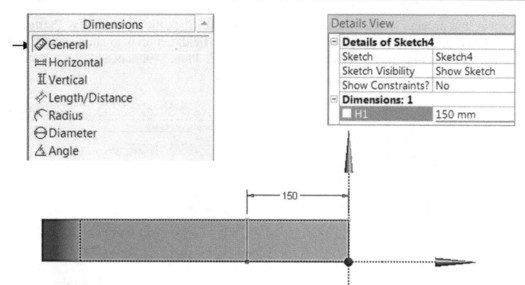

Click **Modeling** > **Extrude**. Select **Imprint Faces** from the **Operation** field. Click **Generate**.

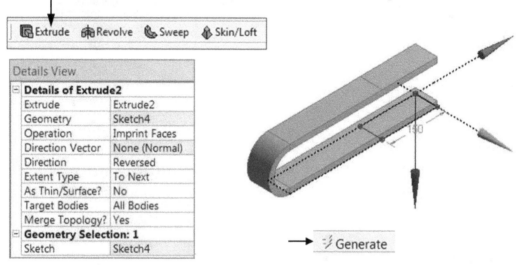

Upon completing the process of adding 2 imprinted surfaces to the clamp model, we need to close Design Modeler. Click **File** > Close **Design Modeler**, and go back to **Project Schematic**. Click **Save Project**.

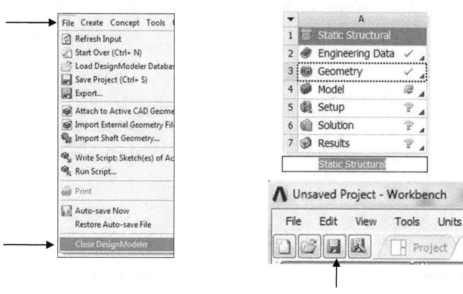

Step 5: Activate the Aluminum Alloy in the Engineering Data Sources

In the Static Structural panel, double click Engineering Data. We enter the Engineering Data Mode.

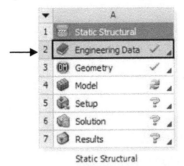

Static Structural

Click the icon of **Engineering Data Sources**. Click **General Materials**. We are able to locate Aluminum Alloy. Click the plus symbol nearby Aluminum Alloy to activate it. Afterwards, click the icon of Engineering Data Sources and click the icon of Project to go back to Project Schematic.

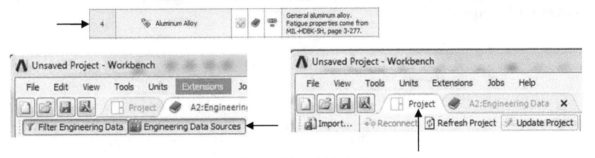

Step 6: Assign Aluminum Alloy to the 3D solid Model and Mesh Generation

Double-click the icon of Model. We enter the Mechanical Mode. Check the unit system. Select Metric (mm, t, N, s, mV, mA).

Static Structural

Now let us do the material assignment. From the Project tree, expand Geometry. Highlight Solid. In Details of Solid, click material assignment, pick Aluminum Alloy to replace Structural Steel, which is the default material assignment.

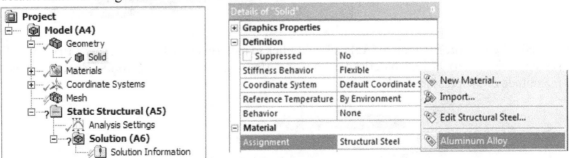

From the project model tree, highlight **Mesh.** Set Resolution to 6, and right-click to pick **Generate Mesh**.

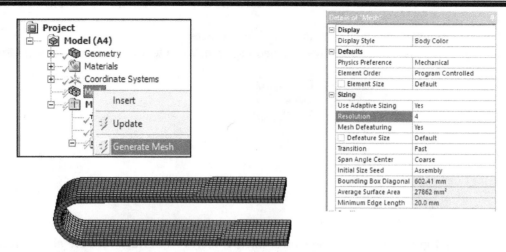

Define the fixed support constraint, highlight **Static Structural (A5)** and right-click to pick **Insert**. Afterwards, select **Fixed Support**. From the screen, pick the surface region created on the bottom surface, as shown. Click **Apply**.

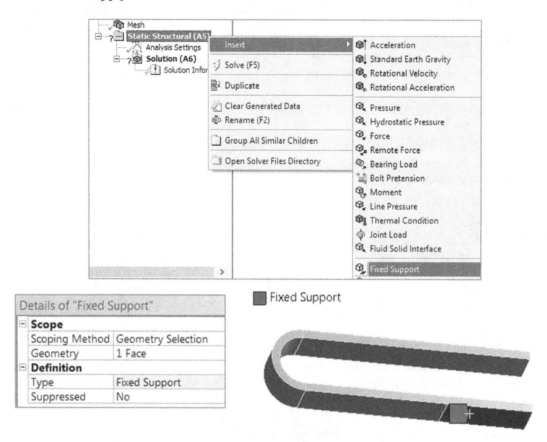

Highlight **Static Structural (A5)** and right-click to pick **Insert**. Afterwards, select **Pressure**. From the screen, pick the surface region defined on the top surface, as shown. Click **Apply**. Specify 0.1 MPa as the magnitude of the pressure load.

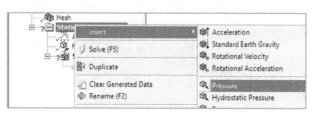

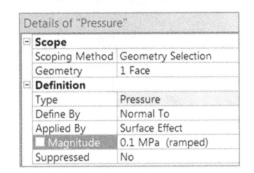

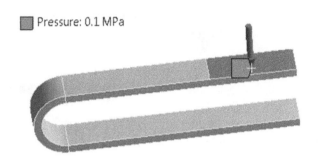

Highlight Solution (**A6**) and right-click to pick **Insert > Deformation > Total.**

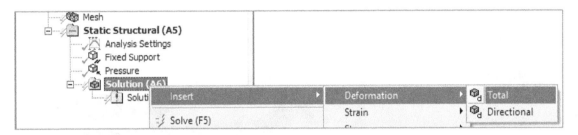

Highlight Solution (**A6**) and right-click to pick **Insert >Stress> Equivalent (von Mises).**

Highlight Solution (**A6**) and right-click to pick **Solve**.

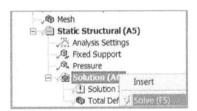

To plot the deformation, highlight **Total Deformation**. The maximum value is 50.7 mm. The required distance is 60 mm. Therefore, let us increase the pressure value from 0.1 MPa to 0.2 MPa.

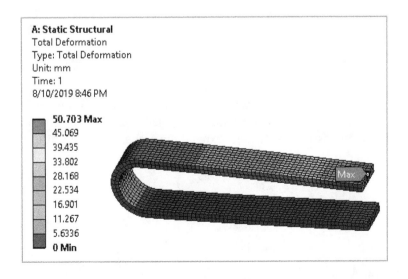

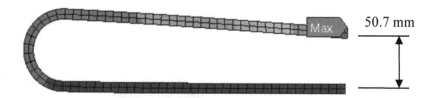

Now highlight Equivalent Stress. The maximum value is 192 MPa, which falls in the yield strength range of 90 – 280 MPa for aluminum alloys.

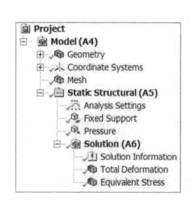

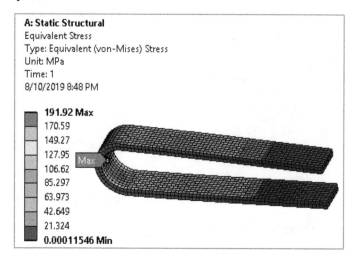

Now let us highlight Pressure listed under Static Structural (A5) and modify the pressure value from 0.1 MPa to 0.12 MPa. From the top menu, click **Solve** to run FEA. A Warning message is on display, indicating that a large deflection is present and is beyond the capability of linear simulation.

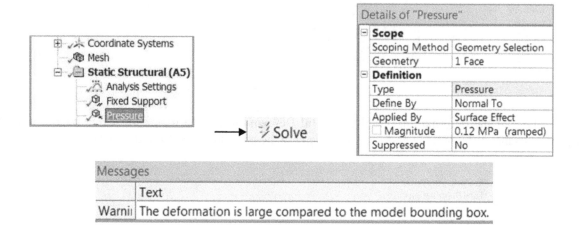

To deal with large deflection cases, Workbench requires users to turn on the Large Deflection module. To do so, users need to highlight Analysis Settings listed under Static Structural (A5). In the system default settings, this module is set to **Off**. Just expand the box and switch **Off** to **On**. Afterwards, click **Solve** to run FEA.

Details of "Analysis Settings"	
Step Controls	
Number Of Steps	1.
Current Step Number	1.
Step End Time	1. s
Auto Time Stepping	Program Controlled
Solver Controls	
Solver Type	Program Controlled
Weak Springs	Off
Solver Pivot Checking	Program Controlled
Large Deflection	Off
Inertia Relief	Off

→ Solve

Details of "Analysis Settings"	
Step Controls	
Number Of Steps	1.
Current Step Number	1.
Step End Time	1. s
Auto Time Stepping	Program Controlled
Solver Controls	
Solver Type	Program Controlled
Weak Springs	Off
Solver Pivot Checking	Program Controlled
Large Deflection	On
Inertia Relief	Off

Now highlight **Total Deformation**. The maximum value is 59.359 mm. This result is close enough to the required distance of 60 mm.

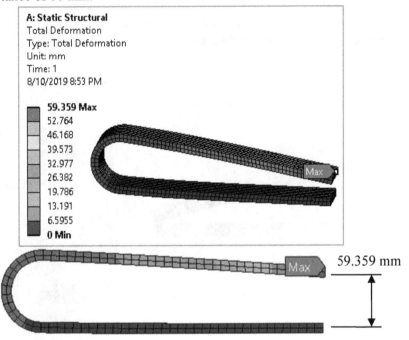

A: Static Structural
Total Deformation
Type: Total Deformation
Unit: mm
Time: 1
8/10/2019 8:53 PM

59.359 Max
52.764
46.168
39.573
32.977
26.382
19.786
13.191
6.5955
0 Min

59.359 mm

Now highlight **Equivalent Stress**. The maximum value is 228 MPa, which still falls in the yield strength range of 90 – 280 MPa for aluminum alloys, indicating that the clamp device will recover from the deflection when the pressure load is removed.

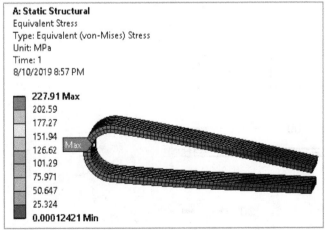

A: Static Structural
Equivalent Stress
Type: Equivalent (von-Mises) Stress
Unit: MPa
Time: 1
8/10/2019 8:57 PM

227.91 Max
202.59
177.27
151.94
126.62
101.29
75.971
50.647
25.324
0.00012421 Min

From the top menu, click **File > Close Mechanical**. In this way, we return to the main page of Workbench, or Project Schematic. From the top menu, click the icon of **Save Project.** From the top menu, click **File > Exit**.

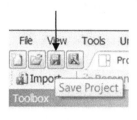

7.3 Clamp Device with Large Deflection (Non-Linear Analysis)

A clamp device was used in the previous case study where turning on the large deflection module was required. In the current case study, our major interest shifts to what the pressure value should be if the two ends meet without penetration.

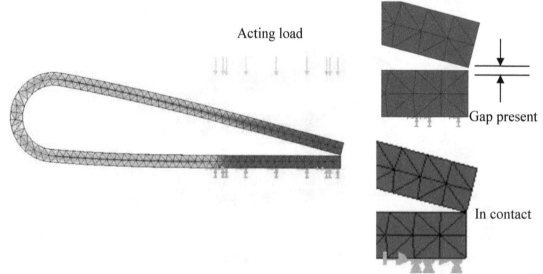

For contact analysis, an assembly model has to be used even though the geometry shape and dimensions are identical to the previously developed clamp device model. Because of this reason, we create a half model and assemble them together. Afterwards, we add two more surface regions serving as the contact regions, as shown below.

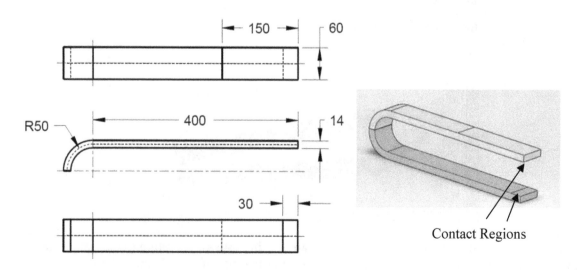

Step 1: Open Workbench.

From the start menu, click **ANSYS 19 > Workbench 19**. Users may directly click **Workbench 19** when the symbol of Workbench 19 is on display.

From the top menu, click the icon of **Save Project**. Specify **contact clamp device** as the file name, and click **Save**.

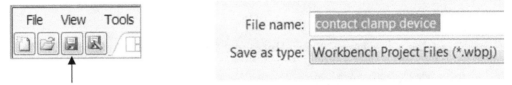

Step 2: From **Toolbox**, highlight **Static Structural** and drag it to the main screen.

Step 3: In Design Modeler, Create 2 Sketches for the Profile and Path, Respectively.

Right-click the **Geometry** cell and pick **New DesignModeler Geometry** to start **Design Modeler**. Users may double click **Geometry** to directly enter **Design Modeler** if **New SpaceClaim Geometry** is not present. Click the **Units** drop down on the top menu, and select **Millimeter**.

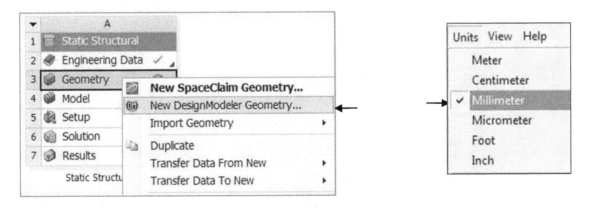

In **Tree Outline**, select **XYPlane** as the sketching plane. Click **Look At** to orient the selected sketching plane.

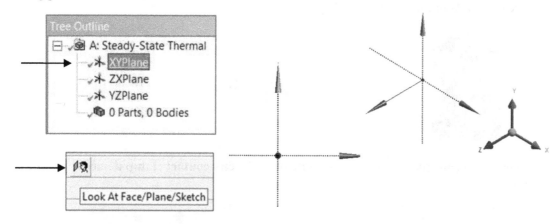

Click **Sketching**. Click **Draw**. Select **Rectangle** and sketch a rectangle, as shown.

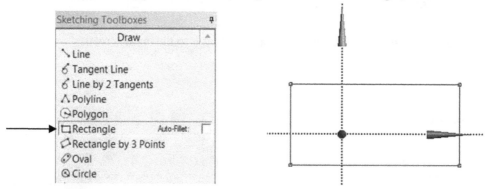

Click **Constraints**. Click **Symmetry**. Make a left click on the Y axis (vertical axis) and make a left click on each of the two vertical lines for symmetry. On the screen, right-click to pick **Select new symmetry axis**. Make a left click on the X axis (vertical axis) and make a left click on each of the two horizontal lines for symmetry.

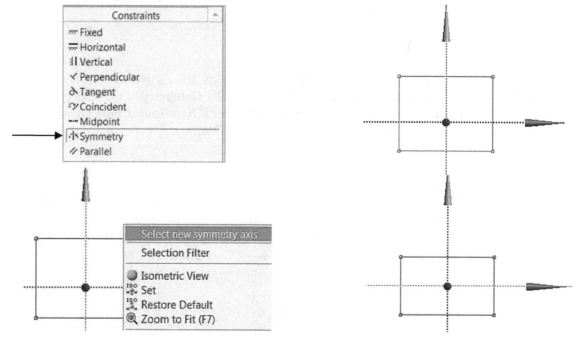

Click **Dimensions. S**elect **General**. Specify 60 and 14 as the 2 size dimensions of the rectangle. At this moment, we have completed the first sketch (profile).

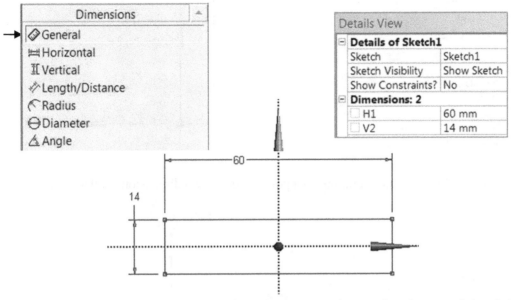

Upon completing this sketch (profile), click **Modeling** for creating the second sketch (path). In the **Tree Outline**, select ZYPlane. Click the icon of **New Sketch** as the sketching plane. Click **Look At** to orient the selected sketching plane.

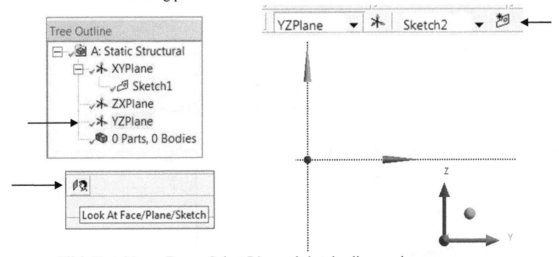

Click **Sketching** > Draw. Select **Line** and sketch a line, as shown.

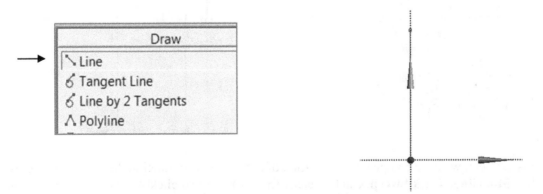

Click **Dimensions. S**elect **General**. Specify 400 as its length value.

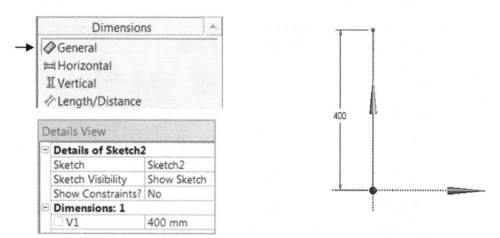

Click Draw > Construction Point. Sketch a point. Specify two dimensions 400 and 50, as shown

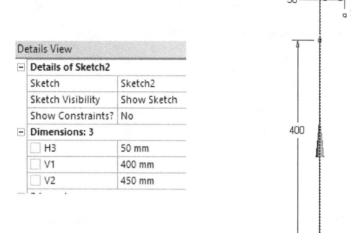

Click **Draw > Arc by 3 Points.** Sketch an arc, as shown.

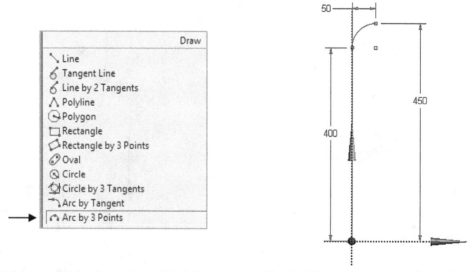

At this moment, we have completed the creation of path. Upon completing the 2 sketches (profile and path), click **Modeling. C**lick **Sweep**. Click Sketch1 (profile) first and click **Apply.** Afterwards, click Sketch2 (path) and **Apply.** Click **Generate** to obtain the 3D solid model of the half clamp, as shown.

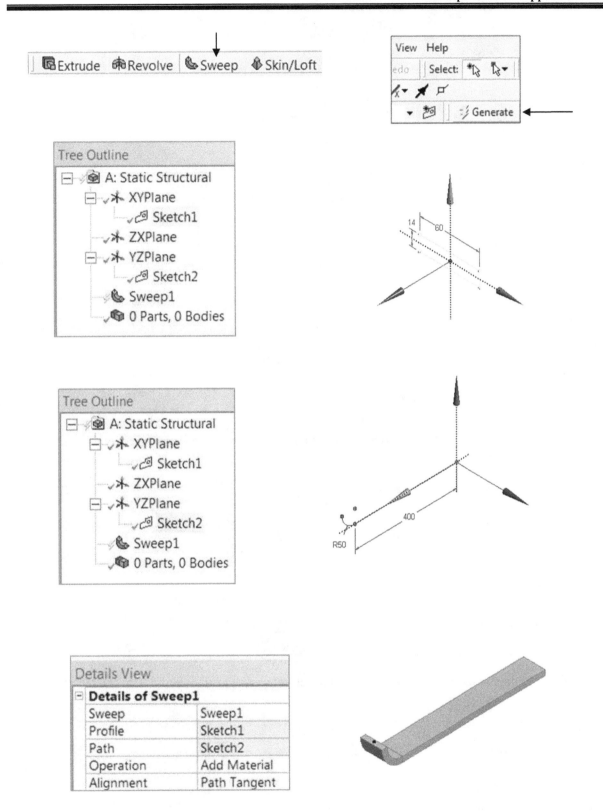

Step 4: Create 2 rectangular surface regions through the use of imprinting 2 rectangular sketches to accommodate the fixed support constraint and define one of the two contact regions.

First pick XZPlane from Tree Outline. Click the icon of **New Plane**. In Details View, select From Face in the Type field. In the Base Plane field, left click to show **Apply** and pick the inner surface of the half clamp model. Click **Apply > Generate**. Plane4 is created.

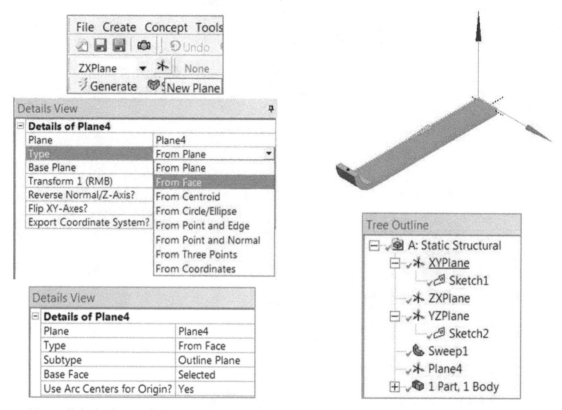

Now click the icon of **New Sketch > Look At**. Click **Sketching > Draw** and select **Rectangle** to sketch a corner rectangle, as shown.

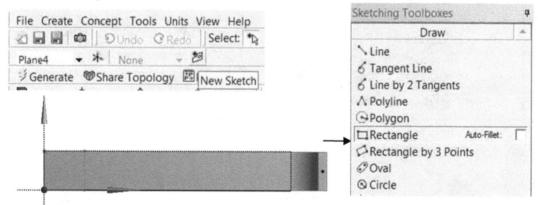

Click **Dimensions.** From the **Dimensions** panel, select **General**. Specify 30 as the length value.

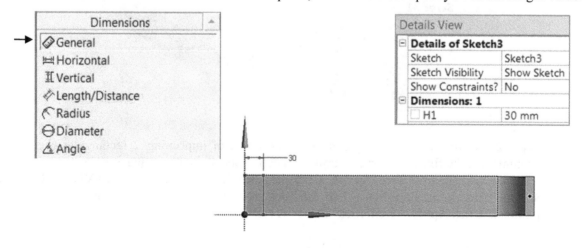

Click **Modeling** > **Extrude**. Select **Imprint Faces** from the **Operation** field. Click **Generate**.

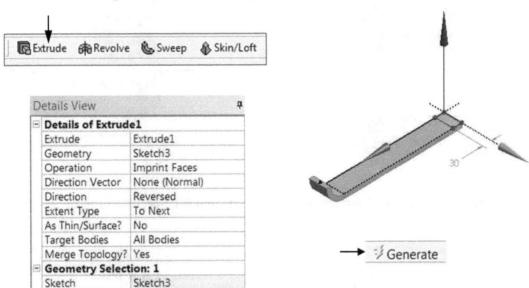

Repeat this process to create a second surface region. Click the icon of **New Plane**. In Details View, select From Face in the Type field. In the Base Plane field, left click to show **Apply** and pick the outer surface of the clamp model. Click **Apply** > **Generate**. Plane5 is created.

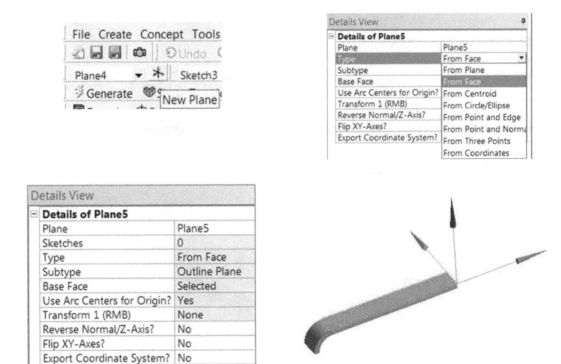

Now click the icon of **New Sketch** > **Look At**. Click **Sketching** > **Draw**, and select **Rectangle** to sketch a corner rectangle, as shown.

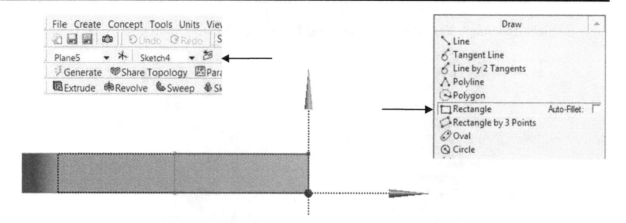

Click **Dimensions.** From the **Dimensions** panel, select **General**. Specify 150 as the length value.

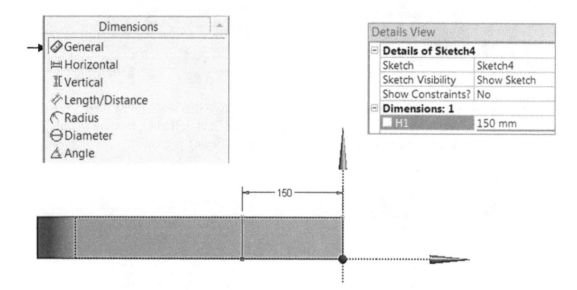

Click **Modeling > Extrude**. Select **Imprint Faces** from the **Operation** field. Click **Generate**.

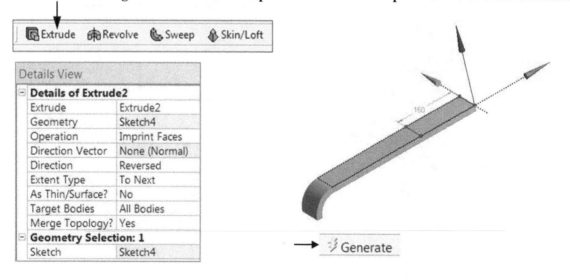

Upon completing the process of adding 2 imprinted surfaces to the half clamp model, we need to freeze this half clamp model so that the second half clamp model to be created will be independent of this half clamp model. From the top menu, click Tools > Freeze. Click **Generate**.

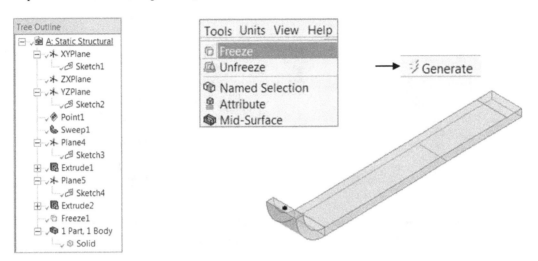

Now let us create the second half clamp model. In **Tree Outline**, select **XYPlane** as the sketching plane. Click the icon of New Sketch. Click **Look At** to orient the selected sketching plane.

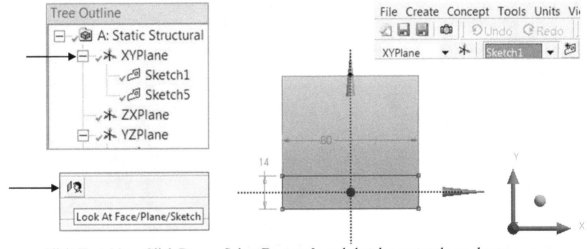

Click **Sketching**. Click **Draw**. Select **Rectangle** and sketch a rectangle, as shown.

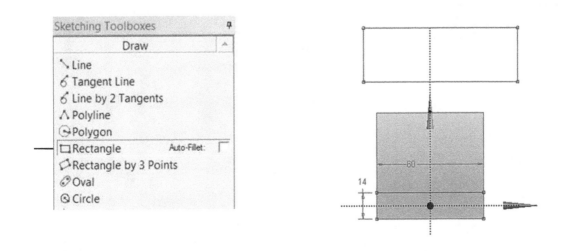

Click **Constraints**. Click **Symmetry**. Make a left click on the Y axis (vertical axis) and make a left click on each of the two vertical lines for symmetry.

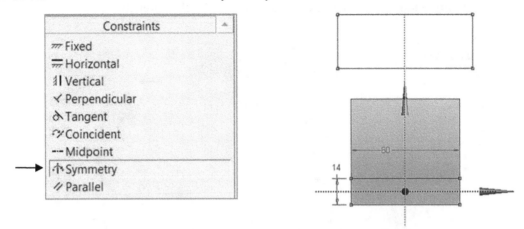

Click **Dimensions. S**elect **General**. Specify 60 and 14 as the 2 size dimensions of the rectangle. Specify 93 as the position dimension. At this moment, we have completed the sketch as the profile for the second half clamp model.

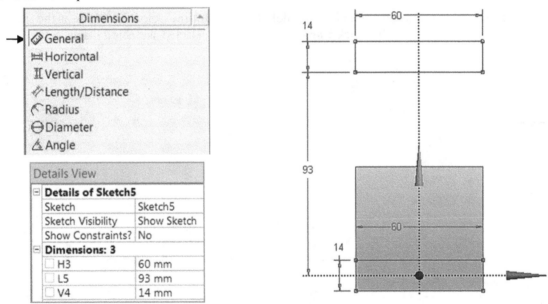

Upon completing this sketch (profile), click **Modeling** for creating the sketch for path. In the **Tree Outline**, select ZYPlane. Click the icon of **New Sketch** as the sketching plane. Click **Look At** to orient the selected sketching plane.

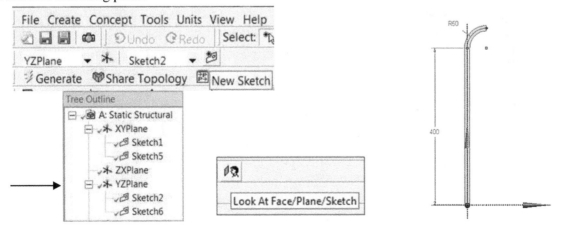

Click **Sketching** > Draw. Select **Line** and sketch a line, as shown.

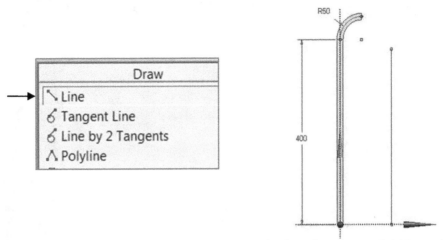

Click **Dimensions.** Select **General**. Specify 400 as its length value and 100 as the position dimension, as shown.

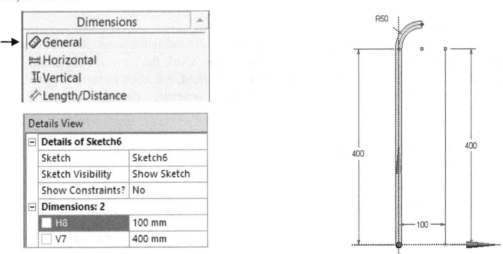

To sketch an arc, click **Draw > Arc by Center.** Sketch an arc, as shown. No dimension is needed.

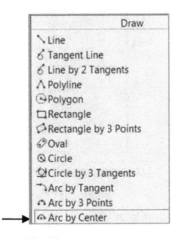

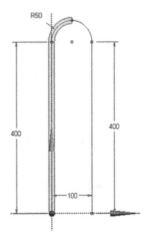

At this moment, we have completed the creation of path. Upon completing the 2 sketches (profile and path), click **Modeling.** Click **Sweep**. Click Sketch5 (profile) first and click **Apply**. Afterwards, click Sketch6 (path) and **Apply.** Click **Generate** to obtain the 3D solid model of the second half clamp, as shown. The first and second half models are assembled together.

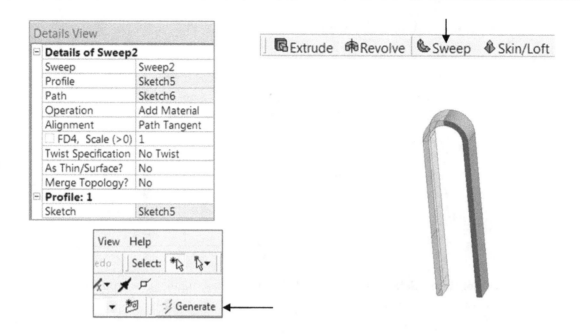

Now we need to create 2 rectangular surface regions for defining the pressure load and defining the second contact region. First pick XZPlane from Tree Outline. Click the icon of **New Plane**. In Details View, select From Face in the Type field. In the Base Plane field, left click to show **Apply** and pick the inner surface of the second half clamp model. Click **Apply > Generate**. Plane6 is created.

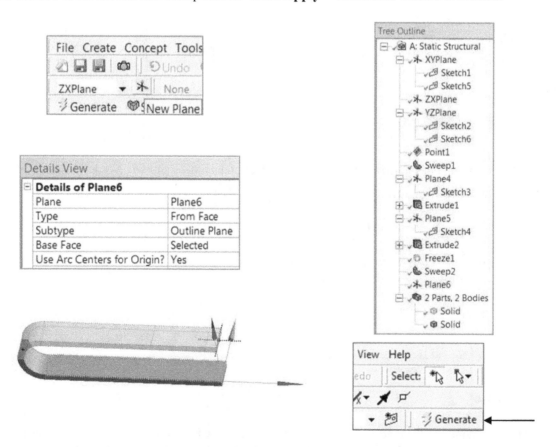

Now click the icon of **New Sketch > Look At**. Click **Sketching > Draw** and select Rectangle to sketch a corner rectangle, as shown. Be sure to align with the previous sketched rectangle (30 mm). In this way, no dimension is needed for this rectangle.

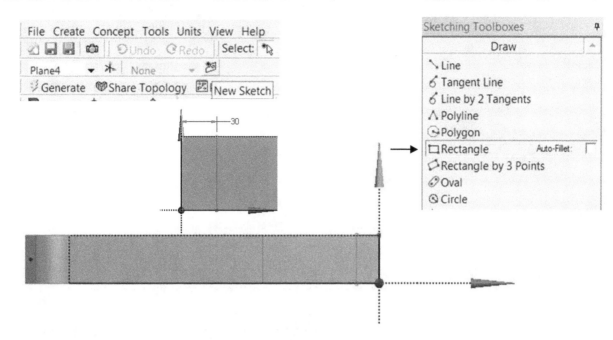

Click **Modeling** > **Extrude**. Select **Imprint Faces** from the **Operation** field. Click **Generate**.

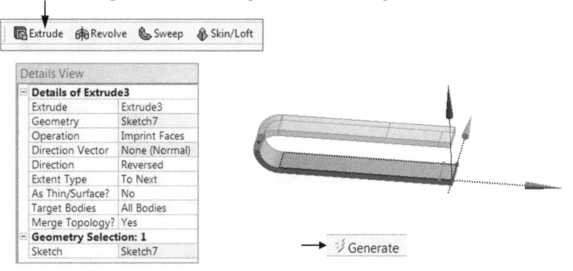

Repeat this process to create a second surface region. Click the icon of **New Plane**. In Details View, select From Face in the Type field. In the Base Plane field, left click to show **Apply** and pick the outer surface of the second clamp model. Click **Apply** > **Generate**. Plane7 is created.

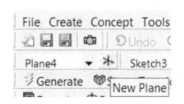

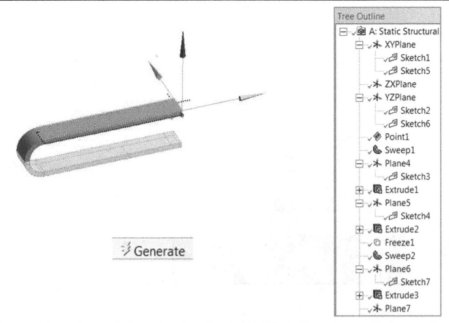

Now click the icon of **New Sketch > Look At**. Click **Sketching > Draw**, and select **Rectangle** to sketch a corner rectangle, as shown.

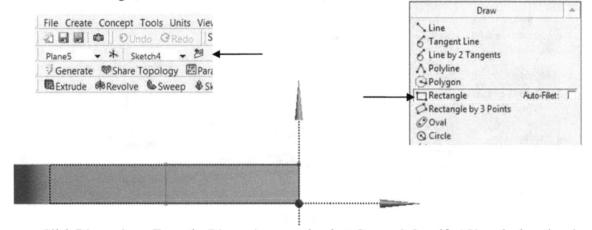

Click **Dimensions.** From the **Dimensions** panel, select **General**. Specify 150 as the length value.

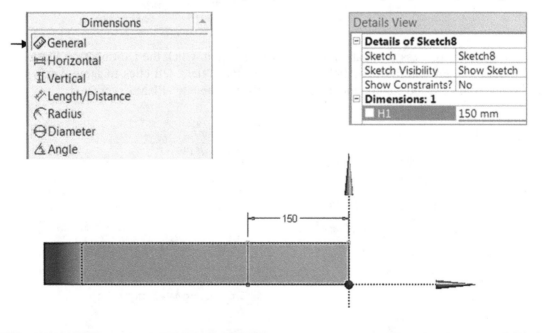

Click **Modeling** > **Extrude**. Select **Imprint Faces** from the **Operation** field. Click **Generate**.

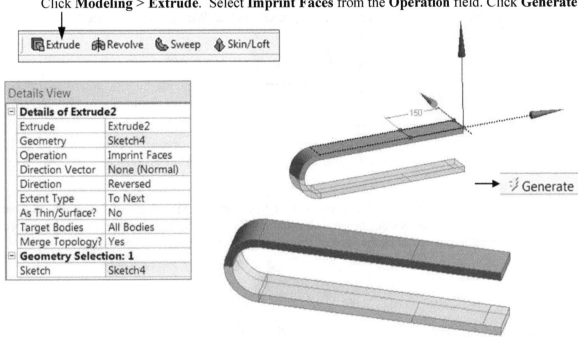

Now let us rename the two Solids. Under 2 Parts, 2 Bodies, highlight the first listed Solid and right-click to pick **Rename**. Change Solid to Lower Half. Do the same, change the second listed Solid to Upper Half.

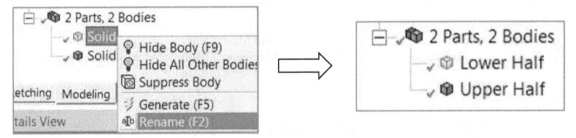

Now let us close Design Modeler. Click **File** > Close **Design Modeler**, and go back to **Project Schematic**. Click **Save Project**.

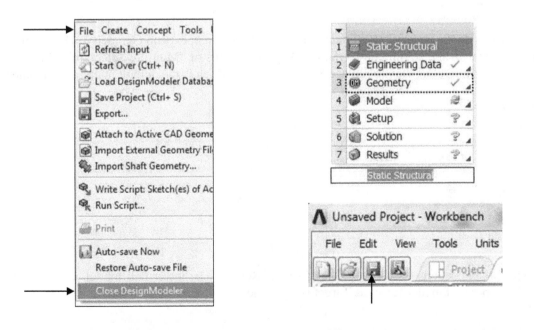

Step 5: Activate the Aluminum Alloy in the Engineering Data Sources
   In the Static Structural panel, double click Engineering Data. We enter the Engineering Data Mode.

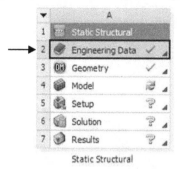

Static Structural

Click the icon of **Engineering Data Sources**. Click **General Materials**. We are able to locate Aluminum Alloy. Click the plus symbol nearby Aluminum Alloy to activate it. Afterwards, click the icon of Engineering Data Sources and click the icon of Project to go back to Project Schematic.

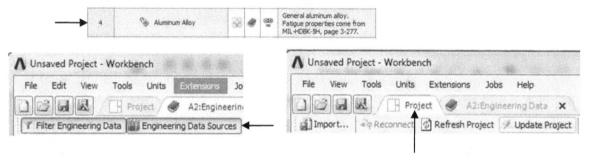

Step 6: Assign Aluminum Alloy to the 3D solid Model and Mesh Generation.
   Double-click the icon of Model. We enter the Mechanical Mode. Check the unit system.

Static Structural

Now let us do the material assignment. From the Project tree, expand Geometry. Highlight Lower Half and Upper Half while holding down the Ctrl key. In Details of "Multiple Selection", click material assignment, pick Aluminum Alloy to replace Structural Steel, which is the default material assignment.

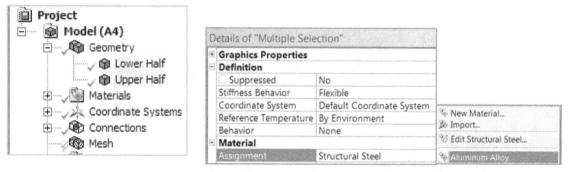

Expand Connections > Contacts. Click Contact Region. The lower half model serves as Contact Body and the upper half model serves as Target Body. The type of contact is Bonded.

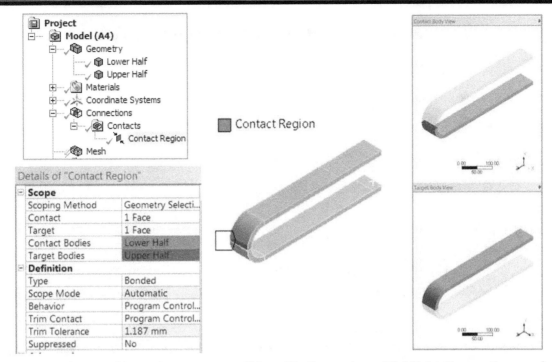

Now let us add a new contact condition: No Separation. Highlight **Connections** and expand **Contact** and select **No Separation**. First select two faces. Pick the surface region from the lower half as Contact. Pick the surface region from the upper half as Target. Afterwards, pick the lower half as Contact Body. Pick the upper half as Target Body.

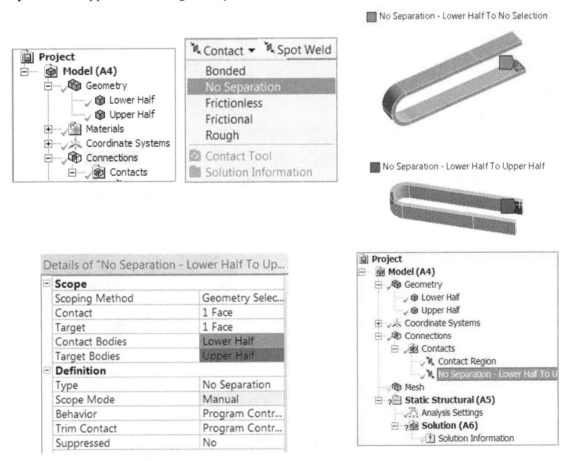

From the project model tree, highlight **Mesh**. Set Resolution to 6. Right click to pick **Generate Mesh**.

Step 7: Define the Pressure Load, fixed Support and Perform FEA

Double-click the icon of **Model**. We enter the **Mechanical Mode**. Check the unit system, and pick Metric (mm, t, N, s, mV, mA).

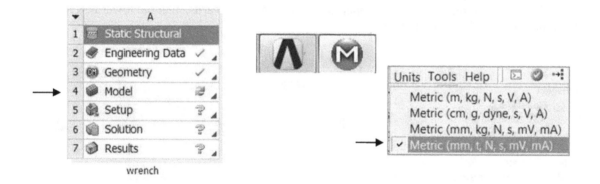

wrench

Define the fixed support constraint, highlight **Static Structural (A5)** and right-click to pick **Insert**. Afterwards, select Fixed Support. From the screen, pick the surface region created on the bottom surface of the lower part model, as shown. Click **Apply**.

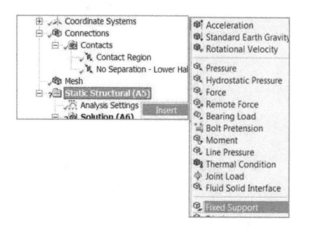

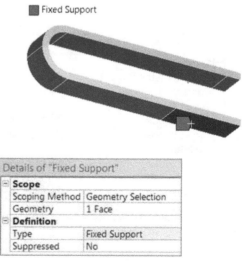

Highlight **Static Structural (A5)** and right-click to pick **Insert**. Afterwards, select **Pressure**. From the screen, pick the surface region defined on the top surface, as shown. Click **Apply**. Specify 0.15 MPa as the magnitude of the pressure load.

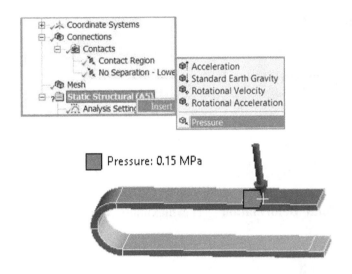

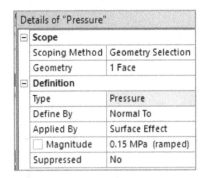

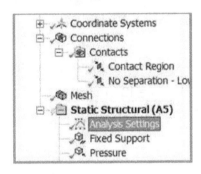

To deal with the large deflection case, we need to turn on the Large Deflection module. Highlight **Analysis Settings** listed under **Static Structural (A5).** Expand the Large Deflection box and switch **Off** to **On**.

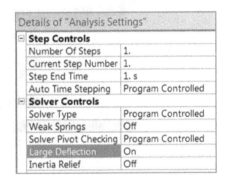

Highlight Solution (**A6**) and right-click to pick **Insert > Deformation > Total.**

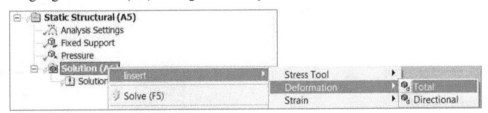

Highlight Solution (**A6**) and right-click to pick **Insert >Stress> Equivalent (von Mises).**

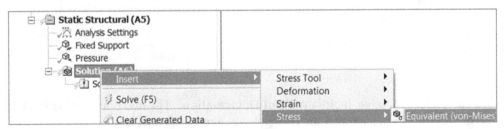

Highlight Solution (**A6**) and right-click to pick **Solve**. Workbench system is running FEA.

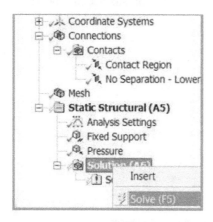

To plot the deformation, highlight **Total Deformation**. The maximum value is 73.79 mm, indicating that the two ends are not in contact.

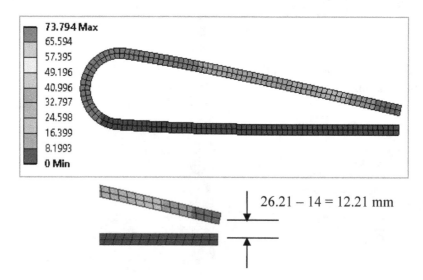

Because the gap value is 12.17 mm. Let us increase the pressure value from 0.15 MPa to 0.17 MPa. Now let us highlight Pressure listed under Static Structural, and modify the pressure value from 0.15 MPa to 0.17 MPa. From the top menu, click **Solve** to run FEA.

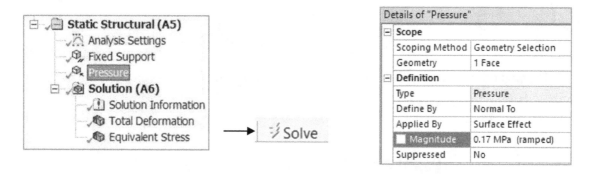

To review the deflection value, highlight **Total Deformation**. The maximum value is 83.25 mm, and the gap value is 2.72 mm, indicating that the two ends are almost in contact.

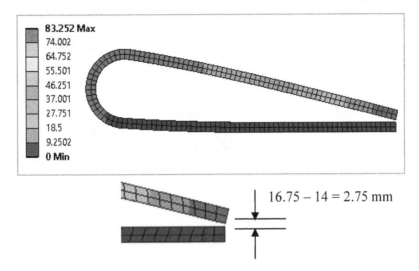

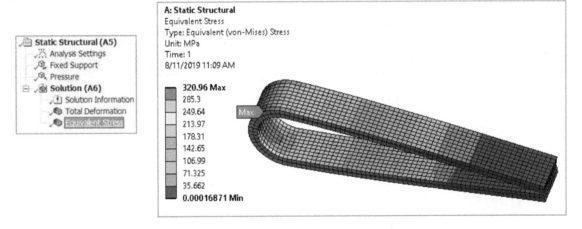

$16.75 - 14 = 2.75$ mm

Now let us check the von Mises stress developed in the clamp device. Highlight Equivalent Stress. The maximum value is 321 MPa, which exceeds the yield strength range of 90 – 280 MPa for aluminum alloys, indicating such a large deflection is permanent. The deformation will remain when the pressure load is removed.

A: Static Structural
Equivalent Stress
Type: Equivalent (von-Mises) Stress
Unit: MPa
Time: 1
8/11/2019 11:09 AM

From the top menu, click **File > Close Mechanical**. In this way, we return to the main page of Workbench, or Project Schematic. From the top menu, click the icon of **Save Project.** From the top menu, click **File > Exit**.

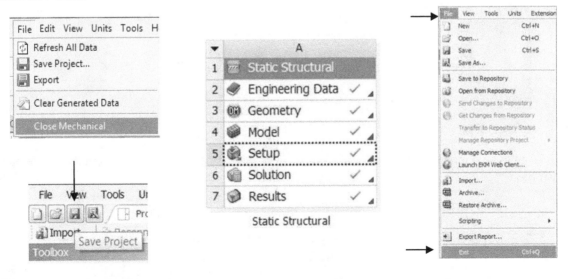

7.4   Contact Analysis for a Support System

The engineering drawing shown below is an assembly of a hanging system, which consists of two support components and a shaft component. To simplify the FEA analysis, we take the right part of the hanging system, as shown. Assume that the material type of support is aluminum alloy. The material type of shaft is structural steel. Note that the diameter of the hole is 35 mm, and the shaft diameter is 34 mm. There is a clearance of 1 mm between the support hole and the shaft. A load of 75000 N is acting on the shaft in the downward direction. As a result, high stress concentration develops in the contact zone between these two components. In this case study, we introduce No Separation configuration for the contact region, instead of using bonded configuration. Separation of faces in contact is not allowed, but small amounts of frictionless sliding can occur along contact faces. We also apply the **Bonded** configuration for the purpose of comparing the difference(s) between the two contact configurations.

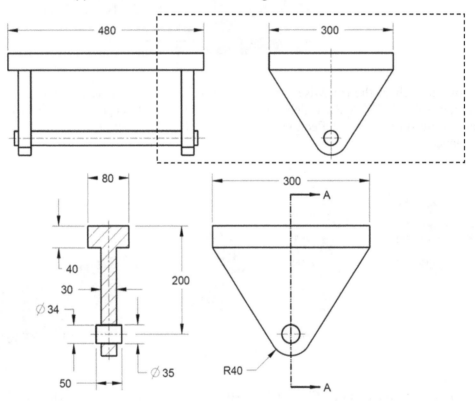

Step 1: Open Workbench.
    From the start menu, click **ANSYS 19** > Workbench 19. Users may directly click Workbench 19 when the symbol of Workbench 19 is on display.

Workbench 19.2

    From the top menu, click the icon of **Save Project**. Specify **contact hang system** as the file name, and click **Save**.

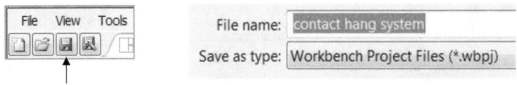

Step 2: From **Toolbox**, highlight **Static Structural** and drag it to the main screen.

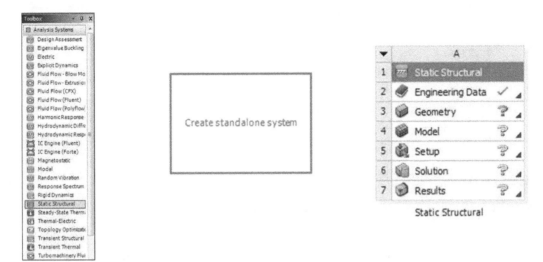

Step 3: In Design Modeler, Create the 2 Parts of Bearing and Shaft.

Right-click the **Geometry** cell and pick **New DesignModeler Geometry** to start **Design Modeler**. Users may double click **Geometry** to directly enter **Design Modeler** if **New SpaceClaim Geometry** is not present. Click the **Units** drop down on the top menu, and select **Millimeter**.

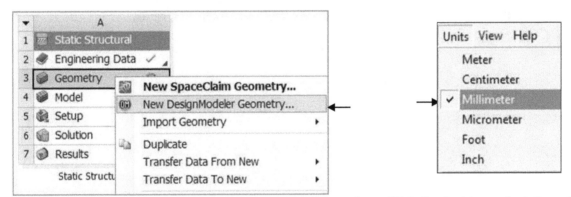

In **Tree Outline**, select **XYPlane** as the sketching plane. Click **Look At** to orient the selected sketching plane.

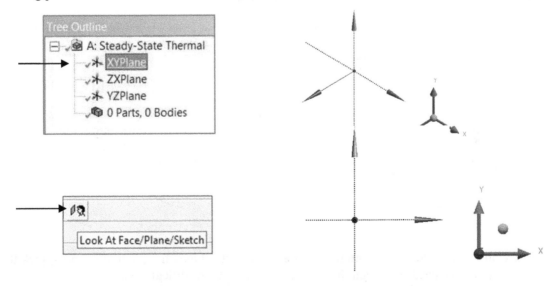

Click **Sketching**. Click **Draw**. Select **Circle** and sketch a circle, as shown.

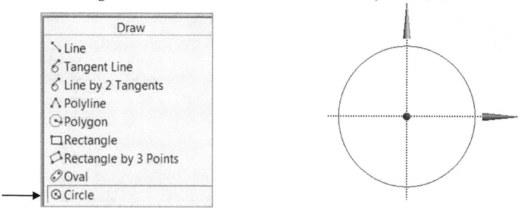

Click **Dimensions. S**elect **Radius**. Specify 40 as the radius value.

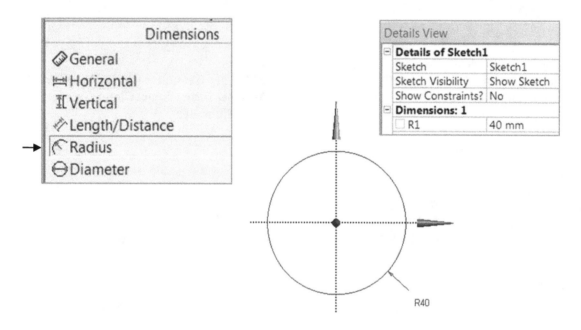

Click **Draw**. Select **Line** and sketch a horizontal line, as shown.

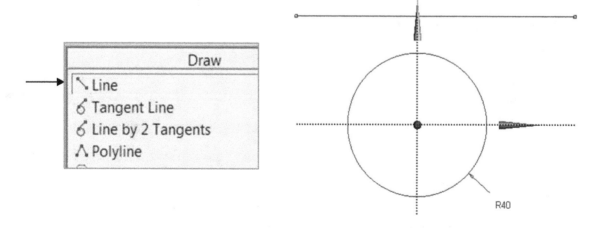

Click **Constraints**. Select **Symmetry** and click the vertical axis first, and afterwards, pick the 2 vertex ends so that the sketched horizontal line is symmetry about the vertical axis.

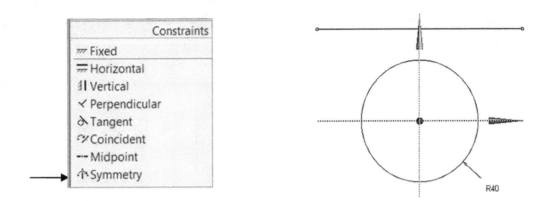

Click **Dimensions.** Select **General**. Specify 300 as the length value. Also specify 160 as the distance value with respect to the horizontal axis.

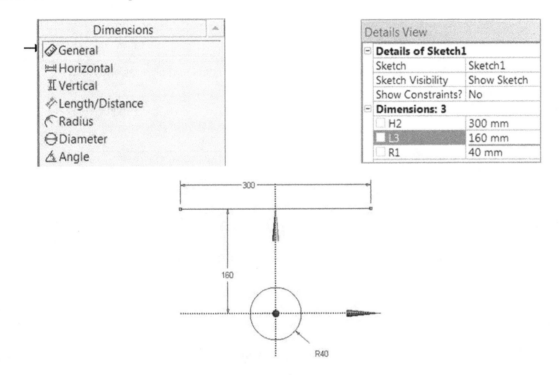

Click **Draw**. Select **Line** and sketch 2 inclined lines starting from the 2 upper corners, as shown.

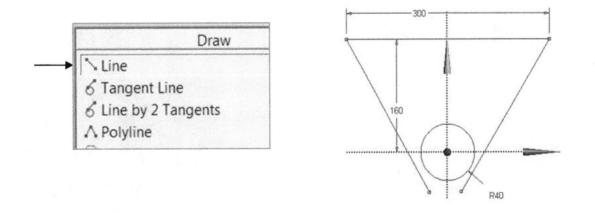

Click **Constraints**. Select **Tangent.** Pick a line and pick the circle. Repeat this process of picking the other line and the circle, as shown.

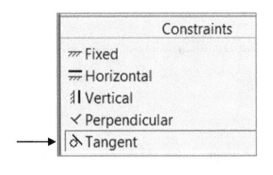

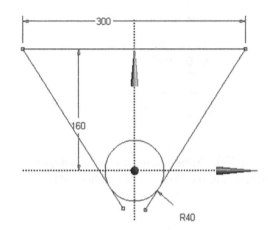

Click **Modify**. Select **Trim.** Delete the unwanted line segments and arcs, as shown.

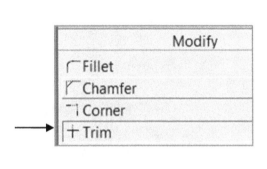

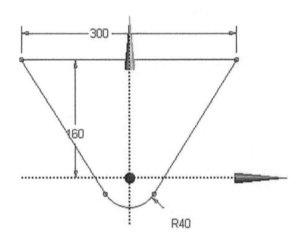

Click **Sketching**. Click **Draw**. Select **Circle** and sketch a circle, as shown.

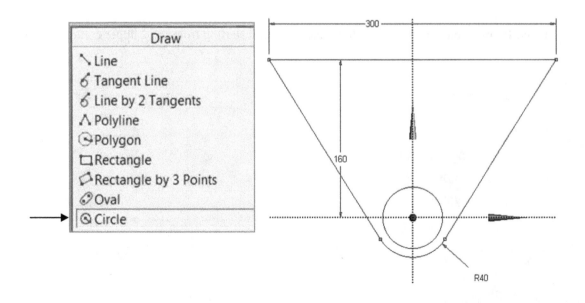

Click **Dimensions. S**elect **General**. Specify 35 as the diameter value.

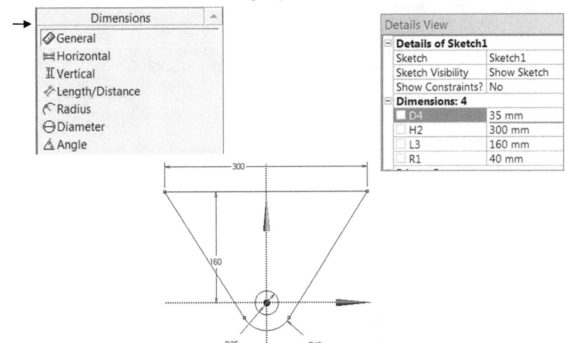

Upon completing Sketch1, Click **Modeling** > **Extrude**. Select **Sketch1**, and click **Apply**. In Direction, select Both – Symmetric and specify 15 as the extrusion distance value. Click **Generate**.

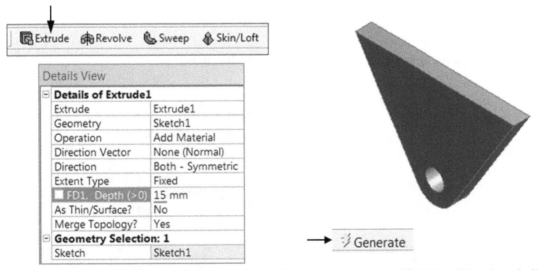

In **Tree Outline**, select **XYPlane,** again, as the sketching plane. Click **New Sketch** and click **Look At** to orient the selected sketching plane.

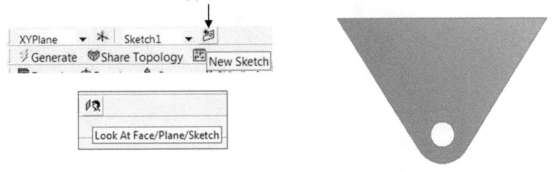

Click **Draw**. Select **Polyline** and sketch a rectangle. Make sure that at the end of sketching, right-click to pick **Closed End**, as shown.

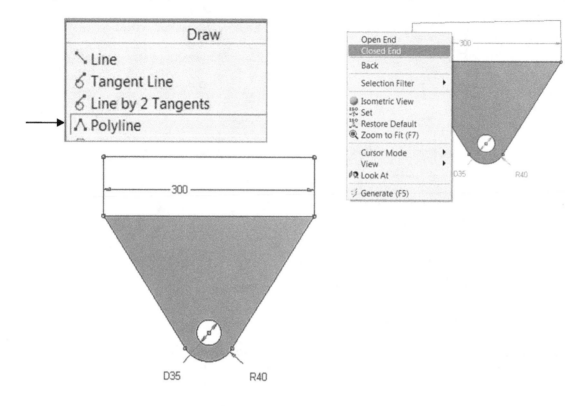

Click **Dimensions.** Select **General**. Specify 40 as the length values.

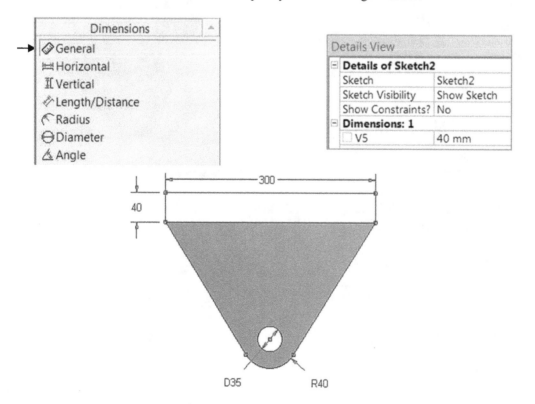

Upon completing Sketch2, Click **Modeling** > **Extrude**. Select Sketch2, and click **Apply**. In Direction, select Both – Symmetric and specify 40 as the extrusion distance value. Click **Generate**.

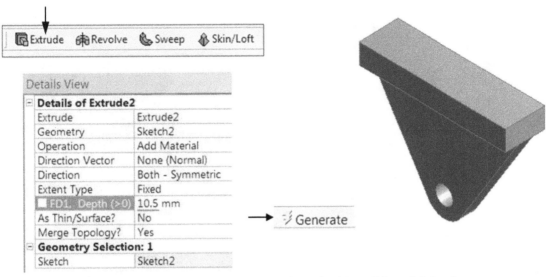

At this moment, we have completed the creation of a 3D solid model for the support component. From the top menu, click the icon of Tools > **Freeze** and click Generate. Highlight Solid listed under 1 Part, 1 Body, and right-click to pick Rename. Change the name to Support.

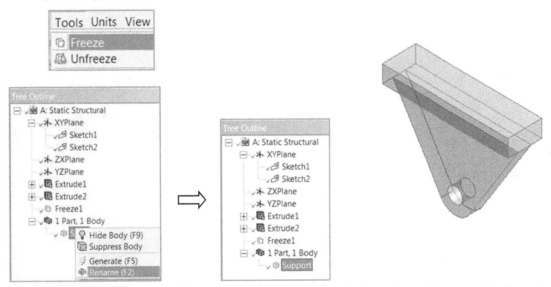

Now let us create a 3D solid model for Shaft, which is independent of Support. In **Tree Outline**, select **XYPlane,** and click New Plane. Afterwards, click Generate. Plane4 is listed in Tree Outline.

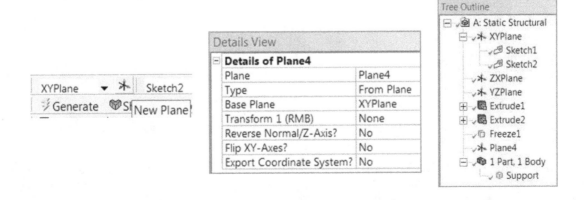

We use Plane4 as the sketching plane. Click New Sketch and click **Look At** to orient the selected sketching plane.

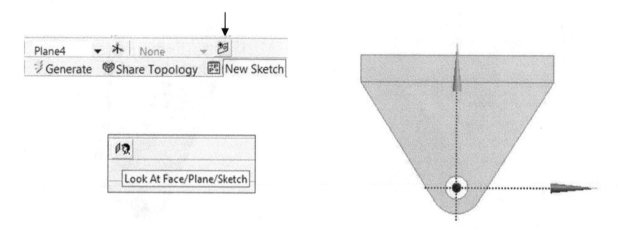

Click **Draw**. Select **Circle** and sketch a circles, as shown.

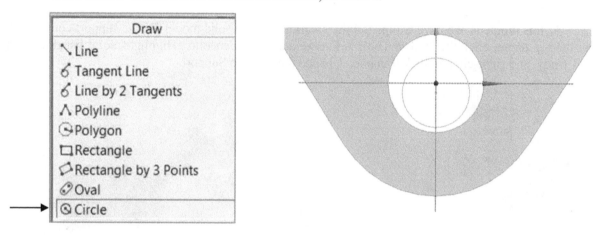

Click **Dimensions. S**elect **General**. Specify 0.5 as the distance value between the center and the horizontal axis, as shown.

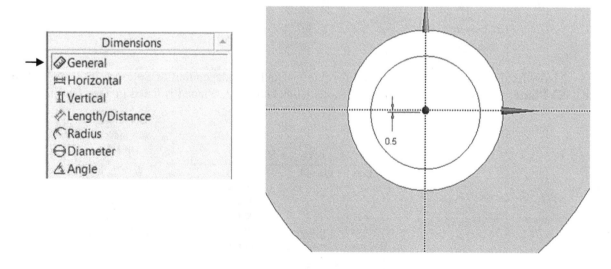

Click **Dimensions. S**elect **General**. Specify 34 as the diameter value of the sketched circle, as shown.

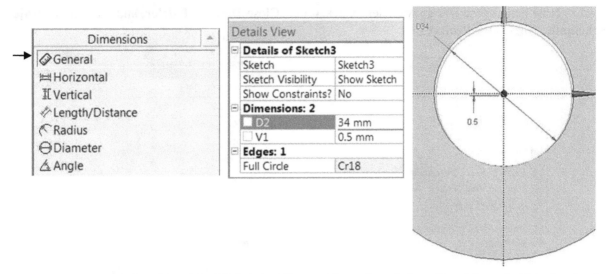

Upon completing Sketch2, Click **Modeling** > **Extrude**. Select Sketch3, and click **Apply**. In Direction, select Both – Symmetric and specify 25 as the extrusion distance value. Click **Generate**.

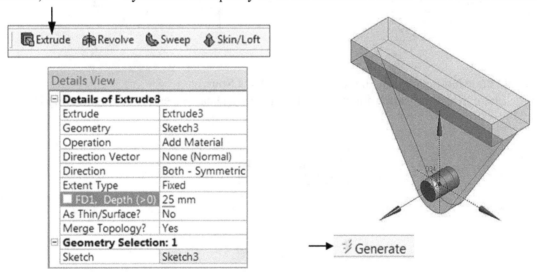

At this moment, we have completed the shaft model. Let us rename Solid listed under 2 Parts, 2 Bodies. Change it to **Shaft**.

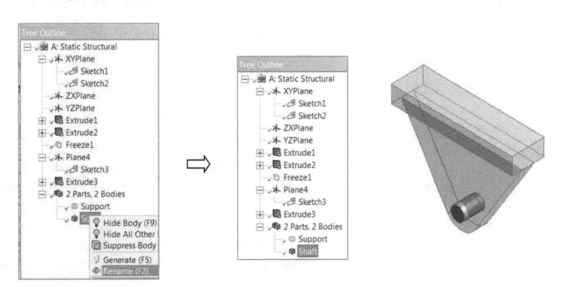

Now let us close Design Modeler. Click **File** > Close **Design Modeler**, and go back to **Project Schematic**. Click **Save Project**.

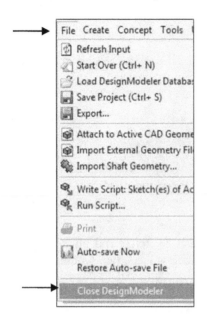

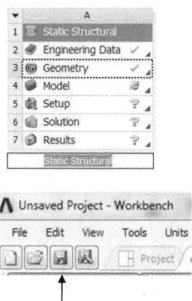

Step 4: Activate the Aluminum Alloy in the Engineering Data Sources

In the Static Structural panel, double click Engineering Data. We enter the Engineering Data Mode.

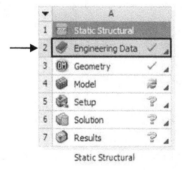

Click the icon of **Engineering Data Sources**. Click **General Materials**. We are able to locate Aluminum Alloy. Click the plus symbol nearby Aluminum Alloy to activate it. Afterwards, click the icon of Engineering Data Sources and click the icon of Project to go back to Project Schematic.

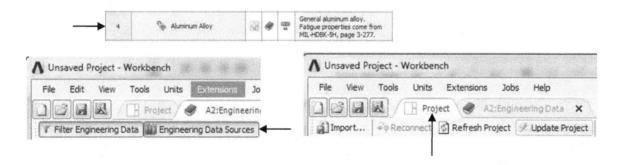

Step 5: Assign Aluminum Alloy to Support and Keep the default Structural Steel for Shaft.

Double-click the icon of Model. We enter the Mechanical Mode. Check the unit system, and pick Metric (mm, t, N, s, mV, mA).

Now let us do the material assignment. From the Project tree, expand Geometry. Highlight Support. In Details of "Support", click material assignment, pick Aluminum Alloy to replace Structural Steel, which is the default material assignment.

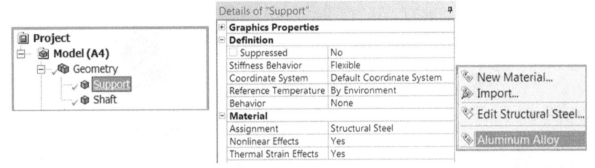

The material type of Structural Steel is assigned to Shaft by the system default assignment.

Step 6: Define No Separation in the Contact Setting.

In the Project tree, Expand Connections > Contacts. Click Contact Region. The Support model serves as Contact Body and the shaft model serves as Target Body. The type of contact is **Bonded**. Let us change it to No Separation. . This contact setting is similar to the bonded case. It only applies to regions of faces. Separation of faces in contact is not allowed, but small amounts of frictionless sliding can occur along contact faces.

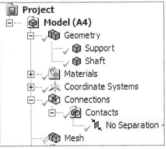

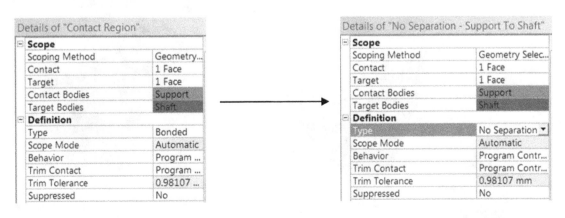

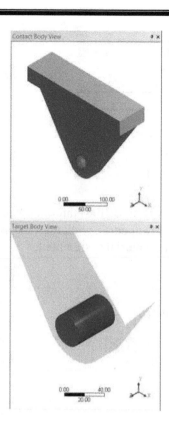

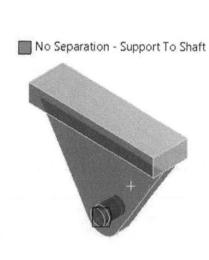

No Separation - Support To Shaft

Step 7: Mesh Generation.

From the project model tree, highlight **Mesh.** Right-click to pick **Insert > Method**. While holding down the **Ctrl** key, pick Support and Shaft and click Apply. In Method, pick Hex Dominant. In Free Face Mesh Type, select Quad/Tri. Expand Sizing, select Fine in Relevance Center. Right-click **Mesh** again to pick **Generate Mesh**.

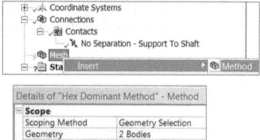

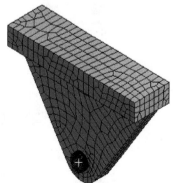

Hex Dominant Method

Step 8: Define the Force Load, Fixed Support and Perform FEA

Define the fixed support constraint, highlight **Static Structural (A5)** and right-click to pick **Insert**. Afterwards, select Fixed Support. From the screen, pick the top surface from Support, as shown. Click **Apply**.

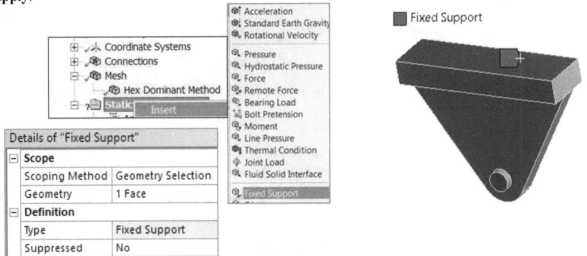

Highlight **Static Structural (A5)** and right-click to pick **Insert**. Afterwards, select **Force**. From the screen, pick the cylindrical surface from Shaft, and click **Apply**. Specify -75000 in Y Component.

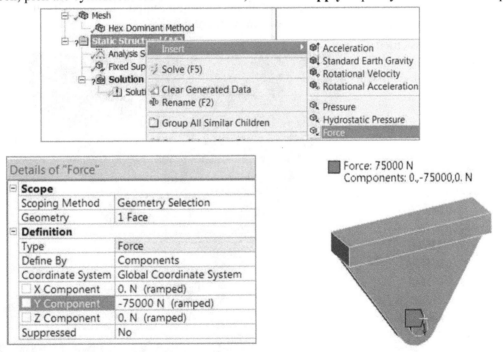

Highlight Solution (**A6**) and right-click to pick **Insert > Deformation > Total.**

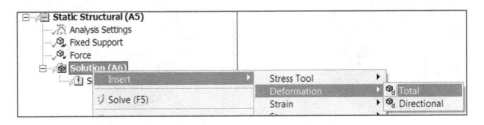

Highlight Solution (**A6**) and right-click to pick **Insert >Stress> Equivalent (von Mises).**

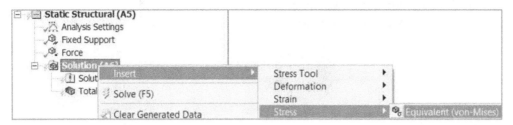

Highlight Solution (**A6**) and right-click to pick **Solve**. Workbench system is running FEA.

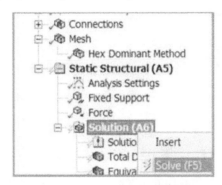

To plot the deformation, highlight **Total Deformation**. The deformation near the contact region is about 0.0568 mm. To show the support component alone, highlight Support in Geometry and right-click to pick Hide All Other Bodies. Do the same to show the shaft component.

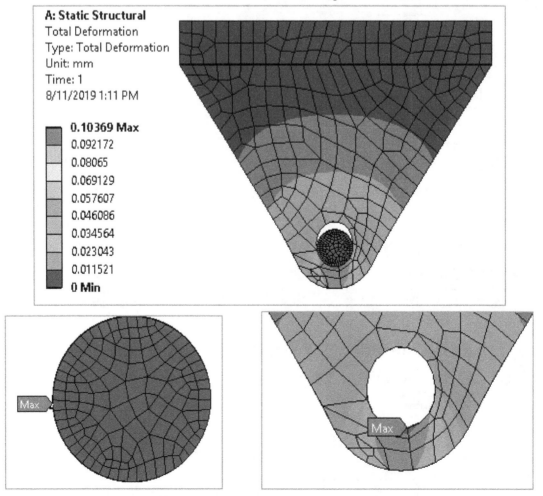

Now let us check the von Mises stress developed. Highlight Equivalent Stress. The maximum value is 164 MPa.

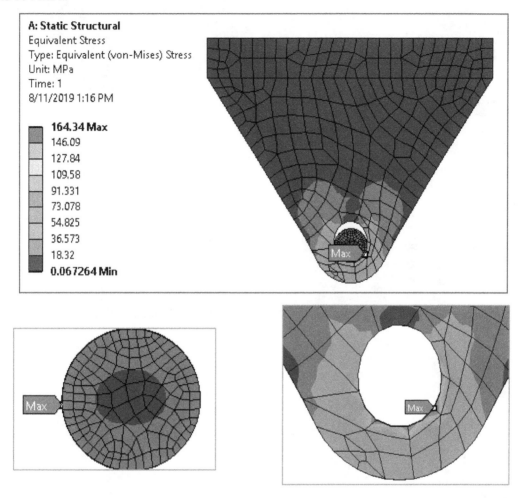

To further examine the stress distribution along a defined path, highlight Model (A4) and right-click to pick Insert > Construction Geometry. Afterwards, highlight Construction Geometry and right-click to pick Insert > Path.

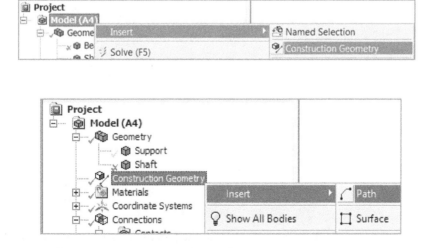

In Details of "Path", specify (42, 0, 0) as the coordinates of the starting point. Specify (-42, 0, 0) as the coordinate of the ending point.

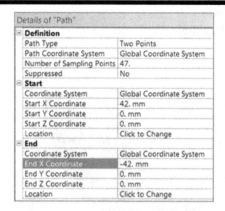

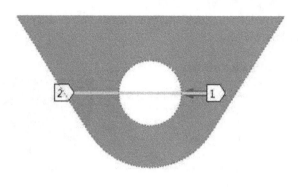

Highlight Solution (A6), and right-click to pick Insert > Linearized Stress > Equivalent (von Mises). In Details of "Linearized Equivalent Stress", input the defined path. Highlight Linearized Equivalent Stress. Right-click to pick Evaluate All Results.

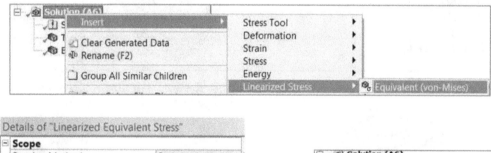

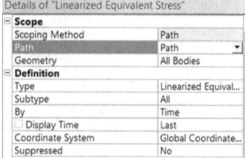

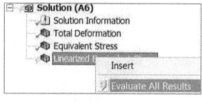

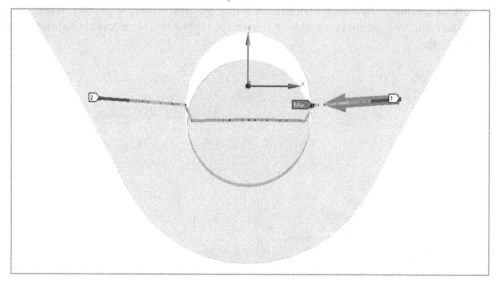

The stress distribution along the path is on display. In the Graph window, the stress distribution is plotted along the path, indicating the location(s) where the maximum stress developed. Also the data are tabulated. Users may save the data in EXCEL to plot.

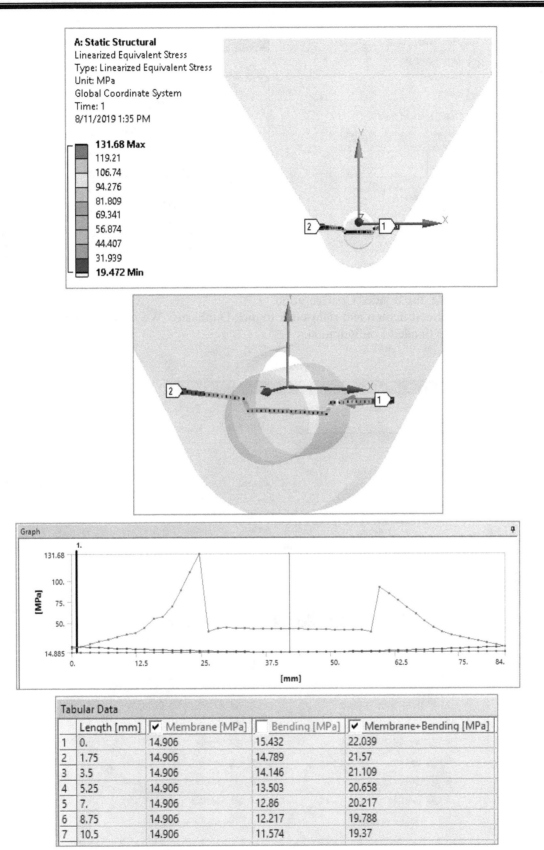

A: Static Structural
Linearized Equivalent Stress
Type: Linearized Equivalent Stress
Unit: MPa
Global Coordinate System
Time: 1
8/11/2019 1:35 PM

131.68 Max
119.21
106.74
94.276
81.809
69.341
56.874
44.407
31.939
19.472 Min

Graph

Tabular Data

	Length [mm]	✔ Membrane [MPa]	☐ Bending [MPa]	✔ Membrane+Bending [MPa]
1	0.	14.906	15.432	22.039
2	1.75	14.906	14.789	21.57
3	3.5	14.906	14.146	21.109
4	5.25	14.906	13.503	20.658
5	7.	14.906	12.86	20.217
6	8.75	14.906	12.217	19.788
7	10.5	14.906	11.574	19.37

Up to this point, we conclude our discussion on applying FEA to perform contact mechanics analysis. From the top menu, click **File** > **Close Mechanical**. In this way, we return to the main page of Workbench, or Project Schematic. From the top menu, click the icon of **Save Project.**

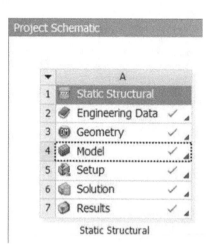

Step 9: Perform FEA under the Bonded Configuration.

Highlight A1 Static Structural and right-click to pick Duplicate. We have a new panel called B. Let us change the name to Bonded Configuration.

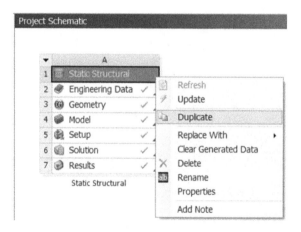

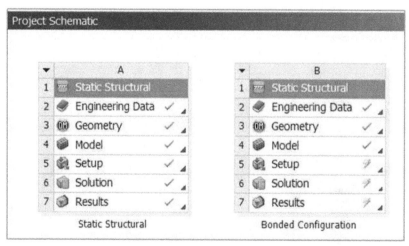

Double-click the icon of Model. We enter the Mechanical Mode. Check the unit system, and pick Metric (mm, t, N, s, mV, mA).

Bonded Configuration

Expand Connections > Contacts. Highlight No Separation. Change No Separation to **Bonded** in Type.

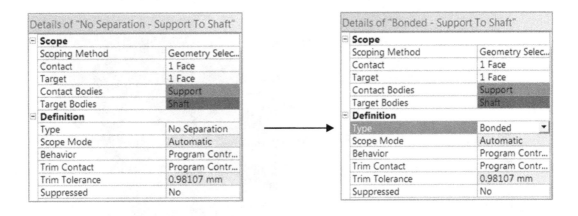

Highlight Solution (**B5**) and right-click to pick **Solve**. Workbench system is running FEA.

To plot the deformation, highlight **Total Deformation**. The deformation near the contact region is about 0.075 mm. There is no gap shown because the two surfaces are bonded.

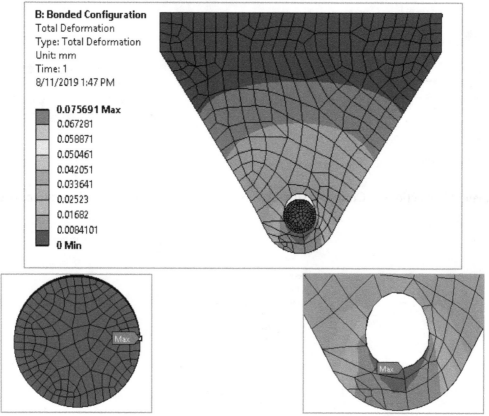

Now let us check the von Mises stress developed. Highlight Equivalent Stress. There is no gap shown because the 2 cylindrical surfaces are bonded. The maximum value is 844 MPa. This value is significantly higher than the value of 164 MPa obtained in the No Separation Configuration.

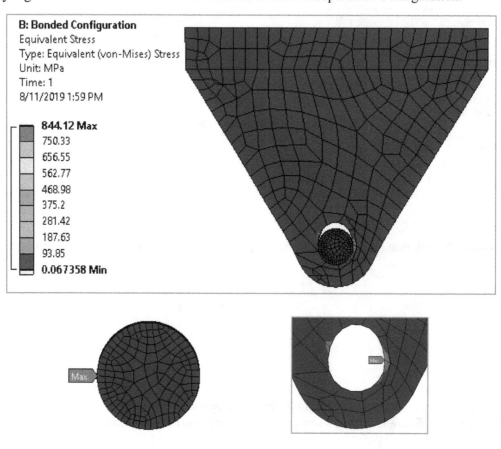

Click **File** > **Close Mechanical**. In this way, we return to the main page of Workbench, or Project Schematic. From the top menu, click the icon of **Save Project.**

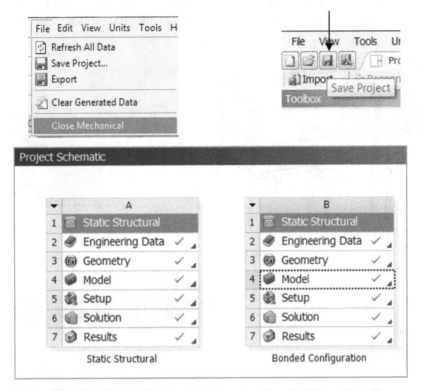

7.5 Case Study: A Solid Model for a Building Structure

In this case study, we use Design Modeler to create a building structure. As shown in the engineering drawing below, the building structure under study is a three (3) story building, and the 3 maximum dimensions of the building structure under study are 720X420X360 inches.

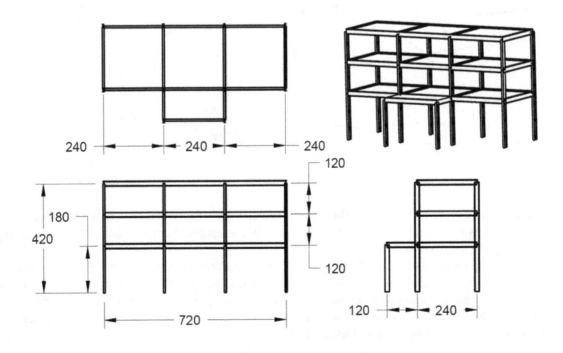

The building structure consists of two parts: building frame and floors. As shown below, the building frame is a combination of horizontal beams and vertical columns. In this case study, we use the standard W18X65 I-Beam to construct the building frame. The section area of W18X65 with dimensions is shown below. Pay attention to the joints where two (2) I-beams intersected. The alignment of the 2 beams is required.

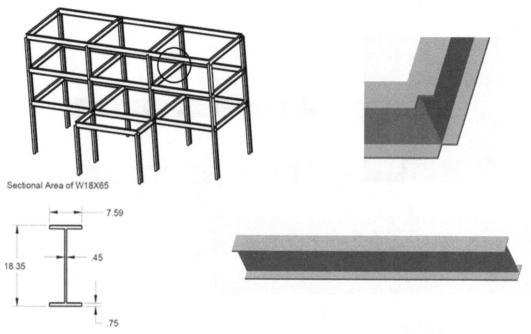

Sectional Area of W18X65

The floors of this building structure are concrete slabs. As shown below, there are 9 pieces of slabs with the 3 dimensions equal to 238X238X8 inches and one piece of slab with the 3 dimensions equal to 238X118X8 inches.

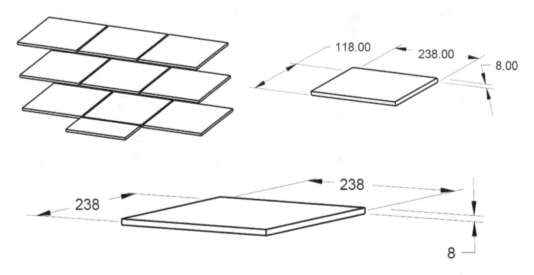

In this case study, we use the DesignModeler to create a Line Body to represent the building frame and 10 Surface Bodies to represents the 10 concrete slabs. We assemble them together to form a building structure. Afterwards, we perform FEA.

The following table lists the coordinates of the 20 construction points to be used in constructing the building frame. We list the coordinates of these 20 construction points in the following two tables. To avoid manually inputting the 3 coordinates for those construction points, we first create a txt file using Notepad. This txt file will include 16 points. We leave point17, point18, point19 and point20 for practicing the procedure of manually adding them in DesignModeler.

Point No	X coordinate	Y coordinate	Z coordinate
1	0	0	0
2	240	0	0
3	480	0	0
4	720	0	0
5	0	0	240
6	240	0	240
7	480	0	240
8	720	0	240
9	0	180	0
10	240	180	0
11	480	180	0
12	720	180	0
13	0	180	240
14	240	180	240
15	480	180	240
16	720	180	240

Point No	X coordinate	Y coordinate	Z coordinate
17	240	0	360
18	480	0	360
19	240	180	360
20	480	180	360

The procedure to create a txt file is listed below, assuming that we are using Notepad, a free basic text editor. There are five (5) columns in the file. The first column is Group No. The second column is Point No. Columns 3, 4 and 5 are the X, Y, and Z coordinates of those construction points. As shown in the following figure, the txt file name is new building points file.

```
new building points file - Notepad
File  Edit  Format  View  Help
1    1    0      0      0
1    2    240    0      0
1    3    480    0      0
1    4    720    0      0
1    5    0      0      240
1    6    240    0      240
1    7    480    0      240
1    8    720    0      240
1    9    0      180    0
1    10   240    180    0
1    11   480    180    0
1    12   720    180    0
1    13   0      180    240
1    14   240    180    240
1    15   480    180    240
1    16   720    180    240
```

After preparing the txt file, we start the design and analysis using Workbench.

Step 1: Open Workbench and Save the File Name.
From the start menu, click **ANSYS 19** > **Workbench 19**. Users may directly click Workbench 19 when the symbol of Workbench 19 is on display.

 Workbench 19.2

From the top menu, click the icon of **Save Project**. Specify **three story building** as the file name, and click **Save**.

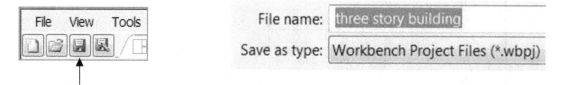

Step 2: From **Toolbox**, highlight **Static Structural** and drag it to the main screen.

Static Structural

Step 3: In Design Modeler, create the first floor frame.

Right-click the **Geometry** cell and pick **New DesignModeler Geometry** to start **Design Modeler**. Users may double click **Geometry** to directly enter **Design Modeler** if **New SpaceClaim Geometry** is not present. Click the **Units** drop down on the top menu, and select U.S. Customary (inch, lbm, lbf, °F, s, V, A).

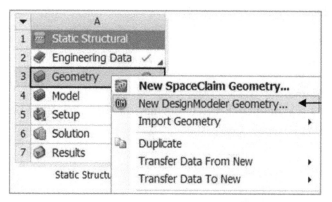

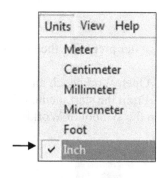

We need to input the key point file to DesignModeler.   From the top menu, click the icon of **Point**. In Details of View, From Coordinates File is selected by the system default. Click the box listed under Coordinate File, and locate the new building points file. Click **Open**. Afterwards, click Generate. The 16 key points are now on display.

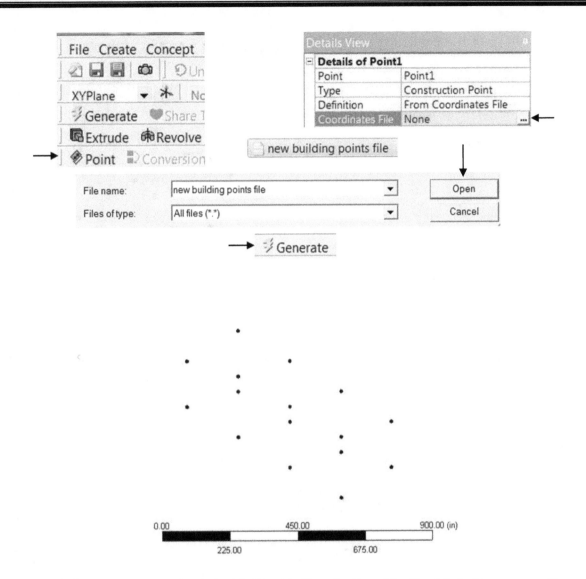

It is important to note that Point1 is listed in Tree Outline. The listed Point1is a group of imported key points, containing 16 construction points.

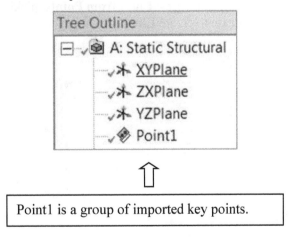

Point1 is a group of imported key points.

Click **Concept** from the top menu and select **Lines from Points**. While holding down the Ctrl key, pick the two points as shown to create Line1. Click **Apply** and click **Generate** to form the first horizon beam, as shown.

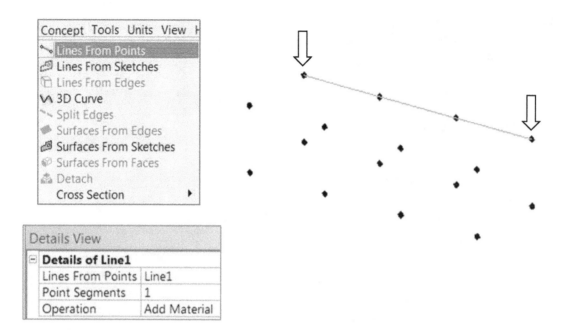

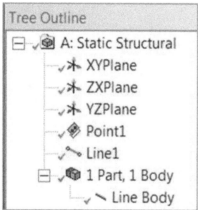

It is important to note that in Tree Outline, Line1 is listed and Line Body is also listed under 1 Part, 1 Body.

Continue this process of clicking **Concept > Lines from Points**, picking two points, while holding down the **Ctrl** key, and clicking **Generate** to create **Line2**, as shown.

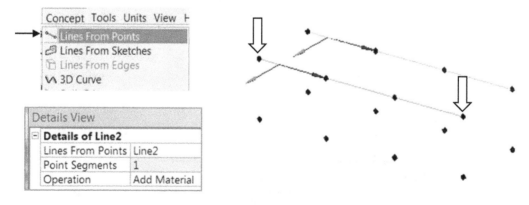

It is important to note that in Tree Outline, Line2 is listed and a second Line Body is also listed under 2 Parts, 1 Bodies because these 2 created Line Bodies do not touch each other. Therefore, they are independent of each other.

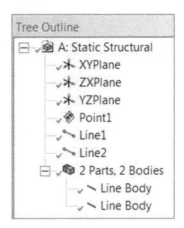

Continue this process of clicking Concept > Lines from Points, picking two points, while holding down the Ctrl key, and clicking Generate to create Line3, as shown.

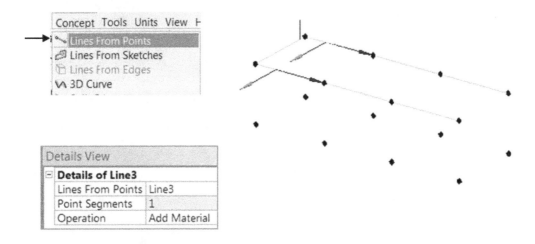

It is important to note that in Tree Outline, Line3 is listed. However, 2 Parts, 2 Bodies is now replaced by 1 Part, 1 Body, under which only one Line Body is listed. This is because Line3 connects Line1 and Line2. The 3 lines form a single Line Body.

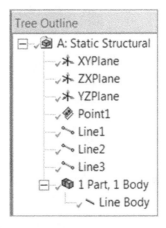

We continue this process to create Line4, Line5, and Line6, as shown. The created 6 lines form a Line Body, representing the first floor frame.

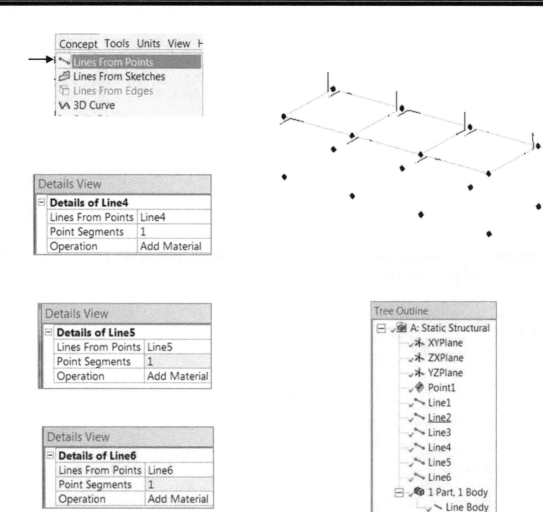

Step 4: To transform the first floor frame at a distance to obtain the second/third floor frame.

To create the second floor frame, we use Transformation. From the top menu, click **View** > clear the box of **Cross Section Alignments**. All the cross section alignment symbols are not on display. The Line Body feature is on display.

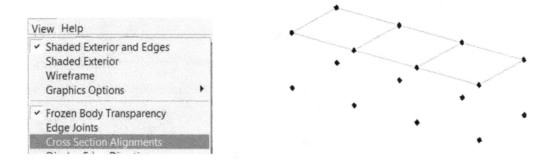

From the top menu, pick **Create** > **Body Transformation** > **Translate**. In Tree Outline, highlight **Line Body**. In **Details View**, select **Yes** to preserve the original **Line Body**, and click **Apply**. In Direction Definition, select Coordinate. Specify 120 as the Y Offset value to transform. Click **Generate** to obtain the second floor frame. Pay attention to Tree Outline, there are 2 Parts, 2 Bodies on the list. Both these 2 parts are Line Bodies.

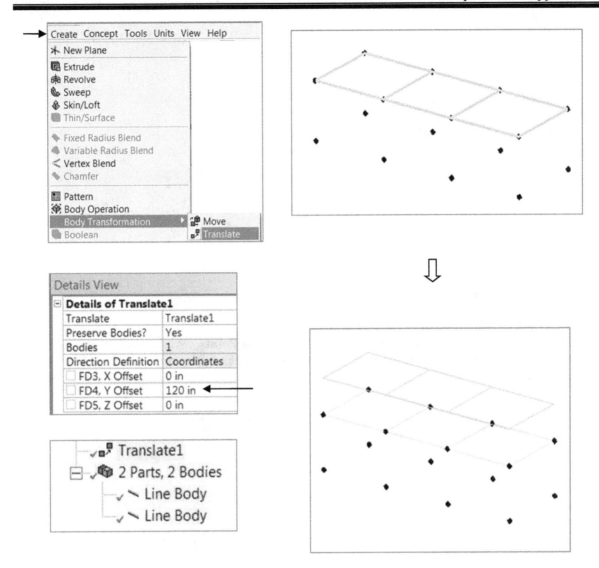

Repeat this process to obtain the third floor frame. From the top menu, pick **Create** > **Body Transformation** > **Translate**.

In Tree Outline, highlight the first **Line Body**. In **Details View**, select **Yes** to preserve the original **Line Body**, and click **Apply**. In Direction Definition, select Coordinate. Specify 240 as the Y Offset value to transform. Click **Generate** to obtain the third floor frame. Pay attention to Tree Outline, there are 3 Parts, 3 Bodies on the list. All 3 parts are Line Bodies.

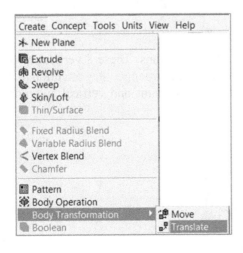

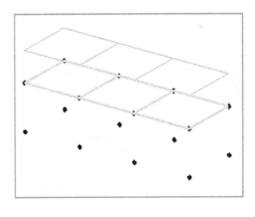

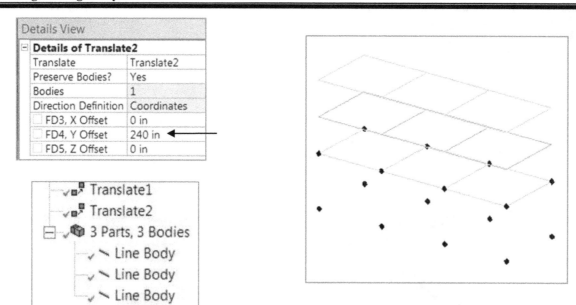

Step 5: Create the Line Body features for the 8 columns.

Click **Concept** from the top menu and select **Lines from Points**. While holding down the **Ctrl** key, pick two points to create a vertical line. Click **Apply** and click **Generate** to form the first vertical column, as shown.

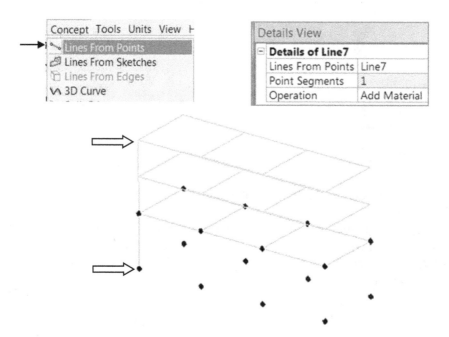

Repeat this procedure 7 more times to create 7 more vertical columns. These 8 vertical columns connecting the first, second and third floor frames to form a building frame structure. Pay attention to **Tree Outline**, there is 1 Part, 1 Body on the list because all these horizontal beams and vertical columns are connected.

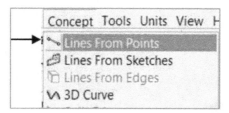

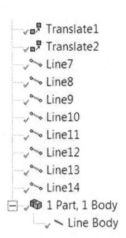

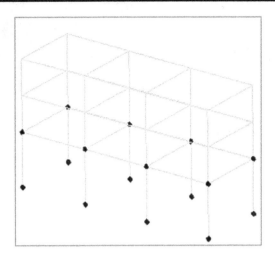

Step 6: Create the Line Body features for the building entrance frame.

First we need to create 4 new construction points. Let us start to create Point2. From the top menu, click the icon of **Point**. In Details of View, select **Manual Input** and specify (240, 0, 360) as the coordinates of Point2. Click **Generate**. (Note: Point1 listed in Tree Outline is a group of construction points)

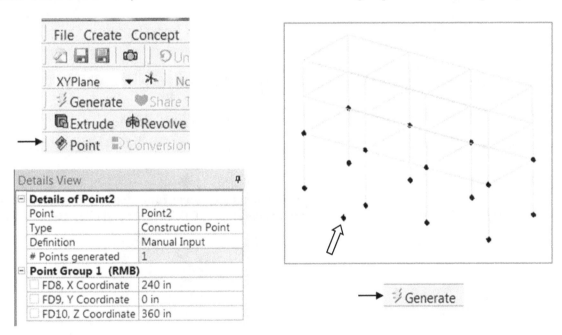

Repeat the above procedure 3 more times to define 3 more construction points: Point3 (480, 0, 360), Point4 (240, 180, 160), Point5 (480, 180, 360), as shown.

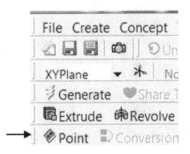

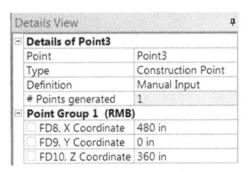

Details View	무
Details of Point4	
Point	Point4
Type	Construction Point
Definition	Manual Input
# Points generated	1
Point Group 1 (RMB)	
☐ FD8, X Coordinate	240 in
☐ FD9, Y Coordinate	180 in
☐ FD10, Z Coordinate	360 in

Details View	무
Details of Point5	
Point	Point5
Type	Construction Point
Definition	Manual Input
# Points generated	1
Point Group 1 (RMB)	
☐ FD8, X Coordinate	480 in
☐ FD9, Y Coordinate	180 in
☐ FD10, Z Coordinate	360 in

⟶ ⚡Generate

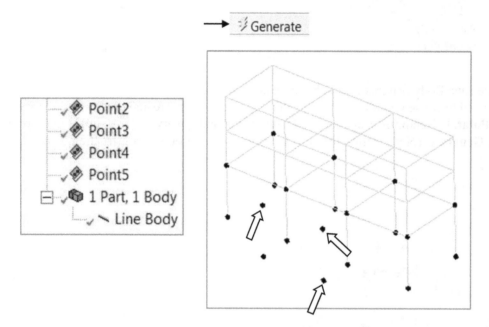

To create 5 lines, click **Concept** from the top menu and select **Lines from Points**. While holding down the Ctrl key, pick the two points as shown to create Line15. Click **Apply** and click **Generate** to form the first horizon beam, as shown.

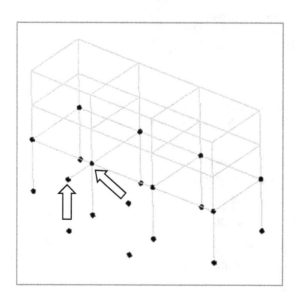

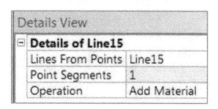

Repeat this process to create Line16, Line17, Line18 and Line19, as shown. Note that there is only one Line Body listed because all the horizontal beam and vertical columns are connected.

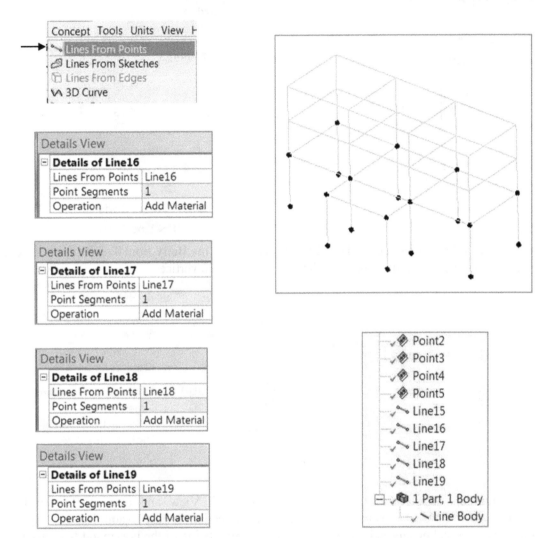

Step 7: Create the Cross Section and Assign it to the Line Body (Building Frame).

Now let us create the cross section for each of the horizontal beams and vertical columns. The shape and size of I-Beam W18X65 are shown below. Click **Concept** from the top menu, and click **Cross Section > I Section**.

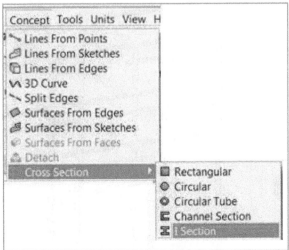

In **Details View**, specify the 6 dimensions, as shown. Cross Section I1 is listed on **Tree Outline**.

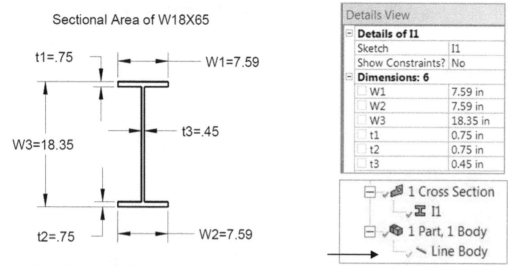

Sectional Area of W18X65

Details View	
□ **Details of I1**	
Sketch	I1
Show Constraints?	No
□ **Dimensions: 6**	
□ W1	7.59 in
□ W2	7.59 in
□ W3	18.35 in
□ t1	0.75 in
□ t2	0.75 in
□ t3	0.45 in

To assign the defined cross-section I1 to **Line Body**, click **Line Body** from Tree Output, select I1 in Cross Section. In Details View, this Line Body has 59 edges and 36 vertices.

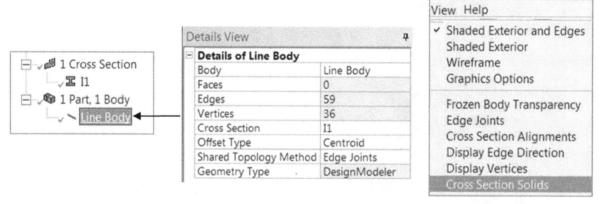

Details View		⊥
□ **Details of Line Body**		
Body	Line Body	
Faces	0	
Edges	59	
Vertices	36	
Cross Section	I1	
Offset Type	Centroid	
Shared Topology Method	Edge Joints	
Geometry Type	DesignModeler	

View Help
- ✓ Shaded Exterior and Edges
- Shaded Exterior
- Wireframe
- Graphics Options

- Frozen Body Transparency
- Edge Joints
- Cross Section Alignments
- Display Edge Direction
- Display Vertices
- Cross Section Solids

To visualize this assignment, from the top menu, click View and turn off Frozen Body Transparency and select Shaded Exterior and Edges and Cross Section Solids. Examining the displayed solid shapes of the horizontal beams and vertical columns, the orientation of the horizontal I-beams along the X direction should be rotated 90 degrees to match the orientation of the horizontal beams along the Z direction on the first, second and third floors.

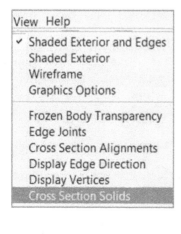

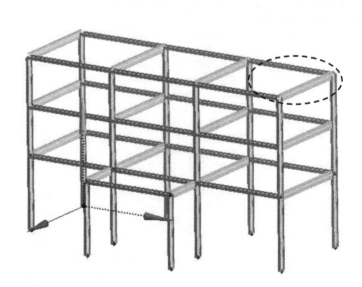

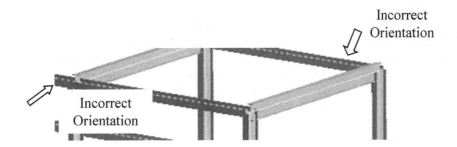

Incorrect
Orientation

Incorrect
Orientation

From the top menu, click Edge Selection, and pick the 18 horizontal edges along the X direction. In Details View, specify 90 degrees in the rotate box.

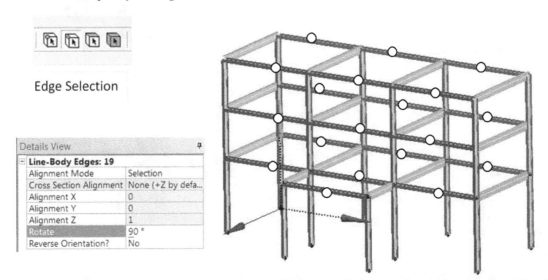

Edge Selection

Details View	╄
Line-Body Edges: 19	
Alignment Mode	Selection
Cross Section Alignment	None (+Z by defa...
Alignment X	0
Alignment Y	0
Alignment Z	1
Rotate	90 °
Reverse Orientation?	No

Now examine the orientation of those 18 horizontal beams, again. At those joints, the alignments are correct.

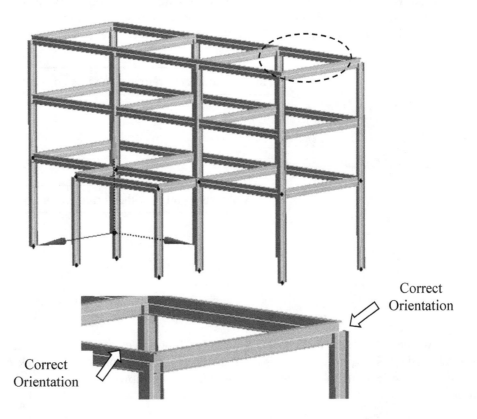

Correct
Orientation

Correct
Orientation

Step 8: Create Surface Bodies for the first, second and third floors.

Click **Concept** from the top menu and select **Surface From Edges**. While holding down the **Ctrl** key, pick the 4 edges, click **Apply**. Specify 8 as the thickness value, and click **Generate**. The first surface body is created. It is very important that 4 is listed as the number of the edges the user has picked. Otherwise, the user has to redo this process before clicking **Generate**.

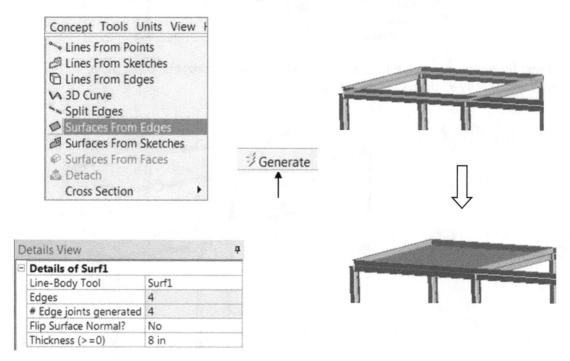

To create a new surface body next to the created surface body, special attention has to be paid for picking the edge, as shown in the following figure. This edge represents a horizontal beam. However, this location is shared with one the 4 edges belonging to the created surface body. Because of this reason, two (2) parallel-shaped selection symbols are on display with two different colors. Make sure you pick the parallel with the color, which is identical to the color of the beam edge you selected.

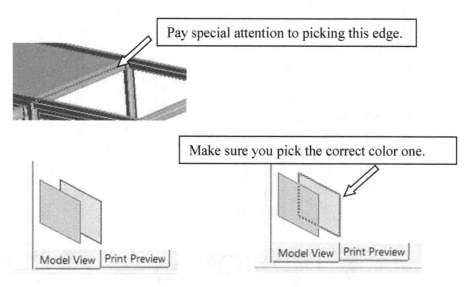

Now we need to repeat this process for 9 more times to create the other 9 concrete slabs. For creating each of the 9 slabs, make sure 4 is the number of the edges picked, and click Apply. Thickness value is 8 defined. Afterwards **Generate**. The following figure shows the slabs created for the second and third floors.

Details View	呂
Details of Surf2	
Line-Body Tool	Surf2
Edges	4
# Edge joints generated	4
Flip Surface Normal?	No
Thickness (>=0)	8 in

Details View	呂
Details of Surf3	
Line-Body Tool	Surf3
Edges	4
# Edge joints generated	4
Flip Surface Normal?	No
Thickness (>=0)	8 in

Details View	呂
Details of Surf4	
Line-Body Tool	Surf4
Edges	4
# Edge joints generated	4
Flip Surface Normal?	No
Thickness (>=0)	8 in

Details View	呂
Details of Surf5	
Line-Body Tool	Surf5
Edges	4
# Edge joints generated	4
Flip Surface Normal?	No
Thickness (>=0)	8 in

Details View	呂
Details of Surf6	
Line-Body Tool	Surf6
Edges	4
# Edge joints generated	4
Flip Surface Normal?	No
Thickness (>=0)	8 in

The concrete slabs created for the first floor are shown below.

Details View	呂
Details of Surf7	
Line-Body Tool	Surf7
Edges	4
# Edge joints generated	4
Flip Surface Normal?	No
Thickness (>=0)	8 in

Details View	呂
Details of Surf8	
Line-Body Tool	Surf8
Edges	4
# Edge joints generated	4
Flip Surface Normal?	No
Thickness (>=0)	8 in

Details View	呂
Details of Surf9	
Line-Body Tool	Surf9
Edges	4
# Edge joints generated	4
Flip Surface Normal?	No
Thickness (>=0)	8 in

Details View	呂
Details of Surf10	
Line-Body Tool	Surf10
Edges	4
# Edge joints generated	4
Flip Surface Normal?	No
Thickness (>=0)	8 in

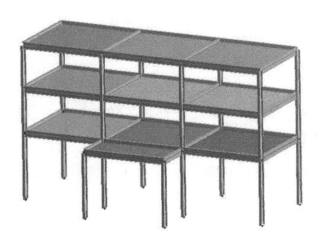

Let us examine the items listed in Tree Outline. There are 10 surfaces created. There are 10 surface bodies listed in Tree Outline. Because these 10 surface bodies are not connected, the total number of parts is 11 Parts, including the Lind Body representing the building frame.

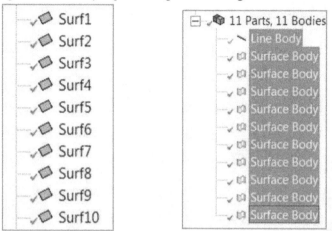

At this moment, we need to check the normal direction of the top surface for each surface body listed in Tree Outline. Common sense dictates a surface has a normal vector presenting its direction. When the direction of the normal vector is upward, this represents the positive direction of the surface, vice versa. For an FEA process, it requires the normal vectors of all surfaces are in the upward direction when a constraint or load condition is applied. To check the normal direction of a surface, from the top menu, click the icon of Surface Selection. In Tree Outline, highlight Surf to inspect the color. If the color is green, the normal vector is upward. Otherwise, click Yes in the box of **Flip Surface Normal** in Details View.

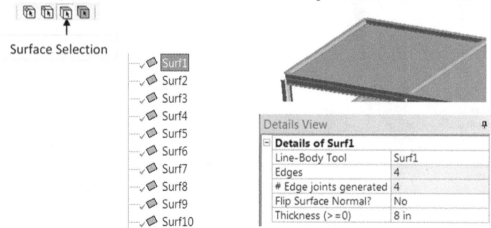

Now let us create a New Part to add a bonded constraint connecting all 11 Parts, 11 Bodies together for performing FEA. While holding down the **Ctrl** key, select all 11 Bodies, and right-click to pick **Form New Part**.

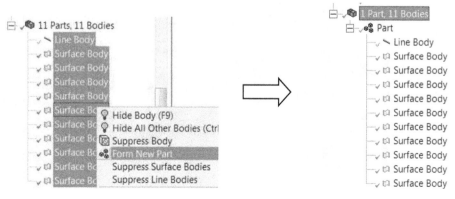

Upon completing the process of creating the surface body, click **File** > Close **Design Modeler**, and go back to **Project Schematic**. Click **Save Project**.

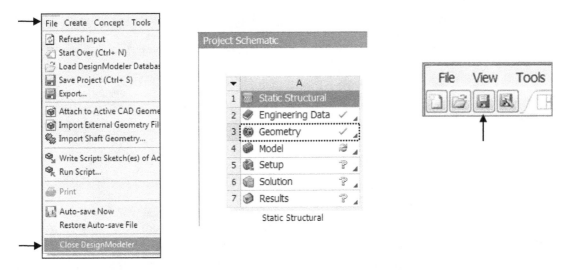

Step 9: Assignments of Material Properties to Beams/Columns & Floors, and **Mesh** Generation.

Double-click **Engineering Data**. Click **Engineering Data Sources**. Click **General Materials** and look for **Concrete**. Click the Plus symbol to activate Concrete Material Properties. Click the icon of Project to go back to Project Schematic page.

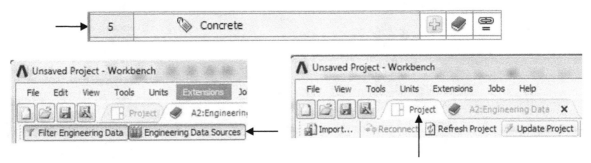

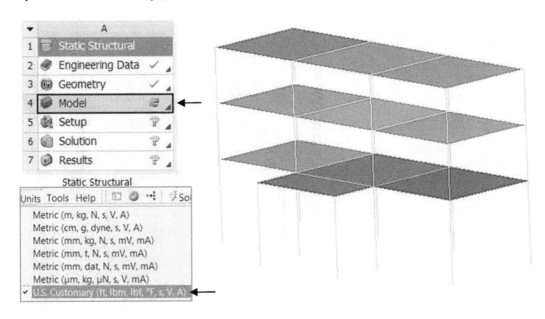

Double-click Model to enter Mechanical. The building structure model is on display. Select the unit system: U.S. Customary (ft, lbm, lbf, °F, s, V, A).

In the Project tree, expand **Geometry** > **Part**. Highlight Line Body. In Details View, check the material assignment. Structural Steel is the assigned material type by the system default.

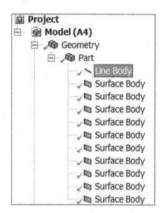

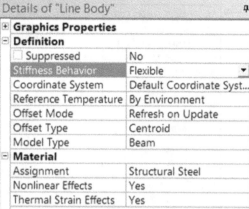

While holding down the **Ctrl** key, highlight the 10 Surface Bodies. In Details View, check the material assignment. Make sure Concrete is the assigned material type.

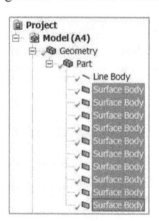

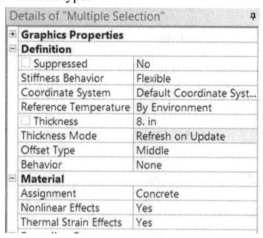

From Project Tree, highlight **Mesh,** and right-click to pick **Generate Mesh**.

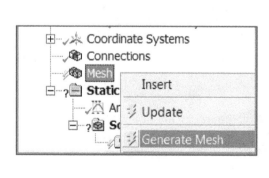

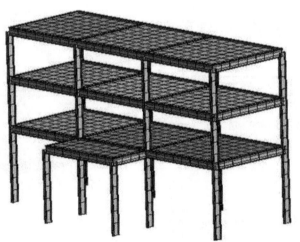

Step 10: Define the load and boundary conditions and perform FEA.

From **Project Tree**, highlight **Static Structural (A5)** and right-click to pick **Insert**. Afterwards, select **Fixed Support**. From the screen, while holding down the **Ctrl** key, pick the 10 construction points from the bottom surface, and click **Apply**, as shown. As a result, this building structure model is fixed to the ground through the 10 construction points of the building structure.

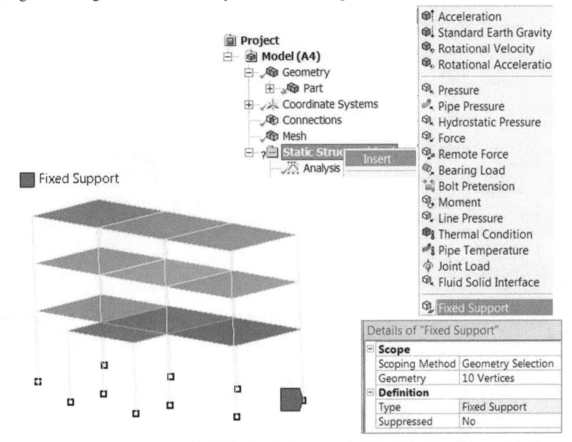

From the Project Tree, highlight **Static Structural (A5)** and right-click to pick **Insert**. Pick **Pressure**. Afterwards, select the 10 floor surfaces (make sure that the positive side is selected from each of the 10 floor surfaces), and click Apply. Specify 0.12 psi as the magnitude.

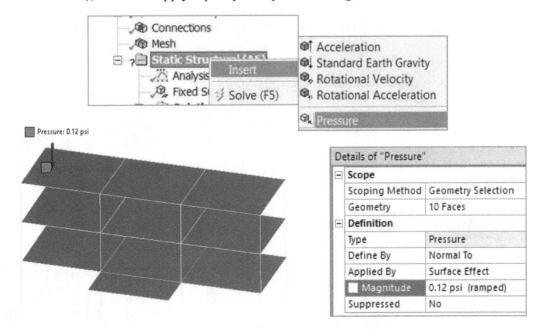

From Project Tree, highlight **Static Structural (A5)** and right-click to pick **Insert**. Select **Standard Earth Gravity**. Select –Y Direction. The gravity constant of 32 ft/s2 is shown as Y Component.

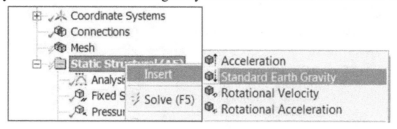

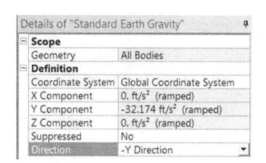

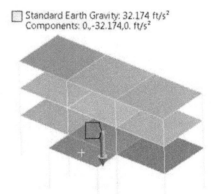

From the project tree, highlight **Solution** (A5) and right-click to pick **Insert**. Select **Deformation** and **Total**.

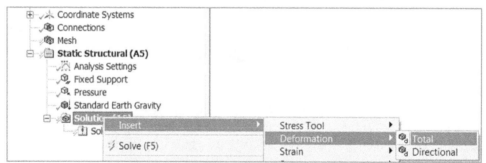

Highlight **Solution** (A5) and right-click to pick **Insert**. Select **Stress** and **Maximum Principal Stress**.

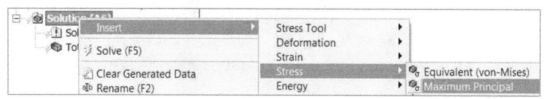

Highlight **Solution** (A5) and right-click to pick **Insert**. Select **Stress** and **Minimum Principal Stress**.

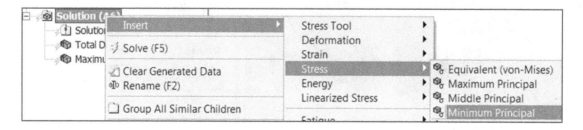

Highlight **Solution** (A5) and right-click to pick **Insert**. Select **Beam Tool** and **Beam Tool**. The 3 stress items (Direct Stress, Minimum Combined Stress, and Maximum Combined Stress are listed.

Highlight **Solution** (A5) and right-click to pick **Solve**.

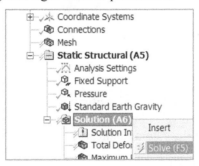

To facilitate the readings, let us change the unit systems back to **Inch,** U.S. Customary (in, lbm, lbf, °F, s, V, A).

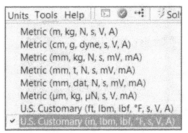

Afterwards, highlight **Total Deformation**. The distribution of the deformation pattern is on display, indicating the maximum value of the deflection at the free-end is 0.13 inch (equivalent to 3.3 mm).

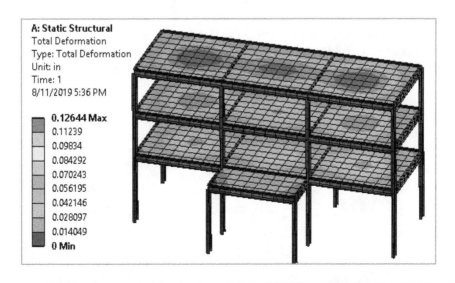

To plot the Minimum Principal Stress, highlight **Solution (A5)** and highlight **Minimum Principal Stress**. The distribution is on display, indicating the maximum compressive stress value is -302 psi (-0.20 MPa) located at the position of the first floor indicated. The building structure is under a compressive stress status.

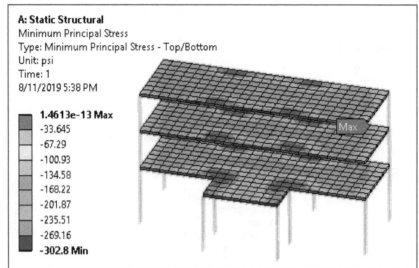

A: Static Structural
Minimum Principal Stress
Type: Minimum Principal Stress - Top/Bottom
Unit: psi
Time: 1
8/11/2019 5:38 PM

1.4613e-13 Max
-33.645
-67.29
-100.93
-134.58
-168.22
-201.87
-235.51
-269.16
-302.8 Min

To plot the Minimum Combined Stress, highlight **Solution (A5)** and highlight **Maximum Combined Stress**. The distribution is on display, indicating the maximum compressive stress value is 3076 psi (21.5 MPa) located at the position of the column indicated.

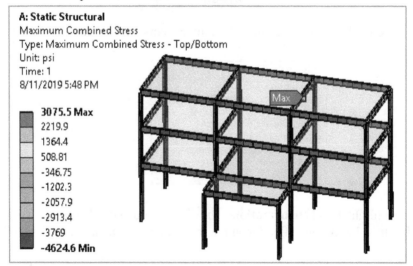

A: Static Structural
Maximum Combined Stress
Type: Maximum Combined Stress - Top/Bottom
Unit: psi
Time: 1
8/11/2019 5:48 PM

3075.5 Max
2219.9
1364.4
508.81
-346.75
-1202.3
-2057.9
-2913.4
-3769
-4624.6 Min

Let us plot the reaction force and moment at the fixed support locations. Highlight **Solution** (A5) and right-click to pick **Insert**. Select **Probe > Force Reaction.** The **Force Reaction** is listed. Highlight it and right-click to pick **Evaluate All Results**. As shown, the reaction force is in the Y direction upward with the magnitude equal to 491×10^5 lbf.

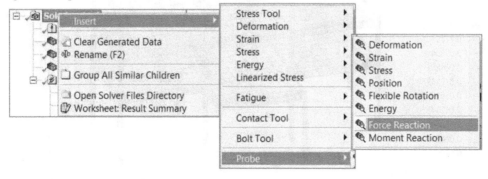

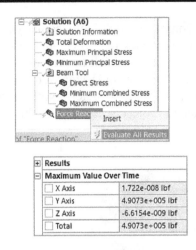

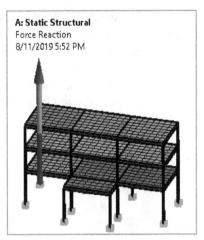

Results	
Maximum Value Over Time	
X Axis	1.722e-008 lbf
Y Axis	4.9073e+005 lbf
Z Axis	-6.6154e-009 lbf
Total	4.9073e+005 lbf

Up to this moment, we complete the evaluation of the building structural design. The obtained information is essential for a future re-design process to improve the quality and safety of this building structure. From the top menu, click **File** > **Close Mechanical**. In this way, we return to the main page of Workbench, or the Project Schematic page. Click **Save Project**.

7.6 CAD Integration

Nowadays engineers in industry use CAD programs, such as Creo Parametric, SolidWorks, Autodesk Inventor, NX and the others, to design because of the capability and efficiency of those CAD programs. The rapid development of digital technology has already equipped those CAD programs with the simulation capability to evaluate the product design and visualize the functionality of the designed product. Engineering With the understanding that all engineering simulation is based on geometry to represent the design, there are two basic methods of accessing CAD models within ANSYS Workbench, depending on the level of integration and the interface between the CAD programs and ANSYS Workbench.

The first basic method requires translation to the intermediate geometry formats. This method does not require the CAD system to be present to import geometry files. However, it is the responsibility of the user to save the CD file to be imported to ANSYS DesignModeler in an appropriate format, such as the IGES format. For example, a user has created a block (100 X 60 X 40 mm) using Creo Parametric or Solidworks or Autodesk Inventor. The user needs to save the block file as .prt under Creo Parametric, or *sldprt, or Autodesk Inventor *ipt. Afterwards, the user needs to save the block.prt file to the block.IGS file under Creo Parametric, or save the block.sldprt file to the block.IGS file under SolidWorks, or save the block.ipt file to the block.IGS under Autodesk Inventor.

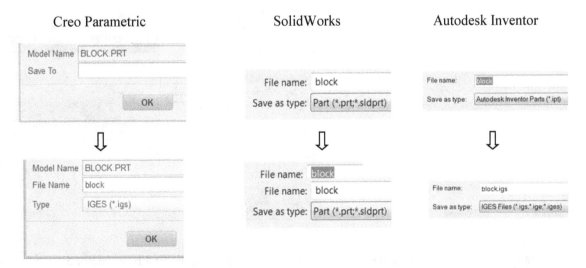

Now the user should launch Workbench 19 from his/her computer. Highlight Geometry and drag it to the Project Schematic screen. Highlight Geometry and right-click to pick Import Geometry > Browse and search for the block.igs file. Afterwards, click Open. The user is now back to Project Schematic. Highlight Geometry and right-click to pick Edit Geometry in DesignModeler. In DesignModeler, the user clicks Generate. The 3D solid model is transferred into the format of DesignModeler. Notes Import1 is listed in Tree Outline. Click File > Close DesignModeler to go back to Project Schematic. At this moment, the geometry file is in DM geometry format.

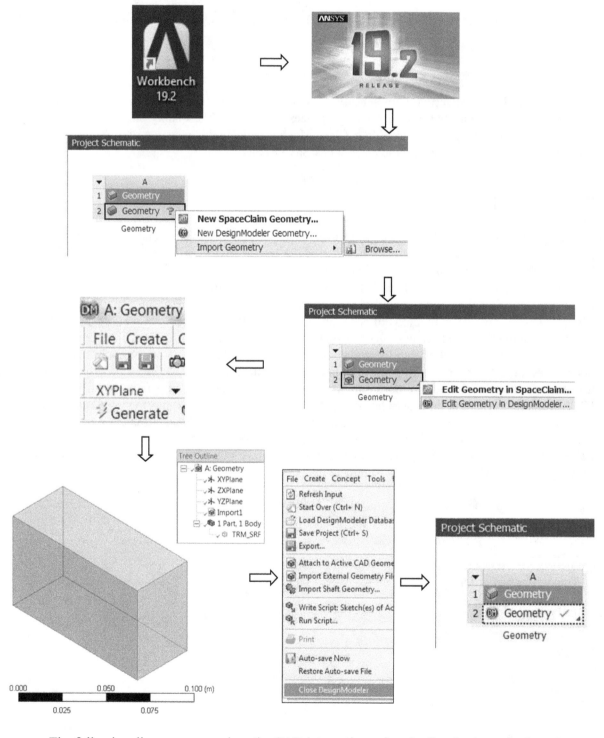

The following diagram summarizes the CAD integration using the first basic method.

Data Intergration

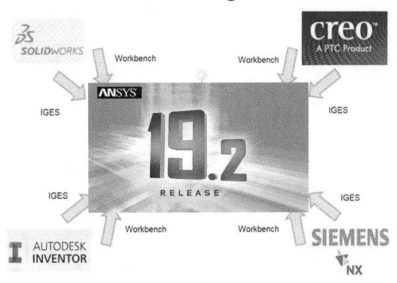

Initial Graphics Exchange Specification (IGES)

The second basic method enables the user directly import his/her CAD geometry model to the DesignModeler Application. Once a CAD file has been edited in DesignModeler, DesignModeler is able to dictate the settings for transferring DesignModeler geometry to a downstream application. Note that the CAD Plug-In property in the image below will appear only when an appropriate plug-in is defined for the selected geometry source. Assume that you were a local administrator for the computer you were using. From the start menu, click All Programs, and look for ANSYS 19. Expand ANSYS 19 > CAD Configuration Manager 19.

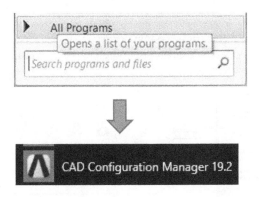

Click the box called CAD Selection. Check the CAD software systems you want to have the connection. For example, Creo Parametric (Pro Engineer), Inventor, SolidWorks. Make sure that Workbench Associative Interface is also checked for each of the three CAD software systems. You have selected. For the next step, you click the box called Creo Parametric. Click Browse and search for Creo Installation Location. For example, C:\Program Files\Creo 5.0\M110\Parametric, which is the location commonly used. It is important to note that this procedure is not required for Autodesk Inventor and not required for SolidWorks. ANSYS system is able to automatically connect to their installation Locations. As a result, Autodesk Inventor and Solid Works are not listed on this page. In order to verify the connections have been successfully made, click the box called CAD Configuration > Configuration Selected CAD Interfaces.

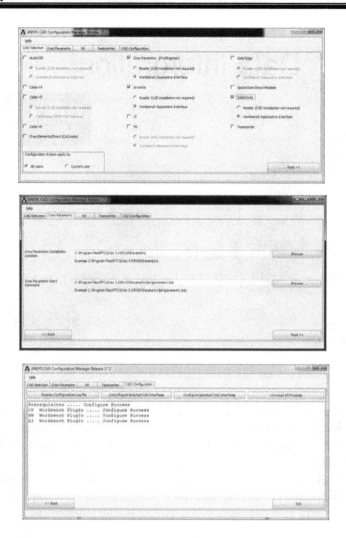

The following figures illustrate the direct connections between Creo Parametric and ANSYS Workbench, between SolidWorks and ANSYS Workbench, and between Autodesk Inventor and ANSYS Workbench.

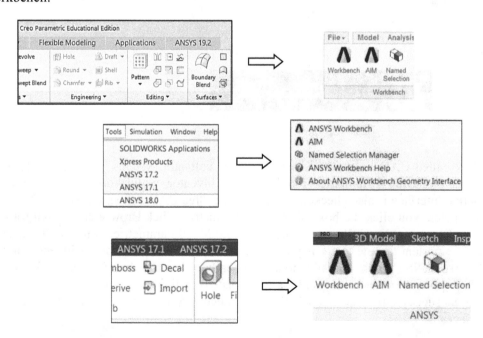

INDEX

CPSIA information can be obtained
at www.ICGtesting.com
Printed in the USA
LVHW011500311219
642052LV00016B/717/P

9 781935 673507